NIKOS D. KYRIAZIS

(Researcher Geometrician)

HARMONIC GEOMETRY - VOL. 2

HARMONIC INVOLUTIONS, HARMONIC MULTILATERALS AND HARMONIC LINES

DEFINITIONS-METHODS-PROPOSITIONS-PROBLEMS-LOCI

(Research-Investigation)

(Our Previous Papers have been honoured from the ACADEMY OF ATHENS).

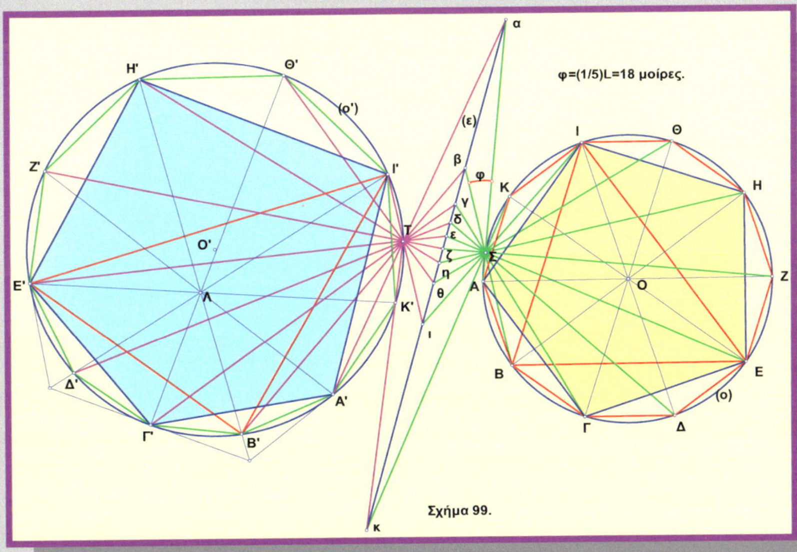

Figure of the Construction of a Harmonic Decagon with the new Method of Harmonic Transformation (Construction 155).

HARMONIC GEOMETRY

Title: Harmonic Geometry, Vol. 2
ISBN SET 978-618-83803-0-1
ISBN 978-618-83803-2-5

Author: Nikos D. Kyriazis

Cover designer: Vassilis Ioannou

Production supervisor: Platon Malliagkas
www.mediterrabooks.com

This work was translated from Greek into English by Ylenia Orfanidou (Mathematician) and Harry Nikolaides (Teacher of English)
August 2nd, 2017

Cover: Implementation of our new "Harmonic Transformation" Method for the Construction of the Harmonic Decagon from a Normal Decagon (Construction 155)

E' HARMONIC COCYCLIC MULTILATERALS

16. In General.

We use the term Harmonic Cocyclic Closed groups of n Series of Points, or Harmonic Cocyclic Multilaterals, or simply Harmonic Multilaterals, to refer to the inscribed in circle Special Multilaterals, whose the exact definition we will give below for each multilateral separately.

Special case of these, are the Closed Collinear groups of n Series of Points, which we mentioned in detail in paragraph 11.

We could not find features, regarding the Harmonic Cocyclic Closed groups of n Series of Points, in bibliography, even though we asked for relevant information at the mathematica.gr website, no such information has been found until today (except for Collinear Harmonic Groups of Four Points and Harmonic Quadrilaterals, that we have forementioned as known to us).

So, at first we approached the matter according to our own view and intuition, while later on we developed and used relevant methods.

16.1. Definitions of Simple Harmonic Multilaterals.

α. Simple Harmonic Quadrilateral.

(Harmonic Closed Cocyclic Group of Four Points).

We use this term to refer to inscribed in circle convex quadrilateral ΑΒΓΔ (Figure 35), whose diagonals are conjugate with repsect to its circumscribed circle (Γεωμετρία [6], § 8-4.10).

Properties of which are mentioned in the known bibliography and specifically in books [1] and [31], where a large amount of properties is given in the form of Criteria of Harmonic Being, among which they have:

$$AB.\Gamma\Delta = B\Gamma.\Delta A \text{ , or } \frac{AB}{B\Gamma} = \frac{A\Delta}{\Delta\Gamma} \{[6]\ \S\ 8\text{-}4.10\ (3)\}. \qquad (1).$$

This relation means that Α, Γ harmonically divide Β, Δ and Β, Δ harmonically divide Α, Γ (See § 11.1a and § 8.61). (Harmonic Quadrilateral is simple and known. We mention it here, though, for generalization purposes, as well as because it is a Fundamental Geometrical Element through the repetition of which any other Cocyclic Harmonic n-lateral is constructed).

Remarks.

(a). To the above group of four we have given the indication «Cocyclic», as there is an equivalent Harmonic «Collinear», Group of Four Points (Harmonic Group of Four § 11.1a).

(b). To the Harmonic Quadrilateral, we will refer later too, only regarding its features that we have not found in bibliography.

(c). Below we will mention some Criteria of harmonic being concerning the Harmonic Quadrilateral, in the form of Propositions, that we did not find in bibliography.

Moreover we will mention new ways of construction of an arbitrary Harmonic Quadrilateral.

(d). Harmonic quadrilaterals are known, many features of them exist in bibliography, mostly in books [1] and [31]. More specifically, in book [1], Complexes of Harmonic Complete Quadrilaterals are mentioned too, and not only that (pages 604 to 763).

β. Simple Harmonic Hexagon.

(Harmonic Cocyclic Closed Group of Six Points).

Definitions.

We consider on circle (o), the group of six ordered points A, B, Γ, Δ, E, Z (Figure 64).

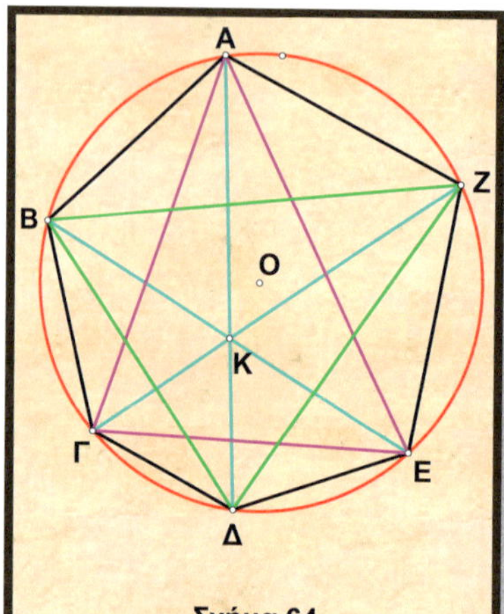

Σχήμα 64.

The inscribed in circle convex hexagon ΑΒΓΔΕΖ will be called Simple Harmonic Hexagon, if all six consecutive quadrilaterals of it below, are harmonic:

ΑΒΓΕ - ΒΓΔΖ - ΓΔΕΑ

- ΔΕΖΒ - ΕΖΑΓ - ΖΑΒΔ. (1).

The above six quadrilaterals of relation (1), we will call «Group of Six Harmonic Quadrilaterals of the Harmonic Hexagon», which do not need to be given that all of them are Harmonic for it to be harmonic, as we will later prove (Criteria 109 and 141).

Remarks

(a). To the group of six above we gave the indication «Cocyclic», since as we saw above (§ 11.1b), an equivalent Harmonic «Collinear» Group of Six Points exists (Harmonic Group of Six).

(b). It is obvious that in the extreme case where the radius of the circumscribed circle of the Simple Harmonic Hexagon is of infinite length or if this circle is of zero curvature, then the group of its six apexes degenerates to a Harmonic «Collinear» Group of Six Points (§ 11.1b).

(c). Below we will mention various properties that every Simple Harmonic Hexagon has, in the form of Criteria of Harmonic being.

Moreover we will refer to the ways of construction of an arbitrary Simple Harmonic Hexagon (Constructions 100, 153).

Comments.

(a). We have developed the Complexes of Complete Harmonic Hexagons, thoroughly in 1990, in our six-volume book [1] (pages 1150 to 1286).

(b). Here, we reached the Simple Harmonic Hexagons through another way irrelevant to the one we used in our book [1].

(c). As we will see below, instead of the above definitions, we ususally use other alternative ones, that we will provide through Criteria later on (for example Criterion 109).

c. Simple Harmonic Octagon.

(Harmonic Cocyclic Closed Group of Eight Points).

Definitions.

We consider on circle (o), the group of eight ordered points Α, Β, Γ, Δ, Ε, Ζ, Η, Θ (Figure 65).

The inscribed in circle convex octagon ΑΒΓΔΕΖΗΘ will be called Simple Harmonic Octagon, if all eight consecutive quadrilaterals of it below, are harmonic: ΑΒΓΖ - ΒΓΔΗ - ΓΔΕΘ - ΔΕΖΑ –ΕΖΗΒ - ΖΗΘΓ - ΗΘΑΔ - ΘΑΒΕ. (1).

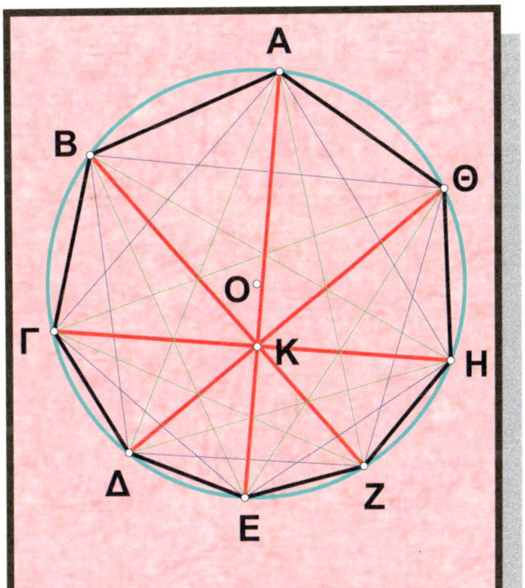

Σχήμα 65.

The above eight quadrilaterals of relation (1), we will call «Group of Eight Harmonic Quadrilaterals of the Harmonic Octagon», which do not need to be given that all of them are Harmonic, as we will later prove (Criterion 110).

The non-regular Asteroid Octagon ΑΔΗΒΕΘΓΖΑ, equivalent to the convex Harmonic Octagon ΑΒΓΔΕΖΗΘ (they share apexes with ΑΒΓΔΕΖΗΘ), is, harmonic, as for this as well, all its eight quadrilaterals are harmonic, which obviously coincide with these in relation (1).

Thus, as we will later see, this matches many main criteria, that a Harmonic Octagon needs to match, but which is not convex. This we would call «Harmonic non Regular Asteroid Octagon», or simply «Harmonic Asteroid Octagon».

Remarks

(a). To the group of eight above we have given the indication «Cocyclic», since as we saw there is an equivalent Harmonic «Collinear» Group of Eight Points (Harmonic Group of Eight) (§ 11.1c).

(b). It is obvious that in the extreme case where the radius of the circumscribed circle of the Simple Harmonic Octagon is of infinite length or if this circle is of zero curvature, then the group of its eight apexes degenerates to a Harmonic «Collinear» Group of Eight Points (§ 11.1d).

(c). Below we will mention various properties that every Simple Harmonic Octagon has, and some Criteria of Harmonic being, in the form of Propositions.

Moreover we will refer to the ways of construction of an arbitrary Simple Harmonic Octagon [Construction 149 (§ 19.64)], 101, 154, etc.

Comments.

(a). We have developed the Complexes of Complete Harmonic Octagons thoroughly in 1991 and they are included in our six-volume book [1] (Volume 4, pages 1352 to 1791).

(b). Here, we reached the Simple Harmonic Octagons through another way irrelevant to the one we used in our book [1].

(c). As we will see below, instead of the above definitions, we ususally use other alternative ones, that we will provide through Criterion 110 (§ 19.18).

(d). As we will later see both its quadrilaterals ΑΓΕΗ and ΒΔΖΘ, are twin harmonic.

d. Simple Harmonic Decagon.

(Harmonic Cocyclic Closed Group of Ten Points).

Definitions.

We consider on circle (o), the group of ten ordered points A, B, Γ, Δ, E, Z, H, Θ, I, K (Figure 66).

The inscribed in circle convex decagon ΑΒΓΔΕΖΗΘΙΚ will be called Simple Harmonic Decagon, if all ten consecutive quadrilaterals of it below, are harmonic:

ΑΒΓΗ - ΒΓΔΘ - ΓΔΕΙ - ΔΕΖΚ - ΕΖΗΑ -
ΖΗΘΒ - ΗΘΙΓ - ΘΙΚΔ - ΙΚΑΕ - ΚΑΒΖ. (1).

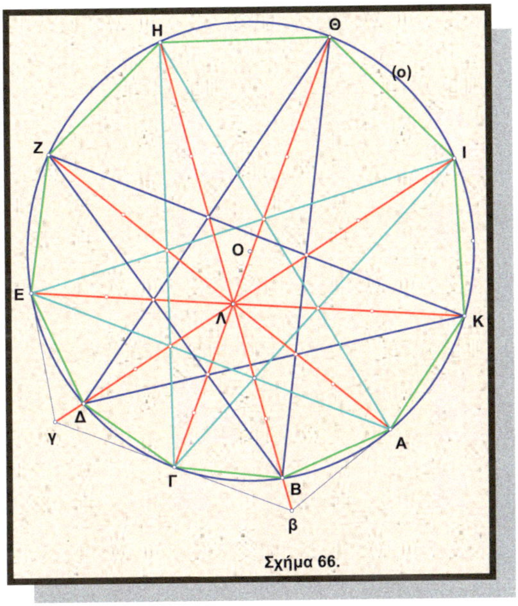

Σχήμα 66.

The above ten quadrilaterals of relation (1), we will call «<u>Group of Ten Harmonic Quadrilaterals of the Harmonic Decagon</u>», which do not need to be given that all of them are Harmonic for it to be harmonic, as we will later prove (Criterion 111).

The non-regular Asteroid Decagon ΑΔΗΚΓΖΙΒΕΘΑ (Figure 73), equivalent to the convex Harmonic Decagon ΑΔΗΚΓΖΙΒΕΘΑ (they share apexes), is, harmonic, as for this as well, all its ten consecutive quadrilaterals are harmonic.

This means that its quadrilaterals ΑΔΗΙ, ΔΗΚΒ, ΗΚΓΕ, ΚΓΖΘ, ΓΖΙΑ, ΖΙΒΔ, ΙΒΕΗ, ΒΕΘΚ, ΕΘΑΓ, ΘΑΔΖ are harmonic (See § 12.8 and Proposition 108 § 19.16). Hence, as we will later see, this matches many criteria, that a Harmonic Decagon should match, but which is not convex. We will call this «<u>Harmonic Non Regular Asteroid Decagon</u>», or simply «<u>Harmonic Asteroid Decagon</u>».

Remarks

(a). To the group of ten above we have given the indication «Cocyclic», since as we saw there is an equivalent Harmonic «Collinear» Group of Ten Points (Harmonic Group of Ten) (§ 11.1d).

(b). It is obvious that in the extreme case where the radius of the circumscribed circle of the Simple Harmonic Decagon is of infinite length or if this circle is of zero curvature, then the group of its Ten apexes degenerates to a Harmonic «Collinear» Group of Ten Points (§ 11.1d).

(c). Below we will mention various properties that every Simple Harmonic Decagon has, in the form of Criteria of Harmonic being.

Moreover we will refer to the ways of construction of an arbitrary Simple Harmonic Decagon (Constructions 102, 155).

(d). As we will see exactly below, both Groups of Five Points A, Γ, E, H, I and B, Δ, Z, Θ, K, constitute apexes of twin Simple Harmonic 5-gons.

Comments.

As we will later see, instead of the definitions above, we usually use other alternative definitions, that we will give in Criteria later and which are useful to us (for example Criterion 111).

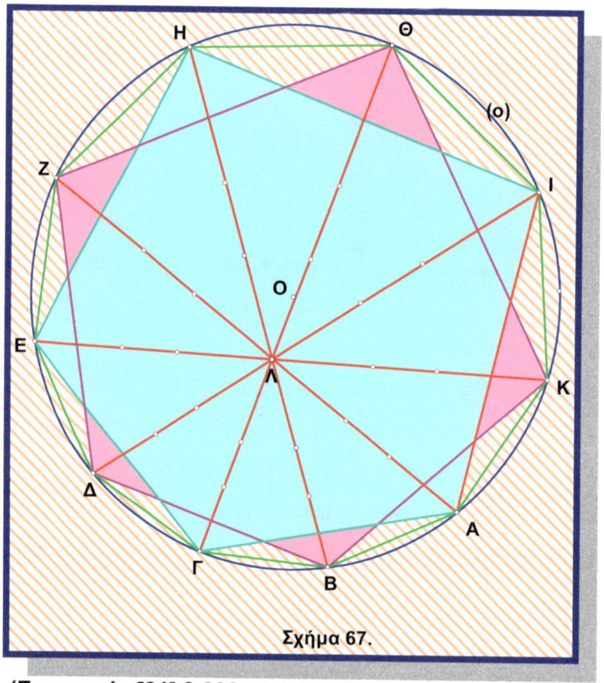

Σχήμα 67.

e. Simple Harmonic Pentagon (Harmonic Cocyclic Closed Group of Five Points).

(a). In General.

As we have mentioned above, it is known that, to construct a Regular Pentagon, we first need to construct the Regular Decagon, from which consequently the Regular Pentagon occurs rather easily (Γεωμετρία [21] § 298 and 299).

The Harmonic Pentagon is linked to the Harmonic Decagon, in the same way the Regular Pentagon is linked to the Regular Decagon.

Hence the definition of the Harmonic Pentagon, is based on the definition of the Harmonic Decagon, we have given above.

The same applies to the construction of the Harmonic Pentagon Construction 156 (§ 22.6), which is carried through with based on the Construction of the Harmonic Decagon.

(b). Definitions.

We consider, the inscribed in circle (o) convex pentagon ΑΓΕΗΙ (Figure 67). We will call this pentagon Harmonic, if between its consecutive apexes, there exists one of the following five points of the circumscribed circle Β, Δ, Ζ, Θ, Κ, respectively, such that the convex Decagon ΑΒΓΔΕΖΗΘΙΚ, would be Harmonic.

As the Regular Pentagon matches to a Regular Asteroid Pentagon, every Harmonic Pentagon matches to a Harmonic Asteroid Pentagon [Here these are ΑΕΙΓΗΑ and ΒΖΚΔΘΒ (figure 66), respectively, of convex ΑΓΕΗΙ and ΒΔΖΘΚ].

The inscribed in circle (o) Asteroid Pentagon ΑΕΙΓΗΑ would be called Harmonic, if between its consecutive apexes, there exists one of the following five points of circle (o) Θ, Β, Ζ, Κ, Δ respectively, such that the Asteroid Decagon ΑΔΗΚΓΖΙΒΕΘΑ, would be harmonic.

Remarks

(a). It is obvious and easily proven, that Pentagon ΒΔΖΘΚ, is Harmonic, as in this Pentagon, there exists one point between its consecutive apexes, from the following five points Α, Γ, Ε, Η, Ι, such that decagon ΑΒΓΔΕΖΗΘΙΚ would be Harmonic (definition).

(b). To the above group of five points Α, Γ, Ε, Η, Ι, we have the indication «Cocyclic» group of five, since as we saw above an equivalent «Collinear» Harmonic Group of Five Points exists (Harmonic degenerate to a line Pentagon) (§ 11.1e).

(c). Given the chance we will refer later to the Twin Harmonic Pentagons ΑΕΙΓΗΑ and ΒΖΚΔΘΒ.

(δ). Later we will refer to various properties that every Harmonic Pentagon has, as well as to some Criteria of its Harmonic being, in the form of Propositions. Additionally we will refer to the ways of construction of an arbitrary Harmonic Pentagon (Constructions 103, 156).

Comment.

As we will later see, instead of the definitions above, we usually use some other alternative definitions, which we will give as Criteria later and which are rather useful to us.

f. Simple Harmonic n-lateral.

(Harmonic Closed Cocyclic Group of n Points).

Working similarly to the above cases of Cocyclic Harmonic 4-laterals, 6-gons, 8-gons, 10-gons, 5-gons, we can provide the definitions of other such Harmonic n-laterals or Harmonic Cocyclic Closed Series of Points, with n sides.

F' HARMONIC CLOSED CENTRAL PENCILS AND THEIR INTERSECTION FROM A LINE AND A CIRCLE.

17. Analysis of the Closed Harmonic Central Pencils and their Intersections from a Line and a Circle.

Closed Harmonic Flat Central Pencils, as we saw above we will also see below, and there are of four, five, six, eight, ten, twelve, etc, beams.

Harmonic Pencils, are rather useful, as we will see from now on, because based on them, we easily construct Harmonic Collinear Groups of n Points, Harmonic Multilaterals, etc and this is why they will interest us later.

We searched in the bibliography online for relevant information, etc, but we were unable to find, apart from some concerning the Cocyclic Harmonic Pencils of Four Beams.

There is a preview regarding Closed Harmonic Flat Central Pencils, in paragraph 12.13 (b) 4/.

17.1. Closed Harmonic Central Pencils of Four Beams.

We consider a Harmonic Pencil Σ(αγβδ) (Figure 68). It is known that this intersected from a line (ε), gives us the Harmonic Group of Four α, γ, β, δ.

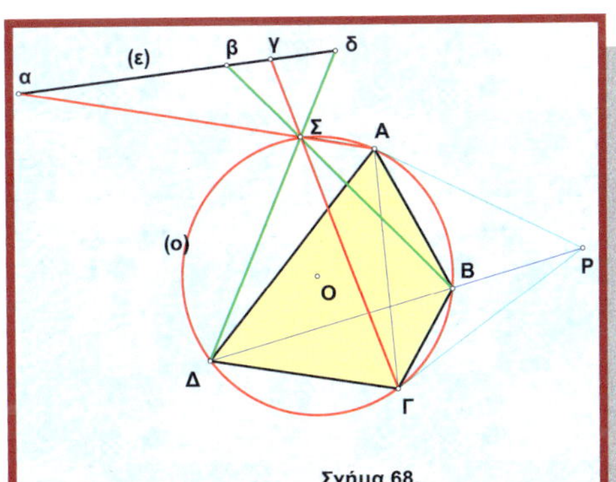

Σχήμα 68.

If we inscribe this pencil in a circle (o), then the beams of this pencil, intersect circle (o), at points A, B, Γ, Δ, which, according to Criterion 35, are apexes to a Harmonic Quadrilateral (§ 16.1a).

These are obviously true when pencil Σ(αβγδ), is Orthooptical and Harmonic, or Regular too (§ 12.1).

17.2. Closed Harmonic Central Pencils of Six Beams.

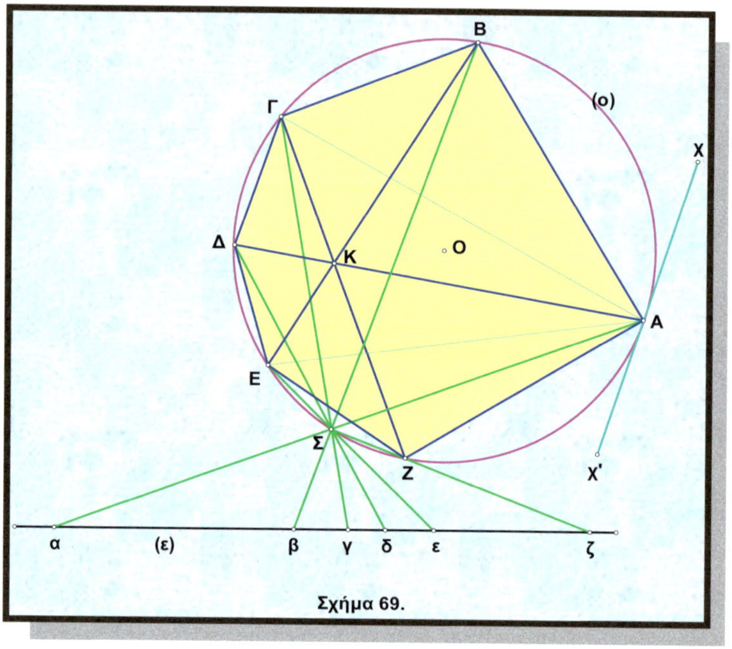

Σχήμα 69.

Definition.

We consider a Pencil Σ(αβγδεζ) (Figure 69). If its six pencils below are harmonic, then this pencil would be called <u>Harmonic</u>:

Σ(αβγε), Σ(βγδζ), Σ(γδεα), Σ(δεζβ), Σ(εζαγ), Σ(ζαβδ). (1).

17.21. Closed Collinear Harmonic Group of Six Points.

It directly occurs that, if we intersect pencil Σ(αβγδεζ), with a line (ε), the <u>Collinear Closed Harmonic Pencil</u> α, β, γ, δ, ε, ζ, occurs (§ 11.1b), as from relation (1) it occurs that its following groups of four points are harmonic:

α, β, γ, ε – β, γ, δ, ζ – γ, δ, ε, α – δ, ε, ζ, β – ε, ζ, α, γ – ζ, α, β, δ. (2).

17.22. Closed Cocyclic Harmonic Group of Six Points or Harmonic Hexagon.

If we Inscribe pencil Σ(αβγδεζ) in circle (ο), then the beams of this pencil, intersect circle (o), at points A, B, Γ, Δ, E, Z, which, according to paragraph 16.1b, are apexes to a Harmonic Hexagon, as from relation (1) it directly occurs that the six pencils below are harmonic:

Σ(ΑΒΓΕ), Σ(ΒΓΔΖ), Σ(ΓΔΕΑ), Σ(ΔΕΖΒ), Σ(ΕΖΑΓ), Σ(ΖΑΒΔ), (3).

Thus from (3) and according to Criterion 35, or paragraph 17.1, it directly occurs that the six quadrilaterals below, are harmonic:

ΑΒΓΕ, ΒΓΔΖ, ΓΔΕΑ, ΔΕΖΒ, ΕΖΑΓ, ΖΑΒΔ. (4).

This, according to the relevant definition (§ 16.1b), means that hexagon ΑΒΓΔΕΖ, is Harmonic.

These are obviously true when pencil Σ(αβγδεζ), is Orthooptical and Harmonic, or Regular, as well (§ 12.4).

17.3. Closed Harmonic Central Pencils of Eight Beams.

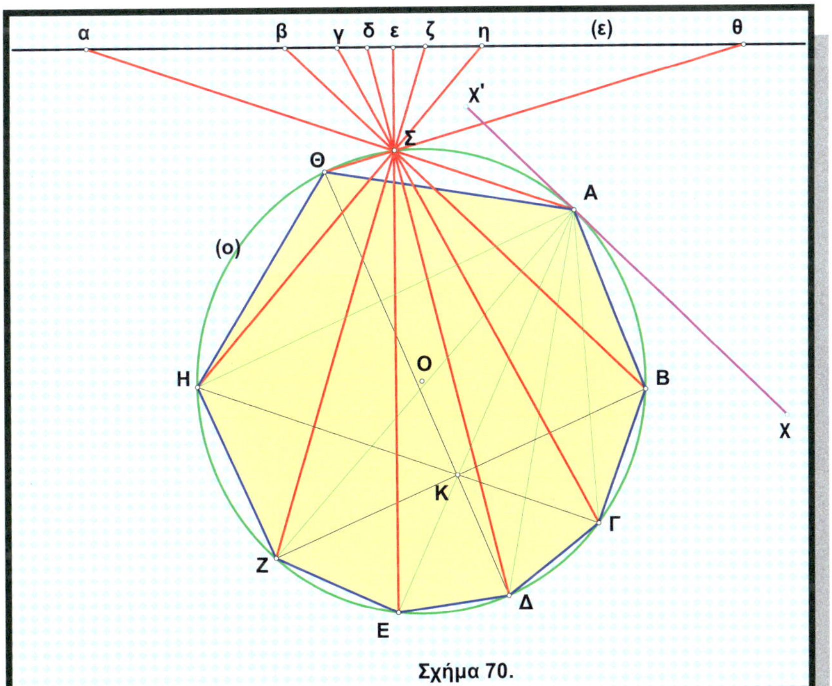

Σχήμα 70.

Definition.

We consider Pencil Σ(αβγδεζηθ) (Figure 70). If for this pencil, its eight pencils below are harmonic, then this pencil would be called Harmonic:

Σ(αβγζ), Σ(βγδη), Σ(γδεθ), Σ(δεζα), Σ(εζηβ), Σ(ζηθγ), Σ(ηθαδ), Σ(θαβε). (5).

17.31. Closed Collinear Harmonic Group of Eight Points.

It directly occurs that, if we intersect pencil Σ(αβγδεζηθ) with line (ε), it occurs that the Collinear Closed Harmonic Group of Eight α, β, γ, δ, ε, ζ, η, θ (§ 11.1c), as it occurs from relation (5) that its groups of four points below would be harmonic:

α, β, γ, ζ – β, γ, δ, η– γ, δ, ε, θ – δ, ε, ζ, α –
ε, ζ, η, β – ζ, η, θ, γ - η, θ, α, δ – θ, α, β, ε. (6).

17.32. Closed Cocyclic Harmonic Group of Eight Points of Harmonic Octagon.

If we inscribe pencil Σ(αβγδεζηθ) in circle (o), then the beams of this pencil, intersect circle (o), at points Α, Β, Γ, Δ, Ε, Ζ, Η, Θ, which, according to paragraph 16.1c, are apexes to a Harmonic Octagon, since from relation (5) it directly occurs that the eight pencils below are harmonic:

Σ(ΑΒΓΖ), Σ(ΒΓΔΗ), Σ(ΓΔΕΘ), Σ(ΔΕΖΑ),
Σ(ΕΖΗΒ), Σ(ΖΗΘΓ), Σ(ΗΘΑΔ), Σ(ΘΑΒΕ), (7).

Hence from (7) and according to Criterion 35, or paragraph 17.1, it directly occurs that the eight quadrilaterals below, are harmonic:

ΑΒΓΖ, ΒΓΔΗ, ΓΔΕΘ, ΔΕΖΑ, ΕΖΗΒ, ΖΗΘΓ, ΗΘΑΔ, ΘΑΒΕ. (8).

This, according to the relevant definition (§ 16.1c), means that Octagon Α-ΒΓΔΕΖΗΘ, is Harmonic.

These are obviously true also when pencil Σ(αβγδεζηθ), is Orthooptical and Harmonic, or Regular (§ 12.6).

17.4. Closed Harmonic Central Pencils of Ten Beams.

Definition.

We consider Pencil Σ(αβγδεζηθικ) (Figure 71). As we have already mentioned before, if the ten pencils of this pencil below are harmonic, then we would call this pencil Harmonic:

Σ(αβγη), Σ(βγδθ), Σ(γδει), Σ(δεζκ), Σ(εζηα),

Σ(ζηθβ), Σ(ηθιγ), Σ(θικδ), Σ(ικαε), Σ(καβζ). (9).

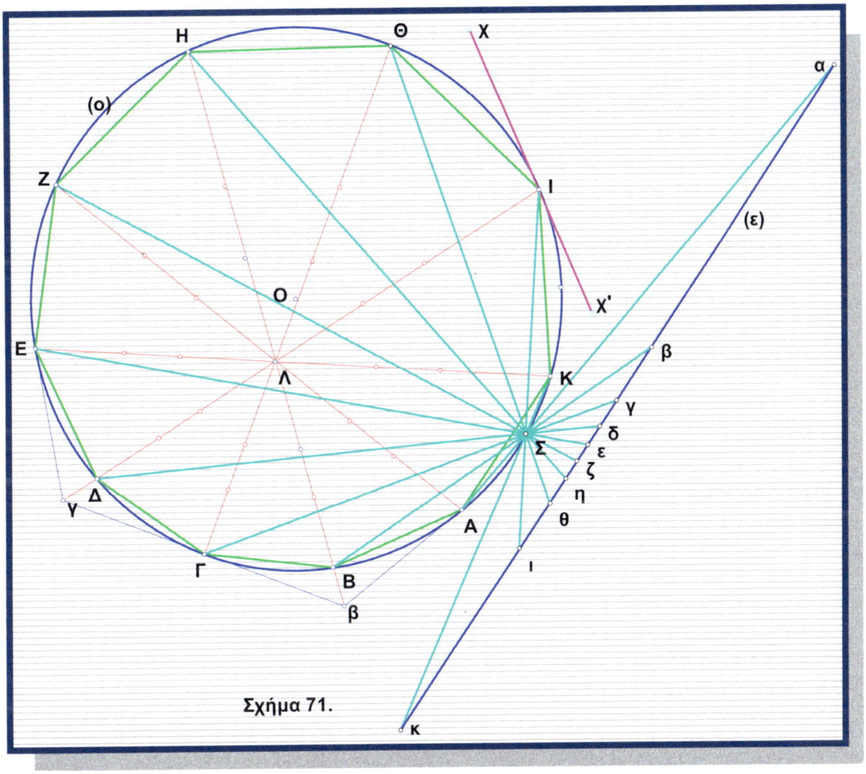

Σχήμα 71.

17.41. Closed Collinear Harmonic Group of Ten Points.

It directly occurs that, if we intersect pencil Σ(αβγδεζηθικ) (Figure 71) with a line (ε), the following Collinear Closed Harmonic Group of Ten α, β, γ, δ, ε, ζ, η, θ, ι, κ occurs (§ 11.1d), since from relation (9) it occurs that its groups of four points below are harmonic:

α, β, γ, η – β, γ, δ, θ– γ, δ, ε, ι – δ, ε, ζ, κ – ε, ζ, η, α –

ζ, η, θ, β – η, θ, ι, γ - θ, ι, κ, δ – ι, κ, α, ε, - κ, α, β, ζ. (10).

17.42. Closed Cocyclic Harmonic Group of Ten Points or Harmonic Decagon.

If we inscribe pencil Σ(αβγδεζηθικ) in circle (o), then its beams, intersect circle (o), at points A, B, Γ, Δ, E, Z, H, Θ, I, K, which, according to paragraph 16.1d, are apexes to a Harmonic Decagon, since from relation (9), it directly occurs that the pencils below are harmonic:

Σ(ΑΒΓΗ), Σ(ΒΓΔΘ), Σ(ΓΔΕΙ), Σ(ΔΕΖΚ), Σ(ΕΖΗΑ),
Σ(ΖΗΘΒ), Σ(ΗΘΙΓ), Σ(ΘΙΚΔ), Σ(ΙΚΑΕ), Σ(ΚΑΒΖ), 11).

hence from (11) and according to Criterion 35, or paragraph 17.1, it directly occurs that the ten quadrilaterals below, are harmonic:

ΑΒΓΗ, ΒΓΔΘ, ΓΔΕΙ, ΔΕΖΚ, ΕΖΗΑ, ΖΗΘΒ, ΗΘΙΓ, ΘΙΚΔ, ΙΚΑΕ, ΚΑΒΖ. (12).

This, according to the relevant definition (§ 16.1d), means that Decagon ΑΒΓΔΕΖΕΖΗΘΙΚ, is actually Harmonic.

These are obviously true as well, when pencil Σ(αβγδεζηθικ), is Orthooptical and Harmonic, or Regular.

If we take into consideration paragraph 12.8, in comparison of course, the above are true for the group of ten α, δ, η. κ, γ, ζ, ι, β, ε, θ, as well and for the Asteroid Decagon ΑΔΗΚΓΖΙΒΕΘΑ.

17.5. Closed Harmonic Central Pencils of Five Beams.

(a). In General.

Here as well, we will define the Harmonic Pencil of Five Beams and the Harmonic Decagon based on the definitions of the Harmonic Pencil of Ten Beams and the Harmonic Decagon, respectively.

(b). Definition.

We consider Pencil Σ(αγεηι) (Figure 71). If between each one of the consecutive beams of this pencil, there are the following five beams Σβ, Σδ, Σζ, Σθ, Σκ, respectively, in such way that pencil Σ(αβγδεζηθικ) would be harmonic, then pencil Σ(αγεηι), would be called Harmonic of Five Beams.

17.51. Closed Collinear Harmonic Group of Five Points.

It directly occurs that, if we intersect pencil Σ(αγεηι) with line (ε), the following Collinear Closed Harmonic Group of Five α, γ, ε, η, ι occurs (§ 11.1e), as obviously pencil Σ(αβγδεζηθικ) is harmonic.

17.52. Closed Cocyclic Harmonic Group of Five Points or Harmonic Pentagon.

If we inscribe pencil Σ(αγεηι) in circle (o) and if the beams of this pencil intersect circle (o) at points A, Γ, E, H, I, respectively, while if the beams of pencil Σ(αβγδεζηθικ), intersect circle (o), at points A, B, Γ, Δ, E, Z, H, Θ, I, K and if these are apexes to a Harmonic Decagon, then this convex pentagon ΑΓΕΗΙ is called Harmonic Pentagon.

17.6. Closed Harmonic Central Pencils of n-Beams and Harmonic n-laterals.

In equivalent ways, that in the above cases 17.1 to 17.52 of Harmonic Pencils and Harmonic Multilaterals, we gave the relevant definitions, we also define the Closed Harmonic Central Pencils of n-Beams and Harmonic n-laterals.

17.7. General Conclusions.

Due to the above (paragraphs 11, 12, 12.13, 16 and 17), the following occur:

(a). Regarding the Closed Harmonic Central Pencils.

1/. The construction of Harmonic Pencils, depends on the ability of construction of the respective Harmonic Collinear groups of n (and theirs, as we have already mentioned, from the respective Regular Multilaterals). This is after all one of the reasons that we named them «Harmonic Pencils».

2/. Apart from the Closed Harmonic Pencils, Open ones exist as well, to which though we will refer later (§ 23).

3/. As it occurs from the aforementioned, from each intersection of a Closed Harmonic Pencil of n beams with an arbitrary line, a Collinear Closed Harmonic Group of n occurs.

4/. If we inscribe these Pencils, in circles, then their beams that intersect these circles, that they are inscribed in, at four, six, eight, ten, five, etc, points which are the apexes to a Harmonic, quadrilateral, hexagon, octagon, decagon, pentagon, etc, respectively (This property is mainly the other reason why we have given them their name as «Harmonic Pencils»).

(b). Regarding the Closed Collinear and Cocyclic Harmonic Groups of n-Points.

1/. It is obvious that the Construction of the Harmonic groups of n (collinear and cocyclic), depends on the ability to construct the respective Regular Pencils, which depend on the ability of construction of the respective regular multilaterals.

2/. Apart from the Closed Harmonic Groups of n (collinear and cocyclic), Open ones exist as well, to which though we will refer later (§ 23).

3/. It is obvious and easily proven that if a Pencil Closed Harmonic of n Beams is on a Closed Collinear Harmonic Group of n Points, whose center is not obligatorily its Orthooptical Point, will have n groups of four of its Beams harmonic, as these will be on n Harmonic Groups of Four Points, respectively. Then this pencil obviously is not always Orthooptical and it is called just «Harmonic Closed Central Pencil of n Beams» (non Regular). These pencils, if we inscribe them in circles, then they intersect them at points that are the apexes to Harmonic Multilaterals, as we will later prove, based on Criterion 35 § 8.72.

(c). Regarding the Inscribed in Circle Harmonic Pencils.

It is obvious that the angles of each Harmonic Pencil of n Beams that are on diagonals (main) of a Harmonic Collinear Group of n, will be on respective diagonals of a Harmonic Cocyclic n-lateral.

(d). Description-Interpretation of Figures 48 and 49).

Based on the above and paragraphs 11, 12, 12.13, 16 and 17, we can now understand the description-interpretation of figures 48 and 49 better, that we mentioned in paragraph 12.13 (d), in which we can see many of the above mentioned and that is why we again advise their thorough study.

Remarks.

(a). We will see and prove the above in Propositions and Problems that will follow, as applications of these.

(b). Later, based on figures 48 and 49 and as we mentioned above, we will refer to the method and the interpretation of the method that a Harmonic Multilateral occurs from a Regular (Multilateral), with equal number of sides, which also consists the solution to a Problem, that troubled us rather often, for at least twenty years (§ 20). We call this New Method, «Method of the Harmonic Transformation » (§ 21).

G'. METHOD OF THE HARMONIC CENTRAL PENCILS.

18. In General.

We referred to the methods that we came up with and apply in Harmonic Geometry in general in Part A (C′ section, paragraphs 4 and 5), while in Part B (C′ section, paragraph 13) we referred to the Method of Regular Central Pencils.

Below we will only develop the method of Closed Harmonic Central Pencils, while to the method of Open Harmonic Central Pencils (§ 24) and the method of Harmonic Transformation (§ 21), we will refer later.

18.1. Method of Closed Harmonic Central Pencils.

It is a general case of Closed Regular Pencils, as Regular Pencils are Harmonic as well, as we mentioned in paragraph 12.13, whose method, tool are the Closed Harmonic Central Pencils (§ 17) and which we apply in Propositions and Problems that concern Collinear and Cocyclic Closed Harmonic Groups of n Series of Points (Collinear Harmonic Multilaterals and mostly in Problems about Cocyclic Harmonic Multilaterals).

Their principles and the way of their application, are similar to the ones of the method of Collinear Symmetrical Pencils (§ 5.3) and this is why we will not mention it here.

To understand this method, many relevant applications will follow, in the form of Propositions and Construstions.

19. APPLICATIONS OF THE METHOD OF HARMONIC PENCILS AND NOT ONLY THAT.

In General.

Consequently we will proceed with applications of the method of Harmonic Pencils, which is mentioned in paragraph 18 and not only that, in the form of Geometrical Propositions and Constructions, with their proofs, as we will need them to depend on them the proofs of other Propositions, etc. Moreover on them and the ones in paragraph 15, the New Method of Harmonic Transformation, will be based, to which we will refer later (§ 21).

Simultaneously to the proofs-solutions that we give with the application of the above, often we will give more, that do not include the application of these, mostly in the purpose of comparison and the proof of the huge value of these.

As the first Criterion of this paragraph, we whould have placed Criterion 35, which though, due to the needs of this book, was preceded and placed in paragraph 8.72.

19.1. First Criterion of Harmonic Beaing of an Inscribed Hexagon.

Extension to Criterion 35 to a Harmonic Hexagon.

Criterion 96 (Figure 69).

Every convex inscribed in circle (o) hexagon ΑΒΓΔΕΖ, is Harmonic, if and only if, pencil Σ(ΑΓΒΔΕΖ) that its center is an arbitrary point Σ of its circumscribed circle (o), is harmonic.

Proof.

(a). Direct.

I.e., if hexagon ΑΒΓΔΕΖ is harmonic, then pencil Σ(ΑΓΒΔΕΖ) is harmonic (Terminology § 16.1b and 17.2).

In fact, since hexagon ΑΓΒΔΕΖ is harmonic, according to the relevant definition of harmonic being (§ 16.1b) of the inscribed hexagon, all its six quadrilaterals below are harmonic:

ΑΒΓΕ, ΒΓΔΖ, ΓΔΕΑ, ΔΕΖΒ, ΕΖΑΓ, ΖΑΒΔ. (1).

Hence, according to the direct of criterion 35 (§ 8.72 or § 17.1), the pencils of four beams below will be harmonic, and the above six quadrilaterals are inscribed as well:

Σ(ΑΒΓΕ), Σ(ΒΓΔΖ), Σ(ΓΔΕΑ), Σ(ΔΕΖΒ), Σ(ΕΖΑΓ), Σ(ΖΑΒΔ). (2).

Thus, if with a line (ε) we intersect pencil Σ(ΑΒΓΔΕΖ) of six beams, at points α, β, γ, δ, ε, ζ, respectively, pencil Σ(αβγδεζ) will occur, for which since (2) are true and according to what we know, the groups of four points below will be harmonic:

α, β, γ, ε – β, γ, δ, ζ – γ, δ, ε, α – δ, ε, ζ, β– ε, ζ, α, γ – ζ, α, β, δ. (3).

Consequently, according to the relevant definition of harmonic being (§ 11.1b), the group of six points α, β, γ, δ, ε, ζ, is harmonic.

This, according to the definition of the harmonic central pencil of six beams (§ 17.21), pencil Σ(αβγδεζ), or pencil Σ(ΑΒΓΔΕΖ), is harmonic.

(b). Converse.

I.e., if pencil Σ(ΑΒΓΔΕΖ), is harmonic, then hexagon ΑΒΓΔΕΖ, is harmonic.

To prove this, we follow the exactly reverse way.

I.e.:

Since pencil Σ(ΑΒΓΔΕΖ), or pencil Σ(αβγδεζ)) is harmonic, if a line (ε) intersects it, according to the relevant definition (§ 17.21), then the collinear group of six points α, β, γ, δ, ε, ζ, will be harmonic as well, thus according to the relevant definition (§ 11.1β) all six groups of collinear points below, will be harmonic:

α, β, γ, ε – β, γ, δ, ζ – γ, δ, ε, α – δ, ε, ζ, β– ε, ζ, α, γ – ζ, α, β, δ. (3).

Thus, due to what we know and because of (3) the pencils below, will be harmonic too:

Σ(ΑΒΓΕ), Σ(ΒΓΔΖ), Σ(ΓΔΕΑ), Σ(ΔΕΖΒ), Σ(ΕΖΑΓ), Σ(ΖΑΒΔ). (2).

Hence, since these six pencils of four beams are harmonic, according to criterion 35 (§ 8.72 or § 17.1), they will intersect circle (o), at six points A, B, Γ, Δ, E, Z, for which, the six groups of four of its cocyclic points below will be harmonic (or that are apexes to six harmonic quadrialterals), as these are also inscribed in circle (o):

A, B, Γ, E – B, Γ, Δ, Z – Γ, Δ, E, A – Δ, E, Z, B – E, Z, A, Γ – Z, A, B, Δ. (1).

This, according to the relevant definition of harmonic being (§ 16.1b), means that hexagon ΑΒΓΔΕΖ will be harmonic.

(c). Investigation.

This criterion is also true in the special case where Σ will coincide with an apex of the hexagon.

In fact, if Σ coincides with for example apex A, then beam ΣA of pencil Σ(ΑΒΓΔΕΖ) will obviously coincide with the tangent χ'Aχ at point A, thus $\angle A\Sigma B = \angle \chi AB$, $\angle B\Sigma\Gamma = \angle BA\Gamma, \ldots$, $\angle \varepsilon\Sigma Z = \angle EAZ$, $\angle Z\Sigma A = \angle ZA\chi'$. This means that, all the angles formed between the consecutive beams of the two pencils Σ(ΑΒΓΔΕΖ) and A(χΒΓΔΕΖ) are equal, one by one. Thus, since the first one of the pencils is harmonic, then the second will be too.

Conslusion.

The Criterion above is significant, as it has many applications, which we will see below and because it allows us to give the alternative definition of the Harmonic Central Pencil of six beams below, apart from the until now known one (§ 17.2):

«A Central Pencil of Six Beams, is called Harmonic, if we inscribe this pencil in a circle and its intersections to it are apexes to a Harmonic Hexagon».

Remarks.

(a). As we will later see, equivalent Criteria, are true for other Harmonic Multilaterals, as the Harmonic Octagon, Decagon, etc.

(b). As we saw in the conclusion above, with this Criterion we extend the definition of the Harmonic Pencils from a line to a circle.

(c). Later we will mention other Criteria and Construction of Harmonic Hexagons.

(d). This Criterion we believe to be new and firstly appear here.

(e). The 2nd, 3rd, 4th, 5th, 6th Criteria of the Harmonic Hexagon, are given in Propositions 109, 112, 139, 140, 141, respectively.

(f). Equivalent Criterion, concering a Regular Hexagon, is Criterion 76.

(g). Relevant Propositions are 35, 76, 97, 98, 109, 112, 139, 140, 141.

19.2. Criterion of Harmonic Being of an Inscribed Octagon.

Extension to Criterion 96 to a Harmonic Octagon.

Criterion 97 (Figure 70).

Every convex inscribed in circle (o) octagon ΑΒΓΔΕΖΗΘ, is Harmonic, if and only if, pencil Σ(ΑΓΒΔΕΖΗΘ) that its center is an arbitrary point Σ of its circumscribed circle (o), is harmonic.

Proof.

(a). Direct.

I.e., if octagon ΑΒΓΔΕΖΗΘ is harmonic, then pencil Σ(ΑΓΒΔΕΖΗΘ) is harmonic (Terminology § 16.1c and 17.3).

In fact, since octagon ΑΓΒΔΕΖΗΘ is harmonic, according to the relevant definition of harmonic being (§ 16.1c) of the inscribed octagon, all its eight quadrilaterals below are harmonic:

ΑΒΓΖ, ΒΓΔΗ, ΓΔΕΘ, ΔΕΖΑ, ΕΖΗΒ, ΖΗΘΓ, ΗΘΑΔ, ΘΑΒΕ. (1).

Hence, according to the direct of criterion 35, or § 17.1, the respective pencils of four beams below will be harmonic too, as the eight quadrilaterals above are inscribed:

Σ(ΑΓΒΖ), Σ(ΒΔΓΗ), Σ(ΓΕΔΘ), Σ(ΔΖΕΑ),
Σ(ΕΗΖΒ), Σ(ΖΘΗΓ), Σ(ΗΑΘΔ), Σ(ΘΒΑΕ). (2).

Thus if we intersect pencil Σ(ΑΒΓΔΕΖΗΘ) of eight beams with line (ε), at points α, β, γ, δ, ε, ζ, η, θ, pencil Σ(αβγδεζηθ) will occur, for which according to what we know and because of relation (2), the groups of four points below will be harmonic:

α, β, γ, ζ – β, γ, δ, η – γ, δ, ε, θ – δ, ε, ζ, α –
ε, ζ, η, β – ζ, η, θ, γ – η, θ, α, δ – θ, α, β, ε. (3).

Therefore, according to the relevant definition in § 12.6, the group of eight points α, β, γ, δ, ε, ζ, η, θ, is harmonic.

This, according to the definition of the harmonic central pencil of eight beams (§ 17.31), pencil Σ(αβγδεζηθ), or pencil Σ(ΑΒΓΔΕΖΗΘ), is harmonic.

(b). Converse.

I.e., if pencil Σ(ΑΒΓΔΕΖΗΘ), is harmonic, then octagon ΑΒΓΔΕΖΗΘ, is harmonic too.

To prove this, we follow the exactly reverse way.

I.e.:

Since pencil Σ(ΑΒΓΔΕΖΗΘ) is harmonic, if a line (ε) intersects it, according to the relevant definition (§ 17.31), then the collinear group of eight points α, β, γ, δ, ε, ζ, η, θ, will be harmonic as well, thus according to the relevant definition in paragraph 12.6 all eight groups of four collinear points below, will be harmonic:

α, β, γ, ζ – β, γ, δ, η – γ, δ, ε, θ – δ, ε, ζ, α –
ε, ζ, η, β – ζ, η, θ, γ – η, θ, α, δ – θ, α, β, ε. (3).

Thus, due to what we know and because of (3) the pencils below, will be harmonic too:

Σ(ΑΓΒΖ), Σ(ΒΔΓΗ), Σ(ΓΕΔΘ), Σ(ΔΖΕΑ),
Σ(ΕΗΖΒ), Σ(ΖΘΗΓ), Σ(ΗΑΘΔ), Σ(ΘΒΑΕ). (2).

Hence, since these eight pencils of four beams are harmonic, according to the converse of criterion 35, or § 17.1 , they will intersect circle (o), at eight points Α, Β, Γ, Δ, Ε, Ζ, Η, Θ, for which, the eight groups of four of its cocyclic points below will be harmonic (or that are apexes to harmonic quadrilaterals), as these are also inscribed in circle (o):

Α, Β, Γ, Ζ – Β, Γ, Δ, Η – Γ, Δ, Ε, Θ – Δ, Ε, Ζ, Α –
Ε, Ζ, Η, Β – Ζ, Η, Θ, Γ – Η, Θ, Α, Δ – Θ, Α, Β, Ε. (1).

This, according to the relevant definition of harmonic being (§ 16.1c), means that octagon ΑΒΓΔΕΖΗΘ will be harmonic.

(c). Investigation.

This criterion is also true in the special case where Σ coincides with an apex of the Octagon.

In fact, if Σ coincides with for example apex A, then beam ΣA of pencil Σ(ΑΒΓΔΕΖΗΘ) will obviously coincide with the tangent χ'Αχ at point A, thus ∠ΑΣΒ = ∠χΑΒ, ∠ΒΣΓ = ∠ΒΑΓ, .., ∠ΗΣΘ = ∠ΗΑΘ, ∠ΘΣα = ∠ΘΗΑ = ∠ΘΑχ'. This means that, all the angles formed between the consecutive beams of the two pencils Σ(ΑΒΓΔΕΖΗΘ) and Α(χΒΓΔΕΖΗΘ) are equal, one by one. Thus, since the first one of the pencils is harmonic, then the second will be too.

Conclusion.

The Criterion above is significant, as it has many applications, which we will see below and because it allows us to give the alternative definition of the Harmonic Central Pencil of eight beams below, apart from the until now known one (§ 17.3):

«A Central Pencil of Eight Beams, is called Harmonic, if we inscribe this pencil in a circle and its intersections to it are apexes to a Harmonic Octagon».

Remarks.

(a). As we will later see, equivalent Criteria, are true for other Harmonic Multilaterals, as the Harmonic Decagon, etc.

(b). As we saw in the conclusion above, with this Criterion we extend the definition of the Harmonic Pencils from a line to a circle.

(c). This Criterion we believe to be new and firstly appear here.

(d). Equivalent Criterion, concering a Regular Octagon, is Criterion 77.

(e). As mentioned and proved in paragraph 16.1c the respective to ΑΒΓΔΕΖΗΘ, asteroid octagon ΑΔΗΒΕΘΓΖΑ, is harmonic as well.

(f). Relevant Propositions are 35, 77, 81, 96, 98, 101.

19.3. Criterion of Harmonic Being of the Inscribed Decagon.

Extension to Criterion 97 to a Harmonic Decagon.

Criterion 98 (Figure 71).

Every convex inscribed in circle (o) decagon ΑΒΓΔΕΖΗΘΙΚ, is Harmonic, if and only if, pencil Σ(ΑΓΒΔΕΖΗΘΙΚ) that its center is an arbitrary point Σ of its circumscribed circle (o), is harmonic.

(Equivalent Proposition is true for an Asteroid Decagon ΑΔΗΚΓΖΗΒΕΘΑ, Proposition 108).

Proof.

1/. Regarding the convex Decagon.

(a). Direct.

I.e., if decagon ΑΒΓΔΕΖΗΘΙΚ is harmonic, then pencil Σ(ΑΓΒΔΕΖΗΘΙΚ) is harmonic (Terminology § 16.1d and 17.4).

In fact, since decagon ΑΓΒΔΕΖΗΘΙΚ is harmonic, according to the relevant definition of harmonic being (§ 16.1d) of the inscribed decagon, all its ten quadrilaterals below are harmonic:

ΑΒΓΗ, ΒΓΔΘ, ΓΔΕΙ, ΔΕΖΚ, ΕΖΗΑ,
ΖΗΘΒ, ΗΘΙΓ, ΘΙΚΔ, ΙΚΑΕ, ΚΑΒΖ. (1).

Hence, according to the direct of criterion 35 (§ 8.72), or § 17.1, the respective ten pencils of four beams in relation (1) below will be harmonic too, as the ten quadrilaterals above are inscribed:

Σ(ΑΒΓΗ), Σ(ΒΓΔΘ), Σ(ΓΔΕΙ),Σ(ΔΕΖΚ), Σ(ΕΖΗΑ),
Σ(ΖΗΘΒ), Σ(ΗΘΙΓ), Σ(ΘΙΚΔ), Σ(ΙΚΑΕ), Σ(ΚΑΒΖ). (2).

Thus if we intersect pencil Σ(ΑΒΓΔΕΖΗΘΙΚ) of ten beams with line (ε), at points α, β, γ, δ, ε, ζ, η, θ, ι, κ, pencil Σ(αβγδεζηθικ) will occur, for which according to what we know and because of relation (2), the ten groups of four points below will be harmonic:

α, β, γ, η – β, γ, δ, θ – γ, δ, ε, ι – δ, ε, ζ, κ– ε, ζ, η, α –
ζ, η, θ, β – η, θ, ι, γ – θ, ι, κ, δ – ι, κ, α, ε – κ, α, β, ζ. (3).

Therefore, according to the relevant definition of harmonic being (§ 11.1d), the group of ten points α, β, γ, δ, ε, ζ, η, θ, ι, κ, will be harmonic.

This, according to the definition of the harmonic central pencil of ten beams (§ 17.41), pencil Σ(αβγδεζηθικ), or pencil Σ(ΑΒΓΔΕΖΗΘΙΚ), is harmonic.

(b). Converse.

I.e., if pencil Σ(ΑΒΓΔΕΖΗΘΙΚ), is harmonic, then decagon ΑΒΓΔΕΖΗΘΙΚ, is harmonic too.

To prove this, we follow the exactly reverse way.

I.e.:

Since pencil Σ(ΑΒΓΔΕΖΗΘΙΚ), or pencil Σ(αβγδεζηθικ) is harmonic, if a line (ε) intersects it, according to the relevant definition (§ 17.41), then the collinear group of ten points α, β, γ, δ, ε, ζ, η, θ, ι, κ, will be harmonic as well, thus according to the relevant definition (§ 11.1d) all ten groups of four collinear points below, will be harmonic:

α, β, γ, η – β, γ, δ, θ – γ, δ, ε, ι – δ, ε, ζ, κ– ε, ζ, η, α –
ζ, η, θ, β – η, θ, ι, γ – θ, ι, κ, δ – ι, κ ,α, ε – κ, α, β, ζ. (3).

Thus, due to what we know and because of (3) the ten pencils below, will be harmonic too:

Σ(ΑΒΓΗ), Σ(ΒΓΔΘ), Σ(ΓΔΕΙ), Σ(ΔΕΖΚ), Σ(ΕΖΗΑ)
Σ(ΖΗΘΒ), Σ(ΗΘΙΓ), Σ(ΘΙΚΔ), Σ(ΙΚΑΕ), Σ(ΚΑΒΖ). (2).

Hence, since these ten pencils of four beams are harmonic, according to the converse of criterion 35 (§ 8.72), or § 17.1 , their beams will intersect circle (o), at ten points Α, Β, Γ, Δ, Ε, Ζ, Η, Θ, Ι, Κ, for which, the ten groups of four of its cocyclic points below will be harmonic (or that are apexes to harmonic quadrialterals), as these are also inscribed in circle (o):

Α, Β, Γ, Η – Β, Γ, Δ, Θ – Γ, Δ, Ε, Ι – Δ, Ε, Ζ, Κ – Ε, Ζ, Η, Α,
Ζ, Η, Θ, Β – Η, Θ, Ι, Γ – Θ, Ι, Κ, Δ – Ι, Κ, Α, Ε – Κ, Α, Β, Ζ. (1).

This, according to the relevant definition of harmonic being (§ 16.1d), means that decagon ΑΒΓΔΕΖΗΘΙΚ, would be harmonic.

(c). Investigation.

This criterion is also true in the special case where Σ coincides with an apex of the decagon.

In fact, if Σ coincides with for example apex I, then beam ΣI of pencil Σ(ΑΒΓΔΕΖΗΘΙΚ) will obviously coincide with the tangent χ'Ιχ at point I, thus:

$$\angle A\Sigma B = \angle AIB,\ \angle B\Sigma\Gamma = \angle BI\Gamma,\ \angle \Gamma\Sigma\Delta = \angle \Gamma I\Delta, \ldots,$$

$$\angle \Theta\Sigma I = \angle \Theta I\chi,\ \angle I\Sigma K = \angle \chi' IK,\ \angle K\Sigma A = \angle KIA, \qquad (4).$$

as quadrilateral ΑΣΚΙ is inscribed.

This means that, all the angles formed between the consecutive beams of the two pencils Σ(ΑΒΓΔΕΖΗΘΙΚ) and Σ(ΑΒΓΔΕΖΗΘxx'K) are equal, one by one. Thus, since the first one of the pencils is harmonic, then the second will be too.

2/. Regarding the Asteroid Decagon.

If the Asteroid Decagon ΑΔΗΚΓΖΙΒΕΘΑ is Harmonic (Proposition 108), we prove as above that pencil Σ(ΑΔΗΚΓΖΙΒΕΘΑ), whose center is an arbitrary point Σ of circle (o) is Harmonic and conversely (the fact that the group of ten α, δ, η, κ, γ, ζ, ι, β, ε, θ, α, is Harmonic, is proved in paragraph 12.8).

Conclusion.

The Criterion above is significant, as it has many applications, which we will see below and because it allows us to give the alternative definition of the Harmonic Central Pencil of ten beams below, apart from the until now known one (§ 17.4):

«A Central Pencil of Ten Beams, is called Harmonic, if we inscribe this pencil in a circle and its intersections to it are apexes to a Harmonic Decagon».

Remarks.

(a). As we will later see, equivalent Criteria, are true for other Harmonic Multilaterals, as the Harmonic Dodecagon, etc.

(b). As we saw in the conclusion above, with this Criterion we extend the definition of the Harmonic Pencils from a line to a circle.

(c). This Criterion we believe to be new and firstly appear here.

(d). Equivalent Criterion, concering a Regular Octagon, is Criterion 78.

(e). Relevant Propositions are 35, 78, 96, 97.

19.4. General Conclusion.

After Criteria 35, 96, 97, 98 above and if we work similarly to their proofs, we reach the general conclusion below:

Equivalent Criterion, to Criteria 35, 96, 97, 98, is true for any other Harmonic n-lateral as well.

19.5. Construction of a Harmonic Quadrilateral from a Harmonic Group of Four and Conversely.

New Easy Way of Construction of a Harmonic Quadrilateral.

Construction 99 (Figure 68).

Construct a Harmonic Quadrilateral, inscribed in a given circle (o), based on a Collinear Harmonic Group of Four Points, which is given to us and conversely, on a given line (ε), construct a Harmonic Group of Four Points, based on a Harmonic Quadrilateral, given to us.

Solution.

(a). Direct.

1/. Analysis.

I.e. we are given on line (ε), the Group of Four Harmonic Points α, β, γ , δ and we are asked, based on this group of four, to construct a Harmonic Quadrilateral ΑΒΓΔ inscribed in given circle (o).

If we take into consideration the converse of Criterion 35 (§ 8.72), we are led to the construction below.

2/. Construction.

On circle (o) we define an arbitrary point Σ and form a pencil Σ(αβγδ), the beams of which, we suppose that they re-intersect circle (o) at points A, B, Γ, Δ, respectively.

We say that quadrilateral ΑΒΓΔ is the desired Harmonic one.

3/. Proof.

Since the Group of Four Points α, β, γ, δ, is Harmonic, as we know, the Central Pencil Σ(αγβδ) or Σ(ΑΓΒΔ) will also be Harmonic.

Hence, according to the converse of Criterion 35 the cocyclic group of points A, B, Γ, Δ, will also be Harmonic (Harmonic Quadrilateral ΑΒΓΔ), thus this quadrilateral is the desired one, as it is also inscribed in circle (o).

(b). Converse.

1/. Analysis.

I.e. we are given on circle (o), the Harmonic Group of Four of its Points A, B, Γ, Δ, and we are asked to construct on a given line (ε) a Harmonic Group of Four Points α, β, γ, δ.

If we take into consideration the direct of Criterion 35, we are led to the Construction below.

2/. Construct.

On circle (o), we set an arbitrary point Σ and we form pencil Σ(ΑΒΓΔ), the beams of which, we suppose they intersect line (ε) at points α, β, γ, δ, respectively.

We say that the desired Harmonic Group of Four Points is the following α, β, γ, δ.

3/. Proof.

Since Quadrilateral ΑΒΓΔ is Harmonic then Pencil Σ(ΑΒΓΔ) or Σ(αβγδ) will be harmonic too (direct of Criterion 35).

Hence since, according to what we know, the Central Pencil Σ(αβγδ) is harmonic, then the Group of Four Points α, β, γ, δ will also be Harmonic. Thus it is the desired one, as it is located on line (ε) as well.

Remarks.

(a). The Problem above and its solution, we believe to be new and firstly appear here.

(b). Problem 99, the way it is given, it is undefined. If we want to define it, then we need to be provided with sufficient data. For example, it becomes specific if we are given the three points on line (ε), or the three points on circle (o), or the given in Construction 134.

(c). In an equivalent way we can construct harmonic groups of six, eight, ten, etc, or in a circle we can construct harmonic hexagons, octagons, decagons, etc, if only their respective Regular Multilaterals are constructible.

(d). Other ways of Construction of a Harmonic Quadrilateral, we will give later (Construction 134).

(e). In the Construction above, we used a special way. This happened because it will help in the better understanding of our new Method of the Harmonic Transformation.

(f). Relevant Propositions are 35, 100-103, 134.

19.6. Construction of a Harmonic Hexagon from a Harmonic Group of Six and Conversely.

New Easy Way of Construction of a Harmonic Hexagon.

Extension to Construction 99 to a Hexagon.

Construction 100 (Figure 69).

Construct a Harmonic Hexagon, inscribed in given circle (o), based on a Collinear Harmonic Group of Six Points, which is given to us and conversely, on given line (ε), construct a Harmonic Group of Six Points, based on a Harmonic Hexagon, given to us.

Solution.

(a). Direct.

1/. Analysis.

I.e. we are given on line (ε), the Group of Six Harmonic Points α, β, γ , δ, ε, ζ and we are asked, based on this group of six, to construct a Harmonic Hexagon ΑΒΓΔΕΖ inscribed in given circle (o).

If we take into consideration the converse of Criterion 96 (§ 19.1) and the definition of a Harmonic Pencil of Six Beams (§ 17.21), we are led to the construction below.

2/. Construction.

On circle (o) we define an arbitrary point Σ and form a pencil Σ(αβγδεζ), the beams of which, we suppose that they re-intersect circle (o) at points A, B, Γ, Δ, E, Z, respectively.

We say that hexagon ΑΒΓΔΕΖ is the desired Harmonic one.

3/. Proof.

Since the Group of Six Points α, β, γ, δ, ε, ζ, is Harmonic, as we know, the Central Pencil Σ(αβγδεζ) or Σ(ΑΒΓΔΕΖ) will also be Harmonic, (definition § 17.21).

Hence, according to the converse of Criterion 96 the cocyclic group of six points Α, Β, Γ, Δ, Ε, Ζ, will also be Harmonic (Harmonic Hexagon ΑΒΓΔΕΖ), thus this hexagon is the desired one, as it is also inscribed in circle (o).

(b). Converse.

1/. Analysis.

I.e. we are given on circle (o), the Harmonic Group of Six of its Points Α, Β, Γ, Δ, Ε, Ζ or the Harmonic Hexagon ΑΒΓΔΕΖ and we are asked to construct on a given line (ε) a Harmonic Group of Six Points α, β, γ, δ, ε, ζ.

If we take into consideration the direct of Criterion 96 (§ 19.1) and the definition of a Harmonic Pencil of Six Beams (§ 17.21), we are led to the Construction below.

2/. Construction.

On circle (o), we set an arbitrary point Σ and we form pencil Σ(ΑΓΒΔΕΖ), the beams of which, we suppose they intersect line (ε) at points α, β, γ, δ, ε, ζ, respectively.

We say that the desired Harmonic Group of Six Points is the following α, β, γ, δ, ε, ζ.

3/. Proof.

Since Hexagon ΑΒΓΔΕΖ is Harmonic then Pencil Σ(ΑΓΒΔΕΖ) or Σ(αβγδεζ) will be harmonic too (direct of Criterion 96).

Hence since, according to the relevant definition (§ 17.21), the Central Pencil Σ(αβγδεζ) is harmonic, then the Group of Six Points α, β, γ, δ, ε, ζ will also be Harmonic. Thus it is the desired one, as it is located on line (ε) as well.

Remarks.

(a). The Problem above and its solution, we believe to be new and firstly appear here.

(b). The Problem, the way it is given, it is undefined. If we want to define it, then we need to be provided with sufficient data.

(c). In an equivalent way we can construct on a line harmonic groups of four, eight, ten, etc, or in a circle we can construct harmonic quadrilaterals, octagons, decagons, etc, if only their respective Regular Multilaterals are constructible.

(d). To other ways of Construction of a Harmonic Hexagon and a Harmonic Group of Six we will refer later.

(e). In the Construction above, we used a special way. This happened because it will help us better understand our new Method of Harmonic Transformation.

(f). Relevant Propositions are 35, 80, 96, 99, 101, 102, 103, 153.

19.7. Construction of a Harmonic Octagon from a Harmonic Group of Eight and Conversely.

New Easy Way of Construction of a Harmonic Octagon.

Extension to Construction 100 to an Octagon.

Construction 101 (Figure 70).

Construct a Harmonic Octagon, inscribed in given circle (o), based on a Collinear Harmonic Group of Eight Points, which is given to us and conversely, on given line (ε), construct a Harmonic Group of Eight Points, based on a Harmonic Octagon, given to us.

Solution,

(a). Direct.

1/. Analysis.

I.e. we are given on line (ε), the Group of Eight Harmonic Points α, β, γ, δ, ε, ζ, η, θ and we are asked, based on this group of eight, to construct a Harmonic Octagon ΑΒΓΔΕΖΗΘ inscribed in given circle (o).

If we take into consideration the converse of Criterion 97 (§ 19.2) and the definition of a Harmonic Pencil of Eight Beams (§ 17.31), we are led to the construction below.

2/. Construction.

On circle (o) we define an arbitrary point Σ and form a pencil Σ(αβγδεζηθ), the beams of which, we suppose that they re-intersect circle (o) at points A, B, Γ, Δ, E, Z, H, Θ, respectively.

We say that octagon ΑΒΓΔΕΖΗΘ is the desired Harmonic one.

3/. Proof.

Since the Group of Eight Points α, β, γ, δ, ε, ζ, η, θ, is Harmonic, as we know, the Central Pencil Σ(αβγδεζηθ), or Σ(ΑΒΓΔΕΖΗΘ) will also be Harmonic, (definition § 17.31).

Hence, according to the converse of Criterion 97 the cocyclic group of eight points A, B, Γ, Δ, E, Z H, Θ, will also be Harmonic, or Harmonic Octagon ΑΒΓΔΕΖΗΘ, thus this octagon is the desired one, as it is also inscribed in circle (o).

(b). Converse.

1/. Analysis.

I.e. we are given on circle (o), the Harmonic Group of Eight of its Points A, B, Γ, Δ, E, Z H, Θ or the Harmonic Octagon ΑΒΓΔΕΖΗΘ and we are asked to construct on a given line (ε) a Harmonic Group of Eight Points α, β, γ, δ, ε, ζ, η, θ.

If we take into consideration the direct of Criterion 97 (§ 19.2) and the definition of a Harmonic Pencil of Eight Beams (§ 17.31), we are led to the Construction below.

2/. Construction.

On circle (o), we set an arbitrary point Σ and we form pencil Σ(ΑΓΒΔΕΖΗΘ), the beams of which, we suppose they intersect line (ε) at points α, β, γ, δ, ε, ζ, η, θ, respectively.

We say that the desired Harmonic Group of Eight Points is the following α, β, γ, δ, ε, ζ, η, θ.

3/. Proof.

Since Octagon ΑΒΓΔΕΖΗΘ is Harmonic then Pencil Σ(ΑΓΒΔΕΖΗΘ) or Σ(αβγδεζηθ) will be harmonic too (direct of Criterion 97).

Hence since, according to the relevant definition (§ 17.31), the Central Pencil Σ(αβγδεζηθ) is harmonic, then the Group of Eight Points α, β, γ, δ, ε, ζ, η, θ, will also be Harmonic. Thus it is the desired one, as it is located on line (ε) as well.

Remarks.

(a). The Problem above and its solution, we believe to be new and firstly appear here.

(b). The Problem, the way it is given, it is undefined. If we want to define it, then we need to be provided with sufficient data.

(c). In an equivalent way we can construct on a line harmonic groups of four, eight, ten, etc, or in a circle we can construct harmonic quadrilaterals, octagons, decagons, etc, if only their respective Regular Multilaterals are constructible.

(d). To other ways of Construction of a Harmonic Octagon and a Harmonic Group of Eight we will refer later.

(e). In the Construction above, we used a special way. This happened because it will help us better understand our new Method of Harmonic Transformation.

(f). Relevant Propositions are 35, 81, 97, 99, 100, 102, 103, 149, 154.

19.8. Construction of a Harmonic Decagon from a Harmonic Group of Ten and Conversely.

New Easy Way of Construction of a Harmonic Decagon.

Extension to Construction 101 to a Decagon.

Construction 102 (Figure 71).

Construct a Harmonic Decagon, inscribed in given circle (o), based on a Collinear Harmonic Group of Ten Points, which is given to us and conversely, on given line (ε), construct a Harmonic Group of Ten Points, based on a Harmonic Decagon, given to us.

Solution.

(a). Direct.

1/. Analysis.

I.e. we are given on line (ε), the Group of Ten Harmonic Points α, β, γ , δ, ε, ζ, η, θ, ι, κ and we are asked, based on this group of ten, to construct a Harmonic Decagon ΑΒΓΔΕΖΗΘΙΚ, inscribed in given circle (o).

If we take into consideration the converse of Criterion 98 (§ 19.3) and the definition of a Harmonic Pencil of Ten Beams (§ 17.41), we are led to the construction below.

2/. Construction.

On circle (o) we define an arbitrary point Σ and form a pencil Σ(αβγδεζηθικ), the beams of which, we suppose that they re-intersect circle (o) at points A, B, Γ, Δ, E, Z, H, Θ, I, K, respectively.

We say that octagon ΑΒΓΔΕΖΗΘΙΚ is the desired Harmonic one.

3/. Proof.

Since the Group of Ten Points α, β, γ, δ, ε, ζ, η, θ, ι, κ, is Harmonic, as we know, the Central Pencil Σ(αβγδεζηθικ), or Σ(ΑΒΓΔΕΖΗΘΙΚ) will also be Harmonic, (definition § 17.41).

Hence, according to the converse of Criterion 98 the cocyclic group of ten points A, B, Γ, Δ, E, Z H, Θ, I, K, will also be Harmonic, or Harmonic Decagon ΑΒΓΔΕΖΗΘΙΚ, thus this decagon is the desired one, as it is also inscribed in circle (o).

(b). Converse.

1/. Analysis.

I.e. we are given on circle (o), the Harmonic Group of Ten of its Points A, B, Γ, Δ, E, Z, H, Θ, I, K or the Harmonic Decagon ΑΒΓΔΕΖΗΘΙΚ and we are asked to construct on a given line (ε) a Harmonic Group of Ten Points α, β, γ, δ, ε, ζ, η, θ, ι, κ.

If we take into consideration the direct of Criterion 98 (§ 19.3) and the definition of a Harmonic Pencil of Ten Beams (§ 17.41), we are led to the Construction below.

2/. Construction.

On circle (o), we set an arbitrary point Σ and we form pencil Σ(ΑΓΒΔΕΖΗΘΙΚ), the beams of which, we suppose they intersect line (ε) at points α, β, γ, δ, ε, ζ, η, θ, ι, κ, respectively.

We say that the desired Harmonic Group of Ten Points is the following α, β, γ, δ, ε, ζ, η, θ, ι, κ.

3/. Proof.

Since Decagon ΑΒΓΔΕΖΗΘΙΚ is Harmonic then Pencil Σ(ΑΓΒΔΕΖΗΘΙΚ) or

Σ(αβγδεζηθικ) will be harmonic too (direct of Criterion 98).

Hence since, according to the relevant definition (§ 17.41), the Central Pencil Σ(αβγδεζηθικ is harmonic, then the Group of Ten Points α, β, γ, δ, ε, ζ, η, θ, ι, κ, will also be Harmonic. Thus it is the desired one, as it is located on line (ε) as well.

Remarks.

(a). The Problem above and its solution, we believe to be new and firstly appear here.

(b). The Problem, the way it is given, it is undefined. If we want to define it, then we need to be provided with sufficient data.

(c). In an equivalent way we can construct on a line harmonic groups of four, six, eight, etc, or in a circle we can construct harmonic quadrilaterals, hexagons, octagons, etc, if only their respective Regular Multilaterals are constructible.

(d). To other ways of Construction of a Harmonic Decagon and a Harmonic Group of Ten we will refer later.

(e).The Construction of an Asteroid Harmonic Decagon can rather easily be made, based on a convex Harmonic Decagon.

(f). In the Construction above, we used a special way. This happened because it will help us better understand our new Method of Harmonic Transformation.

(g). Another way of construction is mentioned in Construction 82.

(h). Relevant Propositions are 35, 82, 98, 99, 100, 101, 103, 155.

19.9. Construction of a Harmonic Pentagon from a Harmonic Group of Five Points and Conversely.

New Easy Way of Construction of a Harmonic Pentagon.

Extension to Construction 102 (§ 19.8) to a Harmonic Pentagon.

Construction 103 (Figure 71).

Construct a Harmonic Pentagon, inscribed in given circle (o), based on a Collinear Harmonic Group of Five Points, which is given to us and conversely, on given line (ε), construct a Harmonic Group of Five Points, based on a Harmonic Pentagon, given to us.

Solution.

(a). Direct.

1/. Analysis.

I.e. we are given on line (ε), the Group of Five Harmonic Points α, γ, ε, η, ι, and we are asked, based on this group of five, to construct a Harmonic Pentagon ΑΓΕΗΙ, inscribed in given circle (o).

If we take into consideration the definitions in paragraphs § 11.1e(b), § 17.41, § 16.1e(b) the converse of Criterion 98 (§19.3), then we are led to the Construction below.

2/. Construction.

On circle (o) we define an arbitrary point Σ and form a pencil Σ(αγεηι), the beams of which, we suppose that they re-intersect circle (o) at points Α, Γ, Ε, Η, Ι, respectively.

We say that pentagon ΑΓΕΗΙ is the desired Harmonic one.

3/. Proof.

Since we are given that the Group of Five Points α, γ, ε, η, ι, is Harmonic, according to the relevant definition [§ 11.1e(b)], it means that there will be let us say the group of five points β, δ, ζ, θ, κ, for which the group of ten points α, β, γ, δ, ε, ζ, η, θ, ι, κ, will be Harmonic, thus according to the relevant definition (§ 17.41), pencil Σ(αβγδεζηθικ) or Σ(ΑΒΓΔΕΖΗΘΙΚ), will be Harmonic and therefore if the beams of this pencil re-intersect circle (o) at points Α, Β, Γ, Δ, Ε, Ζ, Η, Θ, Ι, Κ, respectively, then decagon ΑΒΓΔΕΖΗΘΙΚ, according to the converse of Criterion 98 (§ 19.3), will be Harmonic.

So we see that in fact for the group of five points Α, Γ, Ε, Η, Ι, there is the group of five points Β, Δ, Ζ, Θ, Κ, for which decagon ΑΒΓΔΕΖΗΘΙΚ is Harmonic. This means that the group of five points Α, Γ, Ε, Η, Ι, is Harmonic, or Pentagon ΑΓΕΗΙ is in fact Harmonic [Relevant Definition § 16.1e(b)].

(b). Converse.

1/. Analysis.

I.e. we are given the inscribed in circle (o), Harmonic Pentagon ΑΓΕΗΙ and we are asked, based on that, to construct on given line (ε), a Harmonic Group of Five Points α, γ, ε, η, ι.

If we take into consideration the definitions in paragraphs § 16.1e(b),

§17.42, § 11.1 e(b) and the direct of Criterion 98 (§ 19.3), then we are led to the Construction below.

(b). Construction.

On the circumscribed circle (o) of the given Harmonic Pentagon ΑΓΕΗΙ, we set an arbitrary point Σ.

We draw ΣΑ, ΣΓ, ΣΕ, ΣΗ, ΣΙ, which let us say intersect the given line (ε), at points α, γ, ε, η, ι, respectively.

We say that the group of five point α, γ, ε, η, ι, is the desired Harmonic one.

(c). Proof.

Since we are given that pentagon ΑΓΕΗΙ is Harmonic or the group of five points Α, Γ, Ε, Η, Ι, is Harmonic, according to the relevant definition [§ 16.1e(b)], it means that on circle (o) there will exist a group of five points, let it be the following Β, Δ, Ζ, Θ, Κ, for which the group of ten points Α, Β, Γ, Δ, Ε, Ζ, Η, Θ, Ι, Κ, will be Harmonic or decagon ΑΒΓΔΕΖΗΘΙΚ will be Harmonic, thus according to the relevant definition (§ 17.42) pencil Σ(ΑΒΓΔΕΖΗΘΙΚ) will also be Harmonic and therefore, if the beams of this pencil intersect line (ε) at α, β, γ, δ, ε, ζ, η, θ, ι, κ, then this group of ten, according to the direct of Criterion 98 (§ 19.3), will be Harmonic.

We thus see that in fact for the group of five points α, γ, ε, η, ι, the group of five points β, δ, ζ, θ, κ, exists for which the group of ten points α, β, γ, ..., κ, is Harmonic. This means that the group of five points α, γ, ε, η, ι, is actually Harmonic, [Relevant definition § 11.1e(b)].

Comment.

The Construction above, obviously, the way it was given, it is undefined. If we want to make it more specific, we need to be given sufficient data.

Remarks.

(a). In the same way, as above, we construct and prove that the group of five points β, δ, ζ, θ, κ, is Harmonic as well.

(b). Construction 103 we believe to be new and firstly appear here.

(c). We notice that using the above new special Method, the construction of the Harmonic Group of Five Points, easily occurs, from a Harmonic Pentagon, as it does for a Harmonic Group of Four, Six, Eight, Ten, etc.

(d). In an equivalent way we can construct on a line harmonic groups of four, six, eight, etc, or in a circle we can construct harmonic quadrilaterals, hexagons, octagons, etc, if only their respective Regular Multilaterals are constructible.

(e). To other ways of Construction of a Harmonic Pentagon and a Harmonic Group of Five we will refer later.

(f). In the Construction above, we used a special way. This happened because it will help us better understand our new Method of Harmonic Transformation.

(g). Relevant Propositions are 35, 83, 99, 100, 101, 102, 156.

19.10. General Conclusion.

Due to Criteria 99, 100, 101, 102, 103, above and if we work similarly to their proofs, we reach the following general conclusion:

Equivalent Construction, to Constructions 99 to 103, can be carried through for any other Harmonic n-lateral, as long as its respective Regular n-lateral is constructible.

19.11. Significant Property of the Harmonic Hexagon.

Convergence of Diagonals of Harmonic Hexagons.

Proposition 104 (Figure 72).

Diagonals (main), of Harmonic Hexagons, coincide.

Proof.

1st Way (Based on Auxiliary Proposition 7 in Part E).

We consider a Harmonic Hexagon ΑΒΓΔΕΖ. We will prove that:

$$A\Delta \cap BE \cap \Gamma Z \equiv K. \qquad (1).$$

Since the given hexagon is harmonic (implication), according to the definition (§ 16.1b above), its quadrilaterals for example ΑΒΓΕ and ΔΕΖΒ will be harmonic, thus according to known Theorem [6] [§ 8-4.10 (3)], it will be:

$$\frac{AB}{B\Gamma}\cdot\frac{\Gamma E}{EA}=1 \text{ and } \frac{\Delta E}{EZ}\cdot\frac{ZB}{B\Delta}=1 \Rightarrow \frac{AB}{B\Gamma}\cdot\frac{\Gamma E}{EA}=\frac{\Delta E}{EZ}\cdot\frac{ZB}{B\Delta}. \quad (2).$$

The latter relation, according to Auxiliary Proposition 7 (converse) (§ 2.7) in Part E, means that relation (1) is in fact true.

2nd Way (Based on Auxiliary Proposition 51).

Since the given hexagon is harmonic (implication), according to the definition (§ 16.1b above), its quadrilaterals for example ΑΒΓΕ and ΔΕΖΒ will be harmonic, thus according to the Auxiliary Proposition (Lemma) 51 (§ 9.12) (converse), relation (1) will be true, (but it will also be:

$A Z \cap BE \cap \Gamma\Delta \equiv A_2$, etc).

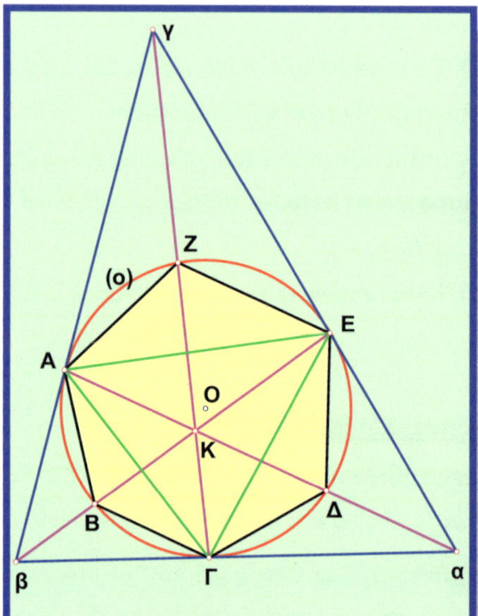

Σχήμα 72.

3rd Way (Based on Poles-Polars and the Gergonne Theorem).

In figure 72, we draw the tangential triangle αβγ (polar converse) of triangle ΑΓΕ. (See paragraph 7 in terminology).

Since the given hexagon ΑΒΓΔΕΖ, is Harmonic, its quadrilaterals ΑΒΓΕ, ΓΔΕΑ, ΕΖΑΓ will be Harmonic (definition).

Hence, since quadrilateral ΑΒΓΕ is Harmonic, BE will run through pole β of ΑΓ, or ΕΒβ is straight.

Additionally, since quadrilateral ΓΔΕΑ is Harmonic, ΑΔ will run through pole α of ΓΕ or ΑΔα is straight.

Moreover, since quadrilateral ΕΖΑΓ is Harmonic too, ΓΖ will run through pole γ of ΕΑ or ΓΖγ is straight.

Therefore, according to Gergonne Theorem, it will be $A\alpha \cap E\beta \cap \Gamma\gamma \equiv K$ or $A\Delta \cap BE \cap \Gamma Z \equiv K$.

(Another way of proof of Proposition 104, is given in the direct of Criterion 109).

Remarks.

(a). Proposition 104 has been recorder in our book [31] [§ 2α(46), where we give five more ways for its proof.

(b). We have also published this Proposition in Magazine ΔΙΑΣΤΑΣΗ of the HMS (issue 3-4 /1998, page 33).

(c). It is obvious that we would reach the same result, if we consider the following pairs of quadrilaterals: ΓΔΕΑ-ΖΑΒΔ or ΕΖΑΓ-ΒΓΔΖ, whose quadrilaterals are harmonic too (definition, § 16.1b above).

(d). It is obvious that for Proposition 104, the converse is not true.

(e). Relevant Propositions are 51, 105, 106, 107.

19.12. Significant Property of the Harmonic Octagon.

Convergence of the Diagonals of Harmonic Octagons. (Extension to Proposition 104).

Proposition 105 (Figure 65).

Diagonals (main), of convex Harmonic Octagons and their respective Harmonic Asteroid Octagons, coincide.

Proof.

(a). Regarding the convex Harmonic Octagon.

1st Way (Based on Auxiliary Proposition 7 in Part E).

We consider a Harmonic Octagon ΑΒΓΔΕΖΗΘ. We will prove that:

$$\text{ΑΕ}\cap\text{ΒΖ}\cap\text{ΓΗ}\cap\text{ΔΘ}\equiv\text{Κ}. \qquad (1).$$

Since the given octagon is harmonic (implication), according to the definition (§ 16.1c above), its quadrilaterals for example ΑΒΓΖ and ΕΖΗΒ will be harmonic, thus according to known Theorem [6] [§ 8-4.10 (3)], it will be:

$$\frac{\text{ΑΒ}}{\text{ΒΓ}}\cdot\frac{\text{ΓΖ}}{\text{ΖΑ}}=1 \text{ and } \frac{\text{ΕΖ}}{\text{ΖΗ}}\cdot\frac{\text{ΗΒ}}{\text{ΒΕ}}=1 \Rightarrow \frac{\text{ΑΒ}}{\text{ΒΓ}}\cdot\frac{\text{ΓΖ}}{\text{ΖΑ}}=\frac{\text{ΕΖ}}{\text{ΖΗ}}\cdot\frac{\text{ΗΒ}}{\text{ΒΕ}}. \qquad (2).$$

The latter relation, according to Auxiliary Proposition 7 (converse) (§ 2.7) in Part E, means that the following relation is true:

$$\text{ΑΕ}\cap\text{ΒΖ}\cap\text{ΓΗ}\equiv\text{Κ}. \qquad (3).$$

Moreover, since from the definition this Octagon, will have its quadrilaterals ΒΓΔΗ and ΖΗΘΓ harmonic too, if we work in the exact same way as bove

we easily find that: ΒΖ∩ΓΗ∩ΔΘ≡Κ. (4).
Hence, from (3), (4), relation (1) directly occurs.

2nd Way (Based on Auxiliary Proposition 51).

Since the given octagon is harmonic (impllication), according to the definition (§ 16.1c above), its quadrilaterals ΑΒΓΖ and ΕΖΗΒ will be harmonic, thus according to Auxiliary Proposition (Lemma) 51 (§ 9.12) (converse), relation (3) will be true.

Moreover, since due to the definition this Octagon, will have its quadrilaterals ΒΓΔΗ and ΖΗΘΓ harmonic too, if we work in the exact same way as bove, we easily find that relation (4) is true.

Thus, from (3), (4), relation (1) directly occurs.

(b). Regarding the Asteroid Harmonic Octagon.

Since the Harmonic Asteroid Octagon ΑΔΗΒΕΘΓΖΑ (definition § 16.1c), shares the same diagonals to ΑΒΓΔΕΖΗΘ, relation (1) is obviously true for it as well.

Remarks.

(a). It is obvious that for Proposition 105, the converse is not true.

(b). This Proposition, we believe, to be new and firstly appear here.

(c). It is obvious that we would reach the same result, if we consider as harmonic two other pairs of consecutive quadrilaterals, instead of the above ones, from the eight quadrialterals of the given Harmonic Octagon.

(d). Untill now we have seen that, apart from the Harmonic Octagon, Harmonic Hexagons have concurrent diagonals as well.

(e). Relevant Propositions are 51, 81, 101, 104, 106, 107.

19.13. Significant Property of the Harmonic Decagon.

Convergence of the Diagonals of Harmonic Decagons.

Proposition 106 (Figure 66).

The diagonals (main), of the Harmonic Decagons and their respective Asteroid Decagons of these, are concurrent.

Proof.

(a). Regarding the convex Harmonic Decagon.

1st Way (Based on Auxiliary Proposition 7 in Part E).

We consider a Harmonic Decagon ΑΒΓΔΕΖΗΘΙΚ. We will prove that:

$$AZ \cap BH \cap \Gamma\Theta \cap \Delta I \cap EK \equiv \Lambda. \qquad (1).$$

Since the given Decagon is Harmonic (implication), according to the definition (§ 16.1d above), its quadrilaterals for example ΑΒΓΗ and ΖΗΘΒ will be harmonic, thus according to known Theorem,

it will be: $\dfrac{AB}{B\Gamma}\cdot\dfrac{\Gamma H}{HA}=1$ and $\dfrac{ZH}{H\Theta}\cdot\dfrac{\Theta B}{BZ}=1 \Rightarrow \dfrac{AB}{B\Gamma}\cdot\dfrac{\Gamma H}{HA}=\dfrac{ZH}{H\Theta}\cdot\dfrac{\Theta B}{BZ}$. (2).

Thus, relation (2) according to Auxiliary Proposition 7 (converse) (§ 2.7) in Part E, regarding Hexagon ΑΒΓΖΗΘ, means that the following relation is true: $AZ \cap BH \cap \Gamma\Theta \equiv \Lambda$. (3).

Moreover, since from the relevant definition this Decagon, will have its quadrilaterals ΒΓΔΘ and ΗΘΙΓ harmonic too, if we work in the exact same way as above we easily find that $BH \cap \Gamma\Theta \cap \Delta I \equiv \Lambda$. (4).

Additionally, since from the relevant definition this Decagon, will have its quadrilaterals ΓΔΕΙ and ΘΙΚΔ harmonic too, if we work in the exact same way as above we easily find that $\Gamma\Theta \cap \Delta I \cap EK \equiv \Lambda$. (5).

Hence, from (3), (4), (5) relation (1) directly occurs.

2nd Way (Based on Auxiliary Proposition 51).

Since the given Decagon is Harmonic (implication), according to the definition (§ 16.1d above), its quadrilaterals ΑΒΓΗ and ΖΗΘΒ will be harmonic, thus according to Auxiliary Proposition 51 (§ 9.12) (converse),

it will be $AZ \cap BH \cap \Gamma\Theta \equiv \Lambda$. (6).

Moreover, since due to the definition this Decagon, will have its quadrilaterals ΒΓΔΘ and ΗΘΙΓ harmonic too, if we work in the exact same way as bove, we easily find that $BH \cap \Gamma\Theta \cap \Delta I \equiv \Lambda$. (7).

Additionally, since due to the definition this Decagon, will have its quadrilaterals ΓΔΕΙ and ΘΙΚΔ harmonic too, if we work in the exact same way as bove, we easily find that $\Gamma\Theta \cap \Delta I \cap EK \equiv \Lambda$. (8).

Therefore, from (6), (7), (8) relation (1) directly occurs.

(b). Regarding the Asteroid Decagon.

It is obvious that the Asteroid Decagon ΑΔΗΚΓΖΙΒΕΘΑ as well, will share its diagonals with the convex decagon ΑΒΓΔΕΖΗΘΙΚ, hence for this Asteroid decagon as well relation (1) will actually be true (See Proposition 107).

[As it is proven in Proposition 108 below (§ 19.16) the Asteroid Decagon is Harmonic too, as Harmonic is its respective convex Decagon].

Remarks.

(a). Proposition 106, we believe, to be new and firstly appear here.

(b). It is obvious that we would reach the same result, if we consider as harmonic any three other pairs of consecutive quadrilaterals, instead of the above ones, from the quadrialterals of the given Harmonic Decagon.

(c). It is obvious that for Proposition 106, the converse is not true.

(d). Untill now we have seen that, apart from the Harmonic Decagon, Harmonic Hexagons and Octagons have concurrent diagonals as well.

(e). An analytical way of proof of Proposition 106, is given in paragraph 200 (§ 29.1), using the method of Harmonic Transformation.

(f). Relevant Propositions are 51, 104, 105, 107, 108, 200.

19.14. Significant Property of the Asteroid Harmonic Decagon.

Convergence of the Diagonals of the Asteroid Harmonic Decagon.

Proposition 107 (Figure 73).

Diagonals (main), of the Asteroid Harmonic Decagons, are concurrent.

Proofs.

1st Way (Based on Auxiliary Proposition 7 in Part E).

We consider a Harmonic Asteroid Decagon ΑΔΗΚΓΖΙΒΕΘΑ. We will provw that it is: ΑΖ∩ΒΗ∩ΓΘ∩ΔΙ∩ΕΚ≡Λ. (1).

Since the given Decagon is Harmonic (implication), according to the relevant definition (§ 16.1d, above), its quadrilaterals for example ΑΓΕΘ and ΖΘΚΓ will be harmonic, thus according to our known Theorem, it wil be:

$$\frac{\mathrm{A}\Gamma}{\Gamma\mathrm{E}}\cdot\frac{\mathrm{E}\Theta}{\Theta\mathrm{A}}=1 \text{ and } \frac{\mathrm{Z}\Theta}{\Theta\mathrm{K}}\cdot\frac{\mathrm{K}\Gamma}{\Gamma\mathrm{Z}}=1 \Rightarrow \frac{\mathrm{A}\Gamma}{\Gamma\mathrm{E}}\cdot\frac{\mathrm{E}\Theta}{\Theta\mathrm{A}}=\frac{\mathrm{Z}\Theta}{\Theta\mathrm{K}}\cdot\frac{\mathrm{K}\Gamma}{\Gamma\mathrm{Z}}. \quad (2).$$

Hence, from relation (2), according to Auxiliary Proposition 7 (converse) (§ 2.7) in Part E, it means that relation is true:

$$\mathrm{AZ}\cap\Gamma\Theta\cap\mathrm{EK}\equiv\Lambda. \quad (3).$$

Moreover, since from the relevant definition this Decagon, will have its quadrialterals ΓΕΗΚ and ΘΚΒΕ harmonic too, if we work in the exact same way as above, we easily find that $\Gamma\Theta\cap\mathrm{EK}\cap\mathrm{HB}\equiv\Lambda$. (4).

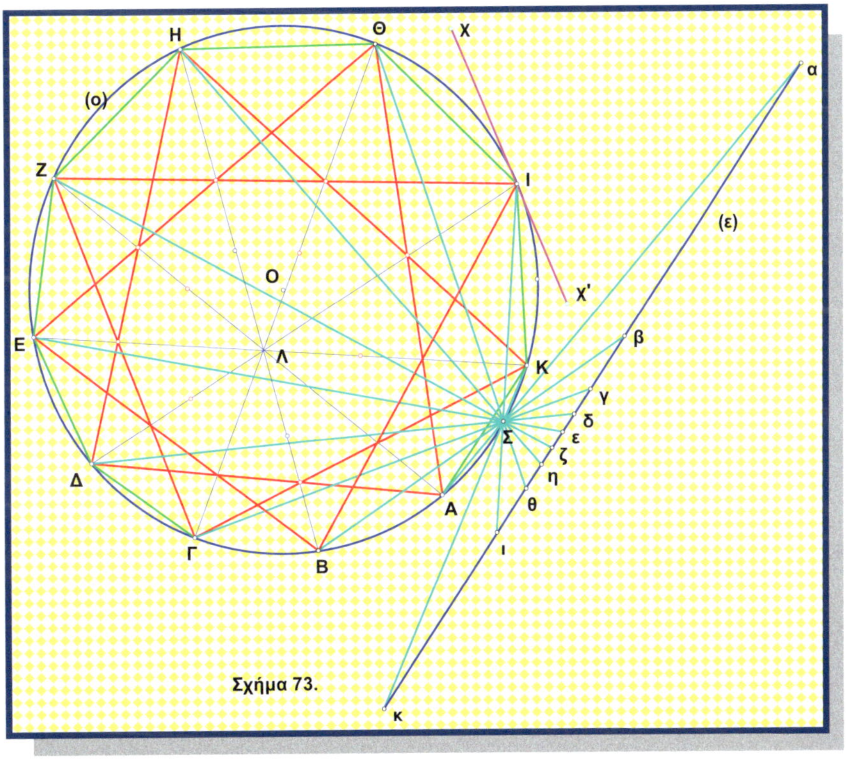

Σχήμα 73.

Additionally, since from the relevant definition this Decagon, will have its quadrialterals ΕΗΙΒ and ΚΒΔΗ harmonic too, if we work in the exact same way as above, we easily find that $\mathrm{EK}\cap\mathrm{HB}\cap\mathrm{I}\Delta\equiv\Lambda$. (5).

Thus, from (3), (4), (5) relation (1) directly occurs.

2nd Way (Based on Auxiliary Proposition 51).

Since the given Decagon is Harmonic (implication), according to the definition (§ 16.1d above), its quadrilaterals for example ΑΓΕΘ and ΖΘΚΓ will be harmonic, thus according to Auxiliary Proposition 51 (§ 9.12, converse), it will be $AZ \cap \Gamma\Theta \cap EK \equiv \Lambda$. (6).

Moreover, since due to the definition this Decagon, will have its quadrilaterals ΓΕΗΚ and ΘΚΒΕ harmonic too, if we work in the exact same way as bove, we easily find that $\Gamma\Theta \cap EK \cap HB \equiv \Lambda$. (7).

Additionally, since due to the definition this Decagon, will have its quadrilaterals ΕΗΙΒ and ΚΒΔΗ harmonic too, if we work in the exact same way as bove, we easily find that $EK \cap HB \cap I\Delta \equiv \Lambda$. (8).

Thus, from (6), (7), (8) relation (1) directly occurs.

Remarks.

(a). Proposition 107, we believe, to be new and firstly appear here.

(b). It is obvious that we would reach the same result, if we consider as harmonic any three other pairs of consecutive quadrilaterals, instead of the above ones, from the quadrilaterals of the given Harmonic Decagon.

(c). It is obvious that for Proposition 107, the converse is not true.

(d). Untill now we have seen that, apart from the Harmonic Asteroid Decagon, Harmonic Hexagons, Octagons, Decagons and Asteroid Harmonic Decagons have concurrent diagonals as well.

(e). Relevant Propositions are 51, 104, 105, 106, 108.

19.15. General Conclusion.

Due to Propositions, 104, 105, 106, 107, above and if we work similarly to their proofs, we reach the follow general conclusion:

Equivalent Proposition, to Propositions 104 to 107, can be applied for any Harmonic 2n-lateral, as long as the respective Regular 2n-lateral is constructible (Convex or Asteroid).

19.16. Property of the Asteroid Harmonic Decagon.

Harmonic being of the Asteroid Harmonic Decagons.

Criterion 108 (Figure 73).

The respective Asteroid Decagon, of each convex Harmonic Decagon, is also Harmonic (§ 16.1d) and conversely.

1/. Direct.

(a). Analysis.

I.e., if convex decagon ΑΒΓΔΕΖΗΘΙΚ is Harmonic inscribed in circle (o), we will prove that decagon ΑΔΗΚΓΖΙΒΕΘΑ is also harmonic.

In fact, if we take into consideration paragraphs § 16.1d, § 12.8, § 17.4 and Criterion 98 (§ 19.3), then we are led to the proof below.

(b). Proof.

Since, Decagon ΑΒΓΔΕΖΗΘΙΚ is Harmonic, according to the direct of Criterion 98, pencil Σ(ΑΒΓΔΕΖΗΘΙΚ) or Σ(αβγδεζηθικ) will also be Harmonic [where Σ is an arbitrary point on circle (o)].

Hence, the group of ten points α, β, γ, δ, ε, ζ, η, θ, ι, κ, that occurs from the intersection of the beams of the pencil above, from an arbitrary line (ε), is harmonic too.

Thus, according to paragraph 12.8, it occurs that the group of ten points α, δ, η, κ, γ, ζ, ι, β, ε, θ will be Harmonic too. This means that pencil Σ(αδηκγζιβεθ) or Σ(ΑΔΗΚΓΖΙΒΕΘ) is Harmonic, hence the Asteroid Decagon ΑΔΗΚΓΖΙΒΕΘΑ, is Harmonic (converse of Criterion 98).

2/. Converse.

To prove the converse, we follow the exact reverse way, of the proof above.

Remarks.

(a). Proposition 108, we believe, to be new and firstly appear here.

(b). Relevant Propositions are, 98, 107.

19.17. Second Criterion of Harmonic Being of a Convex Inscribed Hexagon. Extension to Criterion 67, from a line to a Circle.

Criterion 109 (Figure 64).

If for a convex inscribed in circle hexagon ΑΒΓΔΕΖ its following quadrilaterals are harmonic

ΑΒΓΕ, ΓΔΕΑ, ΕΖΑΓ, or ΒΓΔΖ, ΔΕΖΒ, ΖΑΒΔ, (1).

then and only then this hexagon is harmonic.

Proof.

(1). Direct.

I.e., if the given hexagon ΑΒΓΔΕΖ is convex, inscribed in circle (o) and the quadrilaterals of a group of three of its quadrilaterals in (1) are harmonic, then this hexagon is harmonic.

We suppose that for example this hexagon has harmonic its quadrilaterals ΑΒΓΕ, ΓΔΕΑ, ΕΖΑΓ. We will prove that it is harmonic.

According to known Theorem, regarding harmonic quadrilaterals ΑΒΓΕ, ΓΔΕΑ, ΕΖΑΓ, it will be respectively: $\frac{AB}{B\Gamma}=\frac{AE}{E\Gamma}$, $\frac{\Gamma\Delta}{\Delta E}=\frac{\Gamma A}{AE}$, $\frac{EZ}{ZA}=\frac{E\Gamma}{\Gamma A}$ $\Rightarrow$

$$\frac{AB\cdot\Gamma\Delta\cdot EZ}{B\Gamma\cdot\Delta E\cdot ZA}=\frac{AE\cdot\Gamma A\cdot E\Gamma}{E\Gamma\cdot AE\cdot\Gamma A}=1,\ \text{or}\ \frac{AB\cdot\Gamma\Delta\cdot EZ}{B\Gamma\cdot\Delta E\cdot ZA}=1. \qquad (2).$$

Relation (2), according to the converse of Auxiliary Proposition (Lemma) 5 (§ 2.5) in Part E, means that: $A\Delta\cap BE\cap\Gamma Z\equiv K$. (3).

[Relation (3), can easily be obtained if we work in figure 72, similarly to way 3 in the proof of Proposition 104 (§ 19.11), after we suppose that for the given hexagon ΑΒΓΔΕΖ its quadrilaterals ΑΒΓΕ, ΓΔΕΑ, ΕΖΑΓ are harmonic].

Hence, since (3) is true then quadrilaterals: ΑΒΓΕ, ΓΔΕΑ, ΕΖΑΓ, are harmonic (definition), according to the above Auxiliary Proposition (Lemma) 34 (direct), the respective quadrilaterals ΔΕΖΒ, ΖΑΒΔ, ΒΓΔΖ will also be harmonic. Consequently, all the quadrilaterals in (1), are harmonic, thus according to the relevant definition (§ 16.1β), given hexagon ΑΒΓΔΕΖ is harmonic.

(2). Converse.

We consider a Harmonic Hexagon ΑΒΓΔΕΖ. We will prove that its following quadrilaterals are harmonic: ΑΒΓΕ, ΓΔΕΑ, ΕΖΑΓ, ΒΓΔΖ, ΔΕΖΒ, ΖΑΒΔ. According to definition (§ 16.1 b) of Harmonic Hexagons, all the above quadrilaterals in (1) are harmonic too, thus, it is obvious, that this is true.

Conclusion.

Due to the Proposition about Harmonic Hexagons about, we reach the following conclusion: Every convex hexagon inscribed in circle (o) is harmonic, if the quadrilaterals of at least one of the groups of quadrilaterals in (1) are harmonic.

Consequently, from now on we can use for Harmonic Hexagons, the following Alternative Definition:

«Harmonic hexagon is called each convex inscribed in circle hexagon ΑΒΓΔΕΖ, if the quadrilaterals in at least one of the two groups of three quadrilaterals below are harmonic: ΑΒΓΕ, ΓΔΕΑ, ΕΖΑΓ or ΒΓΔΖ, ΔΕΖΒ, ΖΑΒΔ».

Remarks.

(a). Criterion 109, we believe, that it is new and firstly appears here.

(b). We would obviously reach the same result if we consider that the following quadrilaterals are harmonic ΒΓΔΖ, ΔΕΖΒ, ΖΑΒΔ.

(c). The above Criterion of Harmonic being of Inscribed Hexagons, is rather significant, as it allows us to have a simple alternative DEFINITION, from our initial one, for Harmonic Hexagons.

(d). The 1st, 3rd, 4th, 5th, 6th, Criteria about the Harmonic Hexagon, are given in Propositions 96, 112, 139, 140, 141, respectively.

(e). Extension to Criterion 109, with Criterion 141.

(f). Relevant Propositions are, 94, 96, 110, 111, 112, 139-141.

19.18. Criterion of Harmonic being of a Convex Inscribed Octagon. Extension to Criterion 68, from a line to a Circle.

Criterion 110 (Figure 65).

Every inscribed in circle convex Octagon, is Harmonic, if and only if, it has concurrent diagonals and four of its eight quadrilaterals are harmonic, those that from these four quadrilaterals, if we receive every two of them, they don't share a diagonal.

Proof.

(a). Direct.

Let there be the inscribed in circle (o) convex octagon ΑΒΓΔΕΖΗΘ for which it is: ΑΕ∩ΒΖ∩ΓΗ∩ΔΘ≡Κ, (1).

and four, of its eight quadrilaterals are harmonic, those that no pair of the four quadrilaterals has a shared diagonal with any other of these four quadrilaterals.

We discern the cases below:

1/. During the first case four consecutive quadrilaterals of the octagon are harmonic, as for example if the following ones are harmonic:

ΑΒΓΖ, ΒΓΔΗ, ΓΔΕΘ, ΔΕΖΑ. (2).

In this case, since for example the inscribed hexagon ΑΒΓΕΖΗ it is ΑΕ∩ΒΖ∩ΓΗ≡Κ and its quadrilateral ΑΒΓΖ is harmonic, according to Proposition 34 (§ 8.71) (direct), quadrilateral ΕΖΗΒ will also be harmonic.

We continue with the inscribed hexagon ΒΓΔΖΗΘ, whose quadrilateral ΒΓΔΗ is harmonic and we find, as exactly above, that its quadrilateral ΖΗΘΓ is harmonic too, as (1) is true too.

Moreover, if we continue, as exactly above, we find that quadrilaterals: ΗΘΑΔ, ΘΑΒΕ, are harmonic.

Thus, from the above we that the quadrilaterals in the group of four quadrilaterals below are harmonic:

ΕΖΗΒ, ΖΗΘΓ, ΗΘΑΔ, ΘΑΒΕ. (3).

Thus, since octagon ΑΒΓΔΕΖΗΘ has all its eight quadrilaterals in (2) and (3) harmonic, according to the relevant definition (§ 16.1c), it will be harmonic.

2/. During the <u>second case</u> the four harmonic quadrilaterals belong to one of the two groups of four below, or to another group of four with equivalent order:

ΑΒΓΖ, ΒΓΔΗ, ΗΘΑΔ, ΘΑΒΕ. (4).

ΕΖΗΒ, ΖΗΘΓ, ΓΔΕΘ, ΔΕΖΑ. (5).

Here, if we receive for example the quadrilaterals in (4) as harmonic and we work on hexagon ΑΒΓΕΖΗ, exactly as above, based on Harmonic Quadrilateral ΑΒΓΖ then we find that quadrilateral ΕΖΗΒ is harmonic, since (1) is true as well.

We continue, similarly to above for hexagon ΒΓΔΖΗΘ and based on the harmonic quadrilateral ΒΓΔΗ we find that quadrilateral ΖΗΘΓ is harmonic too, as (1) is true.

Moreover, if we continue as exactly above, we find that quadrilaterals, ΓΔΕΘ, ΔΕΖΑ, too are harmonic, thus all the quadrilaterals in (4) and (5), are harmonic and thus the given decagon is harmonic.

3/. During the <u>third case</u> the four harmonic quadrilaterals belong to one of the two groups of four below, or to another group of four of equivalent order:

ΑΒΓΖ, ΓΔΕΘ, ΖΗΘΓ, ΘΑΒΕ. (6).

ΕΖΗΒ, ΗΘΑΔ, ΒΓΔΗ, ΔΕΖΑ. (7).

Here, if the quadrilaterals in for example (6) are received as harmonic and we work in hexagon ΑΒΓΕΖΗ, as exactly above, based on Harmonic Quadrilateral ΑΒΓΖ, we find that quadrilateral ΕΖΗΒ is harmonic, as (1) is true too.

We continue, as above for hexagon ΓΔΕΗΘΑ as well and based on the harmonic quadrilateral ΓΔΕΘ we find that quadrilateral ΗΘΑΔ is harmonic too, as (1) is true as well.

Moreover, if we continue as exactly above, we find that quadrilaterals ΒΓΔΗ, ΔΕΖΑ, are harmonic too, thus all the quadrilaterals in (6) and (7), are harmonic and hence the given decagon is harmonic. (We work in a similar way for any other order of its four given Harmonic Quadrilaterals).

(b). Converse.

I.e., if the given octagon is harmonic, we will prove that then this will have harmonic, its ten quadrilaterals and (1) will be true. This is obvious, since the given octagon is harmonic, according to the relevant definition (§ 16.1c) and its eight quadrilaterals will be harmonic, while according to Proposition 105, (1) will also be true.

Conclusion.

Due to the Proposition about Harmonic Octagons above (Criterion of Harmonic being of inscribed octagons), we reach the following conclusion:
Every convex octagon inscribed in circle is harmonic, if its diagonals are concurrent and four of its eight quadrilaterals are harmonic, those that from these four quadrilaterals, if we receive each two of them, they don't share a diagonal.
Consequently, from now on we can use for Harmonic Octagons, the following alternatice Definition:

«We call Harmonic Octagon, each convex inscribed in circle octagon, if its diagonals are concurrent and four out of its eight quadrilaterals are harmonic, those that four of these quadrilaterals, if we receive any two of them, they don't share a diagonal ».

Remarks.

(a). The above Criterion of Harmonic being of Octagons, is rather significant, as it allows us to have a simpler alternative DEFINITION, from the initial one, for Harmonic Octagons, but because it also finds many applications, as we will later find out.
(b). We believe that it is a new Criterion and that it firstly appears here.
(c). According to the relevant definition (§ 16.1c), this Criterion is true for the Asteroid Octagon ΑΔΗΒΕΘΓΖΑ as well.
(d). Relevant Propositions are 34, 68, 109, 111.

19.19. Criterion of Harmonic Being of a Convex Inscribed Decagon.

Criterion 111 (Figure 66).

Every inscribed in circle convex Decagon, is Harmonic, if and only if, its diagonals are concurrent and five out of its ten quadrilaterals are harmonic, those that from these five quadrilaterals, if we take each two of them, they don't share a diagonal.

Proof.

(a). Direct.

Let there be the inscribed in circle (o) convex decagon ΑΒΓΔΕΖΗΘΙΚ for which it is: ΑΖ∩ΒΗ∩ΓΘ∩ΔΙ∩ΕΚ≡Λ, (1).

and five, out of its ten quadrilaterals are harmonic, those that no pair from these five quadrialterals has a shared diagonal with any other of these five quadrilaterals.

We will prove that decagon ΑΒΓΔΕΖΗΘΙΚ, is harmonic.

We discern the following cases:

1/. During the first case five harmonic consecutive quadrilaterals of the decagon are harmonic, as for example if the following are harmonic:

ΑΒΓΗ, ΒΓΔΘ, ΓΔΕΙ, ΔΕΖΚ, ΕΖΗΑ. (2).

In this case, since for example in the inscribed hexagon ΑΒΓΖΗΘ it is ΑΖ∩ΒΗ∩ΓΘ≡Λ and its quadrilateral ΑΒΓΗ is harmonic, according to Proposition 34 (§ 8.71) (direct), quadrilateral ΖΗΘΒ will also be harmonic.

We continue with the inscribed hexagon ΒΓΔΗΘΙ, whose quadrilateral ΒΓΔΘ is harmonic and we find, as exactly above, that quadrilateral ΗΘΙΓ is harmonic as well, as relation (1) is true as well.

Moreover, if we continue, as exactly above, we find that quadrilaterals: ΘΙΚΔ, ΙΚΑΕ, ΚΑΒΖ, are harmonic too.

Hence, from the above we see that the quadrilaterals in the group of five quadrilaterals below are harmonic:

ZHΘB, HΘIΓ, ΘIKΔ, IKAE, KABZ. (3).

Thus, since decagon ΑΒΓΔΕΖΗΘΙΚ has all ten quadrilaterals in (2) and (3) harmonic, according to the relevant definition (§ 16.1d), it will be harmonic.

2/. During the second case its five harmonic quadrilaterals are ordered, one by one, thus then we have the two groups of five quadrilaterals below, that at least one of them consists of harmonic quadrilaterals:

ΑΒΓΗ, ΓΔΕΙ, ΕΖΗΑ, ΗΘΙΓ, ΙΚΑΕ, (4).

ΒΓΔΘ, ΔΕΖΚ, ΖΗΘΒ, ΘΙΚΔ, ΚΑΒΖ. (5).

Here, if we consider as harmonic the quadrilaterals in for example (4) and we work in hexagon ΑΒΓΖΗΘ, as exactly above, based on the Harmonic Quadrilateral ΑΒΓΗ, we find that quadrilateral ΖΗΘΒ is harmonic too, as relation (1) is true as well.

We continue, similarly to above for hexagon ΓΔΕΘΙΚ and based on the harmonic quadrilateral ΓΔΕΙ, we find that quadrilateral ΘΙΚΔ is harmonic too, as relation (1) is true.

Additionally, if we continue as exactly above, we find that quadrilaterals ΚΑΒΖ, ΒΓΔΘ, ΔΕΖΚ, are harmonic as well, thus all the quadrilaterals in (4) and (5), are harmonic and thus the given decagon is harmonic.

3/. During the third case its five harmonic quadrilaterals have for example the order: ΚΑΒΖ, ΑΒΓΗ, ΒΓΔΘ, ΔΕΖΚ, ΘΙΚΔ and then working as exactly above we find that the rest five quadrilaterals of the given decagon are harmonic as well, hence this decagon will be harmonic too.

4/. During the fourth case its five harmonic quadrilaterals have the order: ΚΑΒΖ, ΑΒΓΗ, ΓΔΕΙ, ΔΕΖΚ, ΗΘΙΓ and then working as exactly above we find that the rest five quadrilaterals of the given decagon are harmonic as well, hence this decagon will be harmonic too.

(In a similar way, we work for every other order of the given five harmonic quadrilaterals).

(b). Converse.

I.e., if the given decagon is harmonic, we will prove that then this will have harmonic its ten decagons and (1) will be true.

This is obvious, as since the given decagon is harmonic, according to the relevant definition (§ 16.1d) all its ten quadrilaterals will be harmonic, while according to Proposition 106, (1) will be true.

Conclusion.

Due to the Proposition about Harmonic Decagons above (Criterion of Harmonic being of convex inscribed decagons), the following conclusion is reached:

Each convex decagon inscribed in circle is harmonic, if its diagonals are concurrent and five out of its ten quadrilaterals, those that from these five quadrilaterals, if we receive each two of them, they do not share a diagonal.

Consequently, from now on we can use for Harmonic Decagons, the following alternative Definition:

«Harmonic decagon is called, each convex inscribed in circle decagon, if its diagonals are concurrent and five out of its ten quadrilaterals are harmonic, those that from these five quadrilaterals, if we receive each two of them, they do not share a diagonal».

Remarks.

(a). The above Criterion of Harmonic being of Decagons, is rather significant, as it allows us to have a simpler alternative DEFINITION, from the initial one, for Harmonic Decagons, but because it also finds many applications, as we will later find out.

(b). We believe that it is a new Criterion and that it firstly appears here.

(c). According to the relevant definition (§ 16.1d), this Criterion is true for the Asteroid Decagon ΑΔΗΚΓΖΙΒΕΘΑ as well.

(d). Relevant Propositions are 34, 109, 110.

19.20. General Conclusion.

Due to the Propositions, 109, 110, 111, above, and if we work similarly to their proofs, we reach the following general conclusion:

Equivalent Criterion, to Criteria 109 to 111, can be applied for any Harmonic 2n-lateral.

19.21. Third Criterion of Harmonic being of a Convex Inscribed Hexagon.

Extension to known Theorem about the Lemoine Point of a Triangle to a Hexagon.

Does the Harmonic Hexagon have a Lemoine Point?

Criterion 112 (Figure 74).

If convex inscribed in circle (o) hexagon ΑΒΓΔΕΖ, has concurrent diagonals (main) and there exists a point on its plane for which the distances from its sides (the hexagon's sides), are analogous to its sides respectively, then and only then this hexagon is harmonic.

Proof.

(1). Direct.

We consider a Harmonic Hexagon ΑΒΓΔΕΖ, inscribed in circle (o).

We will prove that a point K' exists on its plane, for which it is:

$$\frac{υ_α}{α}=\frac{υ_β}{β}=\frac{υ_γ}{γ}=\frac{υ_δ}{δ}=\frac{υ_ε}{ε}=\frac{υ_ζ}{ζ}, \qquad (1).$$

where $υ_α$, $υ_β$, ..., $υ_ζ$, are the distances of point K', from its sides ΑΒ=α , ΒΓ=β,..., ΖΑ=ζ, respectively.

In fact, since hexagon ΑΒΓΔΕΖ is harmonic, it will also have ΑΔ∩ΒΕ∩ΓΖ≡Κ (Proposition about Harmonic Hexagons 104).

Moreover, according to the relevant definition (§16.1b) of Harmonic Hexagons, in ΑΒΓΔΕΖ quadrilaterals ΑΒΓΕ, ΒΓΔΖ, ..., , ΖΑΒΔ, will be harmonic and therefore ΑΔ, ΒΕ, ΓΖ, will be conjugate to pairs of strings ΒΖ-ΓΕ, ΑΓ-

ΔΖ, ΒΔ-ΕΑ, respectively, thus according to the converse of Auxiliary Proposition (Lemma) 6 in Part E, ΑΔ, ΒΕ, ΓΖ, will be symmedians of the pairs of triangles ΖΑΒ-ΓΔΕ, ΑΒΓ-ΔΕΖ, ΒΓΔ-ΕΖΑ, respectively.

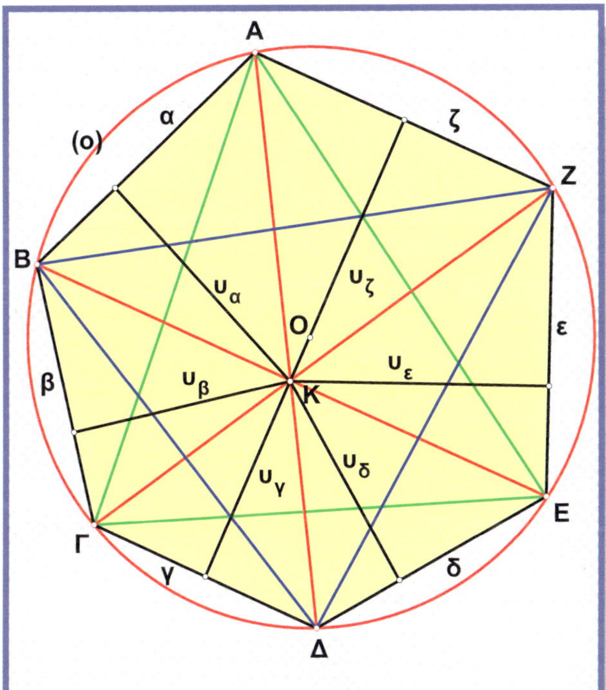

Σχήμα 74.

Therefore, ΑΔ, ΒΕ, ΓΖ, will be the loci, according to which, the distances of each of their points, from the pairs of the respective sides ΖΑ-ΑΒ, ΓΔ-ΔΕ, ΑΒ-ΒΓ, ΔΕ-ΕΖ, ΒΓ-ΓΔ, ΕΖ-ΖΑ, are analogous {[6] § 11-14.2} (direct). This will obviously be true for the shared intersection K of diagonals ΑΔ, ΒΕ, ΓΖ (Propositions 104 about Harmonic Hexagons), as this one belongs to all these three diagonals. This means that point K has all these three properties. Hence for K it will be:

$$\frac{\upsilon_\zeta}{\zeta}=\frac{\upsilon_\alpha}{\alpha},\ \frac{\upsilon_\gamma}{\gamma}=\frac{\upsilon_\delta}{\delta},\ \frac{\upsilon_\alpha}{\alpha}=\frac{\upsilon_\beta}{\beta},\ \frac{\upsilon_\delta}{\delta}=\frac{\upsilon_\varepsilon}{\varepsilon},\ \frac{\upsilon_\beta}{\beta}=\frac{\upsilon_\gamma}{\gamma},\ \frac{\upsilon_\varepsilon}{\varepsilon}=\frac{\upsilon_\zeta}{\zeta}. \qquad (2).$$

Thus, from the equalities in (2), we can easily find relation (1), that is also true for the intersection K≡ΑΔ∩ΒΕ∩ΓΖ and therefore this means that the desired point K', is actually K or K'≡K.

Therefore, the desired point is the point of convergence of the diagonals (main), of the Harmonic Hexagon.

(2). Converse.

If for point K≡AΔ∩BE∩ΓZ, relation (1) is true, we will prove that hexagon ABΓΔEZ is harmonic.

In fact, from relation (1), we also get that for example: $\frac{υ_α}{α}=\frac{υ_ζ}{ζ}$. This means that AK or AΔ is the symmedian of triangle ZAB (converse of the known Proposition {[6] § 11-14.2} and therefore AΔ will be conjugate to ZB with respect to circle (o) [Lemma 6 (direct), in Part E], thus quadrilateral ZABΔ is harmonic (definition § 16.1b).

Similarly we find that quadrilaterals BΓΔZ, ΔEZB, are harmonic too, thus according to the simplified alternative definition of Harmonic Hexagons (Criterion 109 of Harmonic being of Inscribed Hexagons), hexagon ABΓΔEZ will be harmonic.

Remarks.

(a). Two specialized Criteria of Criterion 112, are given in paragraph 8ι(88) (volume 8) in Geometry [31].

(b). The 1st, 2nd, 4th, 5th, 6th Criteria of the Harmonic Hexagon, are given in Propositions 96, 109, 139, 140, 141, respectively.

(c). We believe that Proposition 112 is new and firstly appears here.

(d). Relevant Propositions are 96, 109, 113, 114, 139-141.

Comment.

After Criterion 112 of Harmonic Hexagons, we also notice that:

(a). Point Lemoine in a triangle, has the property, according to which, this point is far from the sides of the triangle, distances analogous to its respective sides (the triangle's).

(b). As it is known and if we depend on the Auxiliary Proposition 6 in Part E, it is easily proven, that the intersection of the diagonals on the harmonic quadrilateral, has the same, as above, property (See also Exercises of Geometry [32] IV, page 54). Hence the distance of this intersection from the sides of the quadrilateral, are analogous, to the respective sides of the quadrilateral. This is why the intersection of the diagonals of the harmonic quadrilateral, was characterized as the Lemoine point of the harmonic quadrilateral.

(c). According to Criterion 112 above in the Harmonic Hexagon, the intersection of its diagonals (main), has the same, as above, property, hence we reach the following conclusion:

Conclusion.

Due to the aforementioned and in equivalence to these, the point of convergence of the diagonals of a Harmonic Hexagon is characterized as the LEMOINE POINT, of this Harmonic Hexagon.

Another formation of Criterion 112.

After the above, Criterion 112 (Criterion of Harmonic being of convex Inscribed Hexagons), can be stated in the following way too:

Every convex Hexagon inscribed in circle, is harmonic, if and only if, it has a LEMOINE POINT. (This coincides with the convergence point of its diagonals).

Alternative Definition of the Harmonic Hexagon.

Therefore, after the above, from now on we can also use for Harmonic Hexagons, the following alternative Definition:

«We call Harmonic Hexagon each convex cyclic Hexagon, if and only if, it has a LEMOINE POINT.

19.22. Criterion of Harmonic being of a Convex Inscribed Octagon.

Extension to Criterion 112 to an Octagon.

Does the Harmonic Octagon have a Lemoine Point?

Criterion 113 (Figure 75).

If convex inscribed in circle (o) octagon ΑΒΓΔΕΖΗΘ, has concurrent diagonals (main) and there exists a point on its plane for which the distances from its sides (the octagon's sides), are analogous to its sides respectively, then and only then this octagon is harmonic.

Proof.

(a). Direct.

We consider a Harmonic Octagon ΑΒΓΔΕΖΗΘ, inscribed in circle (o) (Relevant definition in § 16.1c). We will prove that a point K' exists on its plane, for which it is:

$$\frac{υ_α}{α}=\frac{υ_β}{β}=\frac{υ_γ}{γ}=\frac{υ_δ}{δ}=\frac{υ_ε}{ε}=\frac{υ_ζ}{ζ}=\frac{υ_η}{η}=\frac{υ_θ}{θ}, \qquad (1).$$

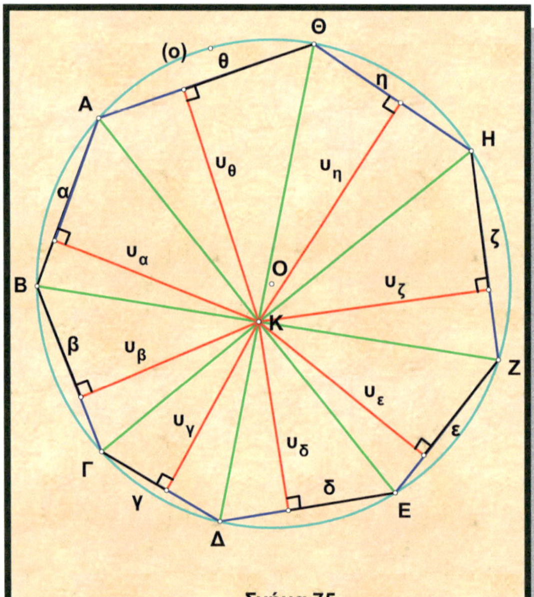

Σχήμα 75.

where $υ_α$, $υ_β$,...., $υ_θ$, are the distances of point K', from its sides AB=α , ΒΓ=β,..., ΘΑ=θ, respectively.

In fact, since Octagon ΑΒΓΔΕΖΗΘ is harmonic, according to Proposition 105 (§ 19.12), it will also have ΑΕ∩ΒΖ∩ΓΗ∩ΔΘ≡Κ, while according to the relevant definition (§16.1c), all its following eight quadrilaterals will be harmonic:

ΑΒΓΖ, ΒΓΔΗ, ΓΔΕΘ ,..., ΘΑΒΕ. (2).

This means that for example for the harmonic quadrilateral ΑΒΓΖ its diagonals ΑΓ, ΒΖ, will be conjugate with respect to circle (o), thus according to Auxiliary Proposition 6 (converse) in Part E, for triangle ΑΒΓ, ΒΖ is its symmedian. Hence, according to the direct of a known Theorem (Γεωμετρία [6] § 11-14.2), each point on BZ will be far from AB, ΒΓ distance analogous to these sides. This will obviously be true for point K of the convergence of diagonals of this Octagon, which belongs to BZ too, thus for point K, it will be:

$$\frac{υ_α}{α}=\frac{υ_β}{β}. \qquad (3).$$

If we work similarly to above for harmonic quadrilateral ΒΓΔΗ and more specifically in triangle ΒΓΔ, we prove that for K it is also:

$$\frac{U_\beta}{\beta}=\frac{U_\gamma}{\gamma}. \qquad (4).$$

Hence, if we continue as above for the rest six triangles ΓΔΕ, ΔΕΖ,...,ΘΑΒ, we find that for K it is also:

$$\frac{U_\gamma}{\gamma}=\frac{U_\delta}{\delta}=,\ldots,=\frac{U_\theta}{\theta}=\frac{U_\alpha}{\alpha}. \qquad (5).$$

Therefore, from relations (3) – (5), we get relation (1), which is true for intersection K as well. This means that the desired point K' is K or K'≡K.
Hence the desired point, is the point where the diagonals of the Harmonic Octagon coincide.

(b). Converse.

I.e., if the convex inscribed in circle (o) Octagon ΑΒΓΔΕΖΗΘ has diagonals concurrent at K and for K relation (1) is true, then this Octagon will be harmonic.
In fact, since (1) is true, for K it will be:

$$\frac{U_\alpha}{\alpha}=\frac{U_\beta}{\beta}. \qquad (3).$$

This, according to the converse of the Theorem in § 11-14.2 in Geometry [6], shows that K belongs to the symmedian matching to the side ΑΓ in triangle ΑΒΓ and thus, according to the direct of the above Auxiliary Proposition 6 in Part E, ΒΖ and ΑΓ are conjugate with respect to circle (o), or quadrilateral ΑΒΓΖ is harmonic.
If we take into consideration, that the rest seven two-part equalities that occur from (1) are true, i.e.:

$$\frac{U_\beta}{\beta}=\frac{U_\gamma}{\gamma},\quad \frac{U_\gamma}{\gamma}=\frac{U_\delta}{\delta},\quad \frac{U_\delta}{\delta}=\frac{U_\varepsilon}{\varepsilon},\ \ldots. \qquad (6).$$

and if we work as above, we find that all three quadrilaterals in (2), ΒΓΔΗ,

ΓΔΕΘ, ΔΕΖΑ, are harmonic, thus, according to Criterion 110 (direct) Octagon ΑΒΓΔΕΖΗΘ (alternative definition) is actually Harmonic, since it also has concurrent diagonals.

Remarks.

(a). Proposition 113 (criterion), we believ to be new and firstly appear here.

(b). Equivalent Proposition (Criterion) we have proved to be true for Harmonic Hexagons as well (Criterion 112).

(c). As we will see next, equivalent Criterion is true for Decagon ΑΒΓΔΕΖΗΘΙΚ, etc.

(d). Relevant Propositions are 112, 114-118.

Comment.

After Criterion 113 of Harmonic Octagons, we also notice that:

(a). Point Lemoine in a triangle, has the property, according to which, this point is far from the sides of the triangle, distances analogous to its respective sides (the triangle's).

(b). As it is known and easily proven {Criterion 8ι(87) in our book [31]} , the intersection of the diagonals on the harmonic quadrilateral, has the same, as above, property. Hence the distance of this intersection from the sides of the quadrilateral, are analogous, to the respective sides of the quadrilateral (See also Γεωμετρία [32] IV, page 54). This is why the intersection of the diagonals of the harmonic quadrilateral, was characterized as the Lemoine point of the harmonic quadrilateral.

(c). According to Criterion 112 above in the Harmonic Hexagon, the intersection of its diagonals (main), has the same, as above, property, and this is why we named it the Lemoine point of the Harmonic Hexagon.

(d). According to Criterion 113 above for every Harmonic Octagon as we saw, the intersection of its diagonals (main), has the same, as above, property, thus we reach the following conclusion:

Conclusion.

After the aforementioned and in equivalence to these, it occurs that:

The point of convergence of the diagonals of the Harmonic Octagon, is a LEMOINE POINT, for the Harmonic Octagon.

Another formation of Criterion 113.

After the above, Criterion 113 (Criterion of Harmonic being of convex Inscribed Octagons), can be stated in the following way as well:

Every convex Octagon inscribed in circle with concurrent diagonals, is harmonic, if and only if, it has a LEMOINE POINT. (This coincides with the convergence point of its diagonals).

Alternative Definition of the Harmonic Octagon.

Therefore, after the above, from now on we can also use for Harmonic Octagons, the following alternative Definition:

«We call Harmonic Octagon each convex cyclic Octagon, if and only if, it has a LEMOINE POINT.

19.23. Criterion of Harmonic being of a Convex Inscribed Decagon.

Extension to Criterion 113 to a Decagon.

Does the Harmonic Decagon have a Lemoine Point?

Criterion 114 (Figure 76).

If a convex inscribed in circle (o) decagon ΑΒΓΔΕΖΗΘΙΚ, has concurrent diagonals (main) and there exists on its plane a point for which the distance from each side (of the decagon), is analogous to the respective side, then and only then, this decagon is harmonic.

Proof.

(a). Direct.

We consider a Harmonic Decagon ΑΒΓΔΕΖΗΘΙΚ, inscribed in circle (o) (Relevant definition in § 16.1d). We will prove that a point Λ' exists on its plane, for which it is:

$$\frac{\alpha'}{\alpha}=\frac{\beta'}{\beta}=\frac{\gamma'}{\gamma}=\frac{\delta'}{\delta}=\frac{\varepsilon'}{\varepsilon}=\frac{\zeta'}{\zeta}=\frac{\eta'}{\eta}=\frac{\theta'}{\theta}=\frac{\iota'}{\iota}=\frac{\kappa'}{\kappa},\qquad (1).$$

where α′, β′,…., κ′, is the distance of point Λ′, from each side AB=α, ΒΓ=β,…, ΚΑ=κ, respectively.

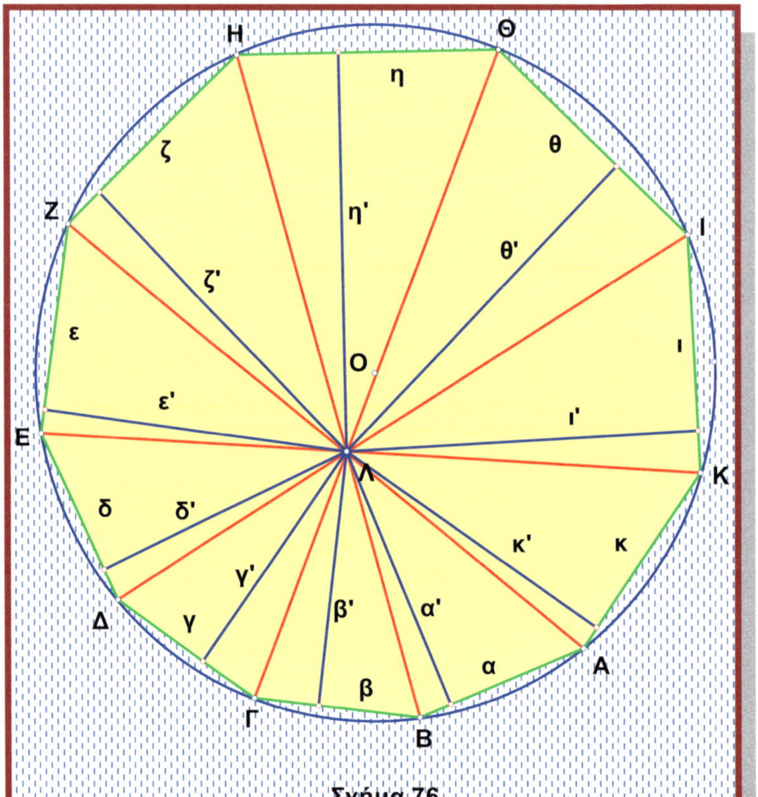

Σχήμα 76.

In fact, since Decagon ΑΒΓΔΕΖΗΘΙΚ is harmonic, according to Proposition 106 (§ 19.13), it will have ΑΖ∩ΒΗ∩ΓΘ∩ΔΙ∩ΕΚ≡Λ, while according to the relevant definition (§ 16.1d), all its ten quadrilaterals will be harmonic:

ΑΒΓΗ, ΒΓΔΘ, ΓΔΕΙ,… , ΚΑΒΖ. (2).

This means that for example for the harmonic quadrilateral ΑΒΓΗ its diagonals ΑΓ, ΒΗ, will be conjugate with respect to circle (o), hence according to the Auxiliary Proposition 6 (converse) in Part E, for triangle ΑΒΓ, ΒΗ is its symmedian. Thus, according to the direct of a known Theorem (Γεωμετρία [6] § 11-14.2), each point on ΒΗ will be far from ΑΒ, ΒΓ distance analogous to these sides. This will obviously be true for the point of convergence of the diagonals of the Decagon Λ, which too belongs to BH, thus for point Λ, it will be:

$$\frac{\alpha'}{\alpha}=\frac{\beta'}{\beta}. \qquad (3).$$

If we work similarly to above for the harmonic quadrilateral ΒΓΔΘ and more specifically in triangle ΒΓΔ, we prove that for Λ it is also:

$$\frac{\beta'}{\beta}=\frac{\gamma'}{\gamma}. \qquad (4).$$

Hence, if we continue as above for the rest eight triangles ΓΔΕ, ΔΕΖ,…,ΚΑΒ, we find that for Λ is also is:

$$\frac{\gamma'}{\gamma}=\frac{\delta'}{\delta},\dots,\frac{\kappa'}{\kappa}=\frac{\alpha'}{\alpha}. \qquad (5).$$

Therefore, from relations (3) – (5), we get relation (1), which is true for intersection Λ. This means that the desired point Λ' is Λ or Λ'≡Λ.

Hence the desired point, is the point where the diagonals of the Decagon coincide.

(b). Converse.

I.e., if the convex inscribed in circle (o) Decagon ΑΒΓΔΕΖΗΘΙΚ has diagonals concurrent at Λ and for Λ relation (1) is true, then this Decagon will be harmonic.

In fact, since (1) is true, for Λ it will be:

$$\frac{\alpha'}{\alpha}=\frac{\beta'}{\beta}. \qquad (3).$$

This, according to the converse of the Theorem in § 11-14.2 in Geometry [6], shows that Λ belongs to the symmedian matching to the side ΑΓ in triangle ΑΒΓ and thus, according to the direct of the above Auxiliary Proposition 6 in Part E, BH and ΑΓ are conjugate with respect to circle (o), or quadrilateral ΑΒΓΗ is harmonic.

If we take into consideration, that the rest nine two-part equalities that occur from (1) are true, i.e:

$$\frac{\beta'}{\beta}=\frac{\gamma'}{\gamma},\quad \frac{\gamma'}{\gamma}=\frac{\delta'}{\delta},\quad \frac{\delta'}{\delta}=\frac{\varepsilon'}{\varepsilon},\quad \frac{\varepsilon'}{\varepsilon}=\frac{\zeta'}{\zeta},\ \ldots \qquad (6).$$

and if we work as above, we find that the rest nine quadrilaterals in (2), ΒΓΔΘ, ΓΔΕΙ, ΔΕΖΚ, ΕΖΗΑ, etc, are harmonic, thus, according to the definition of the Simple Harmonic Decagon (§ 16.1d), then Decagon ΑΒΓΔΕΖΗΘΙΚ is in fact Harmonic.

Remarks.

(a). Proposition 114 (criterion), we believ to be new and firstly appear here.

(b). Equivalent Propositions (Criteria) we have proved to be true for Harmonic Hexagons (Criterion 112) and Octagons (Criterion 113) as well.

(c). As we will see next, equivalent Criterion is true for the Asteroid non regular Decagon ΑΔΗΚΓΖΙΒΕΘΑ, etc. I.e. that this is also harmonic, etc.

(d). Relevant Propositions are 112, 113, 115-118.

Comment.

After Criterion 114 of Harmonic Decagons, we also notice that:

(a). Point Lemoine in a triangle, has the property, according to which, this point is far from the sides of the triangle, distances analogous to its respective sides (the triangle's).

(b). As it is known and easily proven {Criterion 8ι(87) in our book [31] and Exercises [32] page 54}, the intersection of the diagonals on each harmonic quadrilateral, has the same, as above, property. Hence the distance of this intersection from the sides of the quadrilateral, are analogous, to the respective sides of the quadrilateral. This is why the intersection of the diagonals of the harmonic quadrilateral, was characterized as the Lemoine point of the harmonic quadrilateral.

(c). According to Criterion 112 above in the Harmonic Hexagon, the intersection of its diagonals (main), has the same, as above, property, and this is why we named it the Lemoine point of the Harmonic Hexagon.

(d). According to Criterion 113 above for every Harmonic Octagon as we saw, the intersection of its diagonals (main), has the same, as above, property.

(e). According to Criterion 114 above for every Harmonic Decagon, the intersection of its diagonals (main), has the same, as above, property, thus we reach the following conclusion:

Conclusion.

After the aforementioned and in equivalence to these, it occurs that:

The point of convergence of the diagonals of the Harmonic Decagon, is a LEMOINE POINT, for the Harmonic Decagon.

Another formation of Criterion 114.

After the above, Criterion 114 (Criterion of Harmonic being of convex Inscribed Decagons), can be stated in the following way as well:

Every convex Decagon inscribed in circle with concurrent diagonals, is harmonic, if and only if, it has a LEMOINE POINT. (This coincides with the convergence point of its diagonals).

Alternative Definition of the Harmonic Decagon.

Therefore, after the above, from now on we can also use for Harmonic Octagons, the following alternative Definition:

«We call Harmonic Decagon each convex cyclic Octagon, if and only if, it has a LEMOINE POINT.

19.24. Criterion of Harmonic being of an Asteroid Inscribed Octagon.

Extension to Criterion 113 and 114, to aa Asteroid Octagon.

Does the Asteroid Harmonic Octagon have a Lemoine Point?

Criterion 115 (Figure 77).

If an asteroid inscribed in circle (o) octagon ΑΔΗΒΕΘΓΖΑ, has concurrent diagonals (main) and there exists on its plane a point for which the distance from each side (of the Asteroid Octagon), is analogous to the respective side, then and only then, this octagon is harmonic.

Proof.

(a). Direct.

We consider a Harmonic Asteroid Octagon ΑΔΗΒΕΘΓΖΑ, inscribed in circle (o) (Relevant definition in § 16.1c). We will prove that a point K' exists on its plane, for which itis:

$$\frac{\upsilon_\alpha}{\alpha}=\frac{\upsilon_\beta}{\beta}=\frac{\upsilon_\gamma}{\gamma}=\frac{\upsilon_\delta}{\delta}=\frac{\upsilon_\varepsilon}{\varepsilon}=\frac{\upsilon_\zeta}{\zeta}=\frac{\upsilon_\eta}{\eta}=\frac{\upsilon_\theta}{\theta}, \quad (1).$$

where υ_α, υ_β,...., υ_θ, is the distance of point K', from each side ΑΔ=α, ΔΗ=β, ΗΒ=γ,...,ΓΖ=η, ΖΑ=θ, respectively, of the Asteroid Octagon.

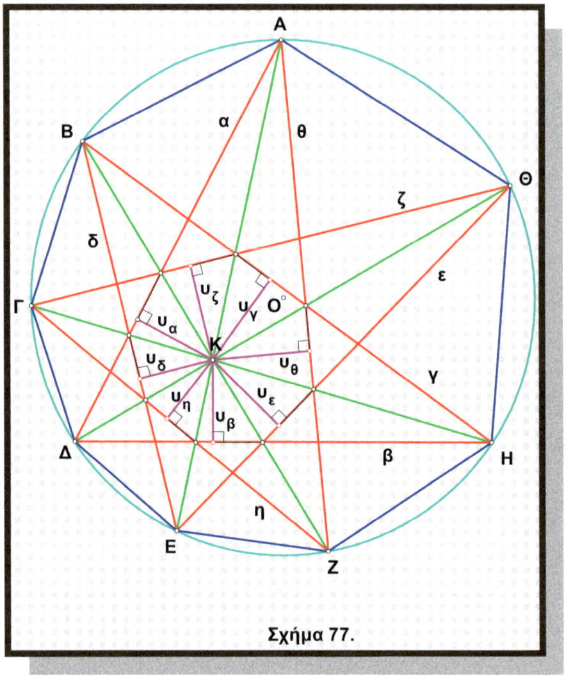

Σχήμα 77.

In fact, since Octagon ΑΔΗΒΕΘΓΖΑis harmonic, according to the relevant definition (§16.1c), all its following eight quadrilaterals will be harmonic:

ΑΔΗΘ, ΔΗΒΓ, ΗΒΕΖ, ΒΕΘΑ, ΕΘΓΔ, ΘΓΖΗ, ΓΖΑΒ, ΖΑΔΕ, (2).

Which are obviously the quadrilaterals of the convex octagon ΑΒΓΔΕΖΗΘ, thus this convex Octagon will also be harmonic. Hence, it will have ΑΕ∩ΒΖ∩ΓΗ∩ΔΘ≡K (Proposition 105, § 19.12).

This means that for example for the harmonic quadrilateral ΑΔΗΘ its diagonals ΑΗ, ΔΘ, will be conjugate with respect to circle (o), thus according to Auxiliary Proposition 6 (converse) in Part E, for triangle ΑΔΗ, ΔΘ is its symmedian. Hence, according to the direct of a known Theorem (Γεωμετρία [6] § 11-14.2), each point on ΔΘ will be far from ΑΔ, ΔΗ distance analogous to these sides. This will obviously be true for point K of the convergence of diagonals of this Octagon, which belongs to ΔΘ too, thus for point K, it will be:

$$\frac{\upsilon_\alpha}{\alpha}=\frac{\upsilon_\beta}{\beta}. \qquad (3).$$

If we work similarly to above for harmonic quadrilateral ΔΗΒΓ and more specifically in triangle ΔΗΒ, we prove that for K it is also:

$$\frac{\upsilon_\beta}{\beta}=\frac{\upsilon_\gamma}{\gamma}. \qquad (4).$$

Hence, if we continue as above for the rest six triangles ΗΒΕ, ΒΕΘ, ΕΘΓ, ΘΓΖ, ΓΖΑ, ΖΑΔ, we find that it is also:

$$\frac{\upsilon_\gamma}{\gamma}=\frac{\upsilon_\delta}{\delta},\ \frac{\upsilon_\delta}{\delta}=\frac{\upsilon_\varepsilon}{\varepsilon},\ \frac{\upsilon_\varepsilon}{\varepsilon}=\frac{\upsilon_\zeta}{\zeta},\ \frac{\upsilon_\zeta}{\zeta}=\frac{\upsilon_\eta}{\eta},\ \frac{\upsilon_\eta}{\eta}=\frac{\upsilon_\theta}{\theta},\ \frac{\upsilon_\theta}{\theta}=\frac{\upsilon_\alpha}{\alpha}. \qquad (5).$$

Therefore, from relations (3) – (5), we get relation (1), which is true for intersection K as well. This means that the desired point K' is K or K'≡K.
Hence the desired point, is the point where the diagonals of the Harmonic Asteroid Octagon coincide, which are the same as the ones for the respective Harmonic Octagon.

(b). Converse.

I.e., if the convex inscribed in circle (o) Asteroid Octagon ΑΔΗΒΕΘΓΖΑ has diagonals concurrent at K and for K relation (1) is true, then this Octagon will be harmonic.
In fact, since (1) is true, for K it will be:

$$\frac{\upsilon_\alpha}{\alpha}=\frac{\upsilon_\beta}{\beta}. \qquad (3).$$

This, according to the converse of the Theorem in § 11-14.2 in Geometry [6], shows that K belongs to the symmedian matching to the side ΑΗ in triangle ΑΔΗ and thus, according to the direct of the above Auxiliary Proposition 6 in Part E, ΔΘ and ΑΗ are conjugate with respect to circle (o), or quadrilateral ΑΔΗΘ is harmonic.

If we take into consideration, that the rest seven two-part equalities that occur from (1) are true, i.e.:

$$\frac{U_\beta}{\beta}=\frac{U_\gamma}{\gamma},\ \frac{U_\gamma}{\gamma}=\frac{U_\delta}{\delta},\ \frac{U_\delta}{\delta}=\frac{U_\varepsilon}{\varepsilon},\ \frac{U_\varepsilon}{\varepsilon}=\frac{U_\zeta}{\zeta},\ \frac{U_\zeta}{\zeta}=\frac{U_\eta}{\eta},\ \frac{U_\eta}{\eta}=\frac{U_\theta}{\theta},\ \frac{U_\theta}{\theta}=\frac{U_\alpha}{\alpha},\quad (6).$$

and if we work as above, we find that the rest seven quadrilaterals in (2), ΔΗΒΓ, ΗΒΕΖ, ΒΕΘΑ, ΕΘΓΔ, ΘΓΖΗ, ΓΖΑΒ, ΖΑΔΕ, are harmonic, thus, according to the definition of the Simple Asteroid Harmonic Octagon (§ 16.1c), Octagon ΑΔΗΒΕΘΓΖΑ is in fact Harmonic, as it is also inscribed in circle (o).

Remarks.

(a). Proposition 115 (criterion), we believe to be new and firstly appear here.

(b). Equivalent Propositions (Criteria) we have proved to be true for Harmonic Hexagons (Criterion 112) and Octagons (Criterion 113), etc.

(c). As we will see next, equivalent Criterion 116 is true for the Asteroid non regular Decagon ΑΔΗΚΓΖΙΒΕΘΑ, etc. I.e. that this is also harmonic, etc.

(d). Relevant Propositions are 112-114, 116-118.

Comment.

After Criterion 115 of Asteroid Harmonic Octagons, we also notice that:

(a). Point Lemoine in a triangle, has the property, according to which, this point is far from the sides of the triangle, distances analogous to its respective sides (the triangle's).

(b). As it is known and easily proven {Criterion 8ι(87) in our book [31]} , the intersection of the diagonals on each harmonic quadrilateral, has the same, as above, property (See also [32] IV, page 54). Hence the distance of this intersection from the sides of the quadrilateral, are analogous, to the respective sides of the quadrilateral. This is why the intersection of the diagonals of the harmonic quadrilateral, was characterized as the Lemoine point of the harmonic quadrilateral.

(c). According to Criterion 112 above in the Harmonic Hexagon, the intersection of its diagonals (main), has the same, as above, property, and this is why we named it the Lemoine point of the Harmonic Hexagon.

(d). According to Criterion 113 above for every Harmonic Octagon as we saw, the intersection of its diagonals (main), has the same, property.

(e). According to Criterion 114 above for every Harmonic Decagon, the intersection of its diagonals (main), has the same, property.

(f). According to Criterion 115 above for every Asteroid Harmonic Octagon, the intersection of its diagonals (main), has the same, as above, property, thus we reach the following conclusion:

Conclusion.

After the aforementioned and in equivalence to these, it occurs that:

The point of convergence of the diagonals of the Asteroid Harmonic Octagon, is a LEMOINE POINT, for the Harmonic Octagon.

Another formation of Criterion 115.

After the above, Criterion 115 (Criterion of Harmonic being of concave Inscribed Asteroid Octagons), can be stated in the following way as well:

Every Asteroid Octagon inscribed in circle with concurrent diagonals, is harmonic, if and only if, it has a LEMOINE POINT. (This coincides with the convergence point of its diagonals).

Alternative Definition of the Harmonic Asteroid Octagon.

Therefore, after the above, from now on we can also use for Asteroid Harmonic Octagons, the following alternative Definition:

««We call Asteroid Harmonic Octagon each asteroid cyclic Octagon, if and only if, it has a LEMOINE POINT».

19.25. Criterion of Harmonic being of an Asteroid Inscribed Decagon.

Extenstion to Criterion 115 to an Asteroid Decagon.

Does the Asteroid Harmonic Decagon have a Lemoine Point?

Criterion 116 (Figure 78).

If an asteroid inscribed in circle (o) decagon ΑΔΗΚΓΖΙΒΕΘΑ, has concurrent diagonals (main) and there exists on its plane a point for which the distance from each side (of the Asteroid Decagon), is analogous to the respective side, then and only then, this decagon is harmonic.

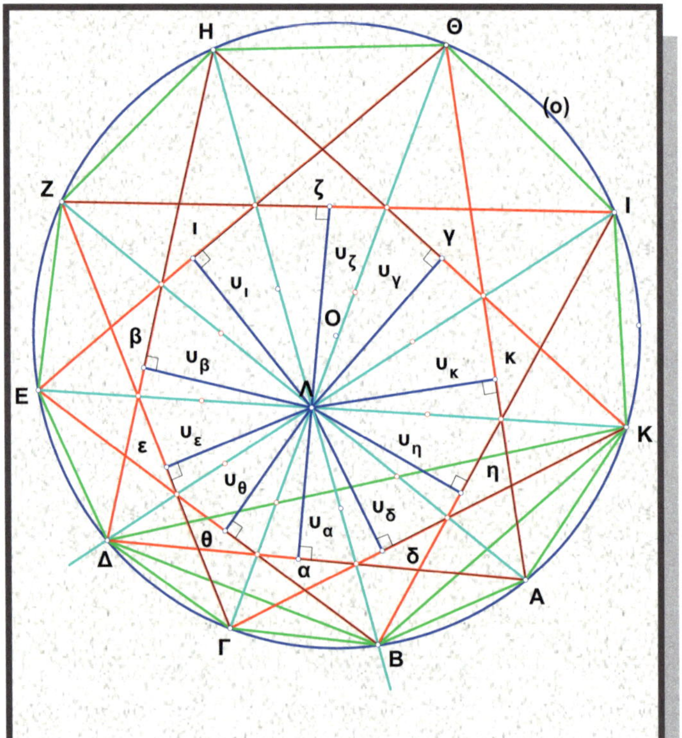

Σχήμα 78.

Proof.

(a). Direct.

We consider a Harmonic Asteroid Decagon ΑΔΗΚΓΖΙΒΕΘΑ, inscribed

in circle (o) (Relevant definition in § 16.1d). We will prove that a point Λ' exists on its plane, for which it is:

$$\frac{u_\alpha}{\alpha}=\frac{u_\beta}{\beta}=\frac{u_\gamma}{\gamma}=\frac{u_\delta}{\delta}=\frac{u_\varepsilon}{\varepsilon}=\frac{u_\zeta}{\zeta}=\frac{u_\eta}{\eta}=\frac{u_\theta}{\theta}=\frac{u_\iota}{\iota}=\frac{u_\kappa}{\kappa}, \qquad (1).$$

where u_α, u_β,...., u_κ, is the distance of point Λ', from each side of the Asteroid Decagon ΑΔ=α, ΔΗ=β, ΗΚ=γ,...,ΕΘ=ι, ΘΑ=κ, respectively.

In fact, since decagon ΑΔΗΚΓΖΙΒΕΘΑ is harmonic, according to Proposition 107 (§ 19.14), it will have ΑΖ∩ΒΗ∩ΓΘ∩ΔΙ∩ΕΚ≡Λ, while according to the relevant definition (§ 16.1d), all its following ten quadrilaterals will be harmonic:

ΑΔΗΙ, ΔΗΚΒ, ΗΚΓΕ, ΚΓΖΘ, ΓΖΙΑ, ΖΙΒΔ, ΙΒΕΗ, ΒΕΘΚ, ΕΘΑΓ, ΘΑΔΖ. (2).

This means that for example for the harmonic quadrilateral ΑΔΗΙ its diagonals ΑΗ, ΔΙ, will be conjugate with respect to circle (o), thus according to Auxiliary Proposition 6 (converse) in Part E, for triangle ΑΔΗ, ΔΙ is its symmedian. Hence, according to the direct of a known Theorem (Γεωμετρία [6] § 11-14.2), each point on ΔΙ will be far from ΑΔ, ΔΗ distance analogous to these sides. This will obviously be true for point Λ of the convergence of diagonals of this Octagon, which belongs to ΔΙ too, thus for point Λ, it will be:

$$\frac{u_\alpha}{\alpha}=\frac{u_\beta}{\beta}. \qquad (3).$$

If we work similarly to above for harmonic quadrilateral ΔΗΚΒ and more specifically in triangle ΔΗΚ, we prove that for Κ it is also:

$$\frac{u_\beta}{\beta}=\frac{u_\gamma}{\gamma}. \qquad (4).$$

Hence, if we continue as above for the rest eight triangles ΗΚΓ, ΚΓΖ, ΓΖΙ, ΖΙΒ, ΙΒΕ, ΒΕΘ, ΕΘΑ, ΘΑΔ, we find that for Λ it is also:

$$\frac{u_\gamma}{\gamma}=\frac{u_\delta}{\delta},\ \frac{u_\delta}{\delta}=\frac{u_\varepsilon}{\varepsilon},\ \frac{u_\varepsilon}{\varepsilon}=\frac{u_\zeta}{\zeta},\ \frac{u_\zeta}{\zeta}=\frac{u_\eta}{\eta},$$

$$\frac{u_\eta}{\eta}=\frac{u_\theta}{\theta},\ \frac{u_\theta}{\theta}=\frac{u_\iota}{\iota},\ \frac{u_\iota}{\iota}=\frac{u_\kappa}{\kappa},\ \frac{u_\kappa}{\kappa}=\frac{u_\alpha}{\alpha}. \qquad (5).$$

Therefore, from relations (3) – (5), we get relation (1), which is true for intersection Λ as well. This means that the desired point Λ' is Λ or Λ'≡Λ.

Hence the desired point, is the point where the diagonals of the Asteroid Decagon coincide.

(b). Converse.

I.e., if the convex inscribed in circle (o) Asteroid Decagon ΑΔΗΚΓΖΙΒΕΘΑ has diagonals concurrent at Λ and for Λ relation (1) is true, then this Decagon will be harmonic.

In fact, since (1) is true, for Λ it will be: $\frac{υ_α}{α}=\frac{υ_β}{β}$. (3).

This, according to the converse of the Theorem in § 11-14.2 in Geometry [6], shows that Λ belongs to the symmedian matching to the side ΑΗ in triangle ΑΔΗ and thus, according to the direct of the above Auxiliary Proposition 6 in Part E, ΔΙ and ΑΗ are conjugate with respect to circle (o), or quadrilateral ΑΔΗΙ is harmonic.

If we take into consideration, that the rest nine two-part equalities that occur from (1) are true, i.e.:

$$\frac{υ_β}{β}=\frac{υ_γ}{γ},\ \frac{υ_γ}{γ}=\frac{υ_δ}{δ},\ \frac{υ_δ}{δ}=\frac{υ_ε}{ε},\ \frac{υ_ε}{ε}=\frac{υ_ζ}{ζ},$$

$$\frac{υ_ζ}{ζ}=\frac{υ_η}{η},\ \frac{υ_η}{η}=\frac{υ_θ}{θ},\ \frac{υ_θ}{θ}=\frac{υ_ι}{ι},\ \frac{υ_ι}{ι}=\frac{υ_κ}{κ},\ \frac{υ_κ}{κ}=\frac{υ_α}{α}. \qquad (6).$$

and if we work as above, we find that the rest nine quadrilaterals in (2), ΔΗΚΒ, ΗΚΓΕ, ΚΓΖΘ, ΓΖΙΑ, ΖΙΒΔ, ΙΒΕΗ, ΒΕΘΚ, ΕΘΑΓ, ΘΑΔΖ, are harmonic, thus, according to the definition of the Simple Asteroid Harmonic Decagon (§ 16.1d), Decagon ΑΔΗΚΓΖΙΒΕΘΑ is in fact Harmonic, as it is also inscribed in circle (o).

[The construction of the Asteriod Harmonic Decagon (§ 16.1d), as we will later see (§ 19,65, Construction 150), is easy. This is carried through with based on the Construction of the convex Harmonic Decagon ΑΒΓΔΕΖΗΘΙΚ and Proposition 108].

Remarks.

(a). Proposition 116 (criterion), we believe to be new and firstly appear here.

(b). Equivalent Propositions (Criteria) we have proved to be true for Harmonic Hexagons (Criterion 112) and Octagons (Criterion 113), etc.

(c). As we will see next, equivalent Criterion is true for the convex Harmonic pentagon ΑΓΕΗΙ and for the Asteroid non regular Pentagon ΑΓΕΗΙΑ, etc.

(d). Relevant Propositions are 108, 112-115, 117, 118, 150.

Comment.

After Criterion 116 of Asteroid Harmonic Decagons, we also notice that:

(a). Point Lemoine in a triangle, has the property, according to which, this point is far from the sides of the triangle, distances analogous to its respective sides (the triangle's).

(b). As it is known and easily proven {Criterion 8ι(87) in our book [31]} , the intersection of the diagonals on each harmonic quadrilateral, has the same, as above, property (See also [32] IV, page 54). Hence the distance of this intersection from the sides of the quadrilateral, are analogous, to the respective sides of the quadrilateral. This is why the intersection of the diagonals of the harmonic quadrilateral, was characterized as the Lemoine point of the harmonic quadrilateral.

(c). According to Criterion 112 above in the Harmonic Hexagon, the intersection of its diagonals (main), has the same, as above, property, and this is why we named it the Lemoine point of the Harmonic Hexagon.

(d). According to Criterion 113 above for every Harmonic Octagon as we saw, the intersection of its diagonals (main), has the same, property.

(e). According to Criterion 114 above for every Harmonic Decagon, the intersection of its diagonals (main), has the same, property.

(f). According to Criterion 115 above for every Asteroid Harmonic Octagon, the intersection of its diagonals (main), has the same, as above, property.

(g). According to Criterion 116 above for every Asteroid Harmonic Decagon, the intersection of its diagonals (main), has the same, as above, property, thus we reach the following conclusion:

Conclusion.

After the aforementioned and in equivalence to these, it occurs that:

The point of convergence of the diagonals of the Asteroid Harmonic Decagon, is a LEMOINE POINT, for the Harmonic Octagon.

Another formation of Criterion 116.

After the above, Criterion 116 (Criterion of Harmonic being of convex Inscribed Asteroid Decagons), can be stated in the following way as well:

Every Asteroid Decagon inscribed in circle with concurrent diagonals, is harmonic, if and only if, it has a LEMOINE POINT. (This coincides with the convergence point of its diagonals).

Alternative Definition of the Harmonic Asteroid Decagon.

Therefore, after the above, from now on we can also use for Asteroid Harmonic Decagons, the following alternative Definition:

««We call Asteroid Harmonic Decagon each asteroid cyclic decagon, if and only if, it has a LEMOINE POINT».

19.26. Criterion of Harmonic being of an Inscribed Pentagon.

Extension to Criterion 116 to a Pentagon.

Does the Harmonic Pentagon have a Lemoine Point?

Criterion 117 (Figure 79).

If for convex inscribed in circle (o) Pentagon ΑΓΕΗΙ, there exists on its plane a point for which its distance from its side (of the Pentagon), is analogous to the respective side, then and only then this Pentagon is Harmonic.

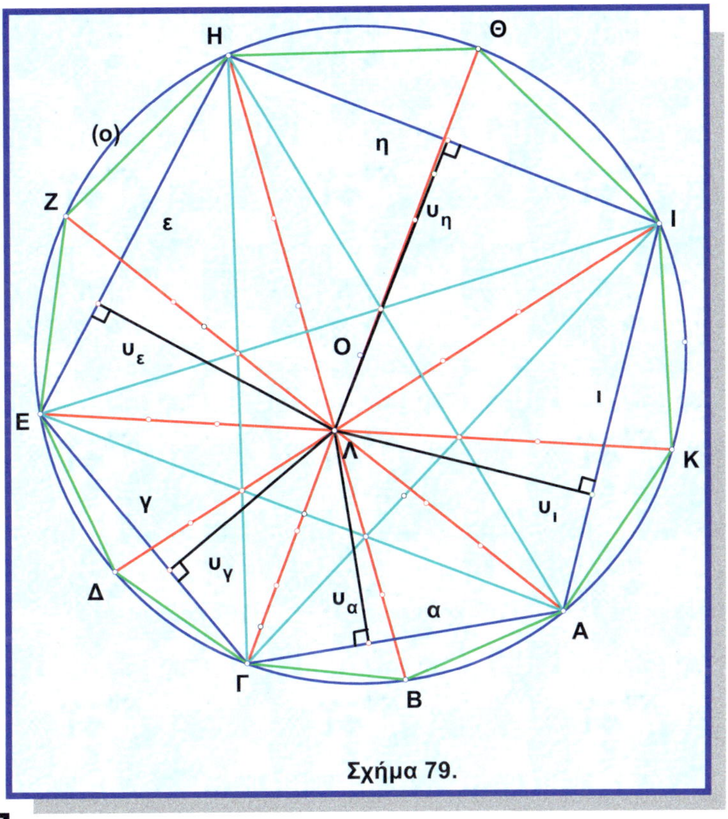

Σχήμα 79.

Proof.

(a). Direct.

We consider a Harmonic Pentagon ΑΓΕΗΙ, inscribed in circle (o) (Relevant terminology in § 16.1 e). We will show that there is a point Λ′ on its plane, for which it is:

$$\frac{υ_α}{α}=\frac{υ_γ}{γ}=\frac{υ_ε}{ε}=\frac{υ_η}{η}=\frac{υ_ι}{ι}, \qquad (1).$$

where $υ_α$, $υ_γ$, $υ_ε$, $υ_η$, $υ_ι$, is the distance of point Λ′, from each of its sides ΑΓ=α , ΓΕ=γ, ΕΗ=ε, ΗΙ=η, ΙΑ=ι, respectively.

In fact, since pentagon ΑΓΕΗΙ is Harmonic, according to the relevant definition (§ 16.1e), there will be points Β, Δ, Ζ, Θ, Κ, on circle (o), such that decagon ΑΒΓΔΕΖΗΘΙΚ, would be Harmonic, thus this will have concurrent diagonals, at point let us say Λ (AZ∩BH∩ΓΘ∩ΔI∩EK≡Λ) ([Proposition 106 (§ 19.13).

Moreover, since the convex decagon ΑΒΓΔΕΖΗΘΙΚ ia Harmonic, according to the direct of Proposition 108 (§ 19.16), the Asteroid Decagon ΑΔΗΚΓΖΙ-ΒΕΘΑ will also be Harmonic, which shares its diagonals with the above convex one, hence they will share point Λ too, thus for this, according to the relevant definition (§ 16,1d), all its ten following quadrilaterals will be Harmonic:

ΑΓΕΘ, ΒΔΖΙ, ΓΕΗΚ, ΔΖΘΑ, ΕΗΙΒ, ΖΘΚΓ, ΗΙΑΔ, ΘΚΒΕ, ΙΑΓΖ, ΚΒΔΗ. (2).

This means that for example quadrilateral EHIB is harmonic, then BH, EI, will be conjugate with respect to circle (o), thus according to Auxiliary Proposition 6 (converse) in Part E, for triangle EHI, HB is its symmedian and hence, according to the direct of a known Theorem (Γεωμετρία [6] § 11-14.2), each point on HB will be far from sides EH, HI, distance analogous to these sides. Hence from point Λ too. Thus for point Λ, it will be:

$$\frac{\upsilon_\varepsilon}{\varepsilon}=\frac{\upsilon_\eta}{\eta}. \qquad (3).$$

Similarly, for Λ and the Harmonic Quadrilaterals ΑΓΕΘ, ΓΕΗΚ, ΗΙΑΔ, ΙΑΓΖ, we find that also, respectively:. $\frac{\upsilon_\alpha}{\alpha}=\frac{\upsilon_\gamma}{\gamma}, \frac{\upsilon_\gamma}{\gamma}=\frac{\upsilon_\varepsilon}{\varepsilon}, \frac{\upsilon_\eta}{\eta}=\frac{\upsilon_\iota}{\iota}, \frac{\upsilon_\iota}{\iota}=\frac{\upsilon_\alpha}{\alpha}$. (4).

Therefore, from (3), (4), relation (1) occurs, which is true for Λ.

This means that the desired point Λ' is Λ or Λ'≡Λ.

Hence the desired point, is the point where the diagonals of the Harmonic Decagon coincide, from which it is produced.

(b). Converse.

I.e., if for the convex inscribed in circle (o) pentagon ΑΓΕΗΙ a point Λ exists, for which relation (1) is true, then we will prove that this pentagon is Harmonic.

We define the intersections Z, Θ, K, B, Δ, of ΑΛ, ΓΛ, ΕΛ, ΗΛ, ΙΛ, with circle (o) respectively.

Hence, since relation (1) is true, for Λ, it will be for example:

$$\frac{U_\varepsilon}{\varepsilon}=\frac{U_\eta}{\eta}. \qquad (3).$$

This, according to the converse of the Theorem in § 11-14.2 in Geometry [6], means that Λ belongs to the symmedian that matches to the side EI of triangle EHI and thus, according to the direct of the above Auxiliary Proposition 6, in Part E, ΛH or BH and EI are conjugate with respect to circle (o) [where B is the intersection of HΛ with circle (o)], or that quadrilateral EHIB is Harmonic.

If we take into to consideration, the fact that the rest four two-part equalities in (1) are true, i.e.:

$$\frac{U_\alpha}{\alpha}=\frac{U_\gamma}{\gamma},\quad \frac{U_\gamma}{\gamma}=\frac{U_\varepsilon}{\varepsilon},\quad \frac{U_\eta}{\eta}=\frac{U_\iota}{\iota},\quad \frac{U_\iota}{\iota}=\frac{U_\alpha}{\alpha}, \qquad (4).$$

and if we work similarly to above, we find that the rest four quadrilaterals in (2), ΑΓΕΘ, ΓΕΗΚ, ΗΙΑΔ, ΙΑΓΖ, are Harmonic.

Hence, since for example quadrilateral ΑΓΕΘ, is Harmonic and ΑΖ∩ΓΘ∩ΕΚ≡Λ, according to Auxiliary Proposition 34, quadrilateral ΖΘΚΓ will also be Harmonic.

In a similar way we find that the following quadrilaterals are harmonic too:

ΒΔΖΙ, ΔΖΘΑ, ΘΚΒΕ, ΚΒΔΗ. (5).

Thus, all the quadrilaterals in (2) are harmonic, hence the Asteroid Decagon ΑΔΗΚΓΖΙΒΕΘΑ (definition § 16.1d), will be harmonic and thus ΑΒΓΔΕΖΗΘΙΚ will also be harmonic (Proposition 108, converse).

Hence, according to the relevant definition (§ 16.1e), Pentagon ΑΓΕΗΙ will in fact be harmonic, as it is also inscribed in circle (o).

Remarks.

(a). Proposition 117 (criterion), we believe to be new and firstly appear here.

(b). Equivalent Propositions (Criteria) we have proved to be true for Harmonic Hexagons (Criterion 112), Octagons (Criterion 113), etc.

(c). As we already saw, equivalent Criterion is true for the Asteroid non regular Decagon ΑΔΗΚΓΖΙΒΕΘΑ [Proposition 116 (§ 19.25)].

(d). As we will later see, equivalent Criterion is also true for the Asteroid non Regular Pentagon ΑΕΙΓΗΑ (Proposition 118), etc.

(e). Relevant Propositions are 112-116, 118.

Comment.

After Criterion 117 of Harmonic Pentagons, we also notice that:

(a). Point Lemoine in a triangle, has the property, according to which, this point is far from the sides of the triangle, distances analogous to its respective sides (the triangle's).

(b). As it is known and easily proven {Criterion 8ı(87) in our book [31]} , the intersection of the diagonals on each harmonic quadrilateral, has the same, as above, property (See also [32] IV, page 54). Hence the distance of this intersection from the sides of the quadrilateral, are analogous, to the respective sides of the quadrilateral. This is why the intersection of the diagonals of the harmonic quadrilateral, was characterized as the Lemoine point of the harmonic quadrilateral.

(c). According to Criterion 112 above in the Harmonic Hexagon, the intersection of its diagonals (main), has the same, as above, property, and this is why we named it the Lemoine point of the Harmonic Hexagon.

(d). According to Criterion 113 above for every Harmonic Octagon as we saw, the intersection of its diagonals (main), has the same, property.

(e). According to Criterion 114 above for every Harmonic Decagon, the intersection of its diagonals (main), has the same, property.

(f). According to Criteria 115 and 116 above for every Asteroid Harmonic Octagon, as well as for every Asteroid Harmonic Decagon the intersection of its diagonals (main), has the same, as above, property.

(g). According to Criterion 117 above for every convex Harmonic Pentagon, the intersection of the diagonals (main) of the decagon from which this is created, has the same, as above, property, thus we reach the following conclusion:

Conclusion.

After the aforementioned and in equivalence to these, it occurs that:

The point of convergence of the diagonals of every Harmonic Decagon, is a LEMOINE POINT for its twin Harmonic Pentagons as well.

Another formation of Criterion 117.

After the above, Criterion 117 (Criterion of Harmonic being of convex Inscribed Pentagons), can be stated in the following way as well:

Every pentagon inscribed in circle, is harmonic, if and only if, it has a LEMOINE POINT. (This coincides with the convergence point of the diagonals of the Harmonic Decagon, from which it is created).

Alternative Definition of the Harmonic Pentagon.

Therefore, after the above, from now on we can also use for Harmonic Pentagons, the following alternative Definition:

«We call Harmonic Pentagon each cyclic pentagon, if and only if, it has a LEMOINE POINT».

19.27. Criterion of Harmonic being of an Inscribed Asteroid Pentagon.

Extension to Criterion 117 to an Asteroid Pentagon.

Does the Asteroid Harmonic Pentagon have a Lemoine Point?

Criterion 118 (Figure 80).

If for convex inscribed in circle (o) Asteroid Pentagon ΑΕΙΓΗΑ, there exists on its plane a point for which its distance from its side (of the Asteroid Pentagon), is analogous to the respective side, then and only then this Pentagon is Harmonic.

Proof.

(a). Direct.

We consider the Asteroid Harmonic Pentagon ΑΕΙΓΗΑ, inscribed in circle (o) (Relevant terminology in § 16.1 e). We will show that there is a point Λ′ on its plane, for which it is:

$$\frac{u_\alpha}{\alpha}=\frac{u_\varepsilon}{\varepsilon}=\frac{u_\iota}{\iota}=\frac{u_\gamma}{\gamma}=\frac{u_\eta}{\eta}, \qquad (1).$$

where $υ_α$, $υ_ε$, $υ_ι$, $υ_γ$, $υ_η$, is the distance of point Λ′, from each of its sides A-E=α, EI=ε, IΓ=ι, ΓH=γ, HA=η, respectively.

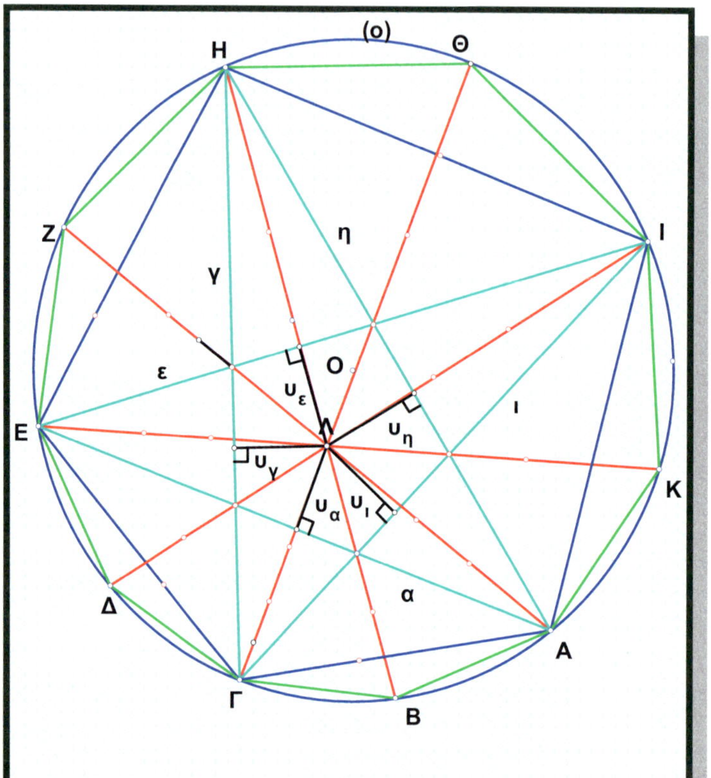

Σχήμα 80.

Since from the implication, the Asteroid Pentagon AEIΓHA is harmonic, according to the relevant definition [§ 16.1 e (b)], between its consecutive apexes, there will be one point of the circle (o), from the group of five points Θ, B, Z, K, Δ, respectively, such that the Asteroid Decagon AΔHKZI-BEΘA, would be harmonic.

Thus, since the Asteroid Decagon AΔHKZIBEΘA, is harmonic, then the convex decagon ABΓΔEZHΘIK will also be harmonic (converse of Proposition 108), while these will also have AZ∩BH∩ΓΘ∩ΔI∩EK≡Λ ([Propositions 106 (§ 19.13) and 107 (§ 19.14).

Thus, since decagon ΑΒΓΔΕΖΗΘΙΚ is Harmonic, according to the relevant definition (§ 16.1d), all its ten quadrilaterals will be harmonic:

ΑΒΓΗ, ΒΓΔΘ, ΓΔΕΙ, ΔΕΖΚ, ΕΖΗΑ, ΖΗΘΒ, ΗΘΙΓ, ΘΙΚΔ, ΙΚΑΕ, ΚΑΒΖ. (2).

Therefore, since for example quadrilateral ΙΚΑΕ is Harmonic, its diagonals ΚΕ, ΙΑ will be conjugate with respect to circle (o). This means that ΕΚ is symmedian of the triangle ΑΕΙ [Auxiliary Proposition 6 (converse) in Part E. Thus all the points on ΕΚ will be far from ΑΕ, ΕΙ analogous distance $υ_α$, $υ_ε$ (Theorem in § 11-14.2 in Geometry [6]). Thus for point Λ as well.

Hence for Λ it will be: $$\frac{υ_α}{α}=\frac{υ_ε}{ε}. \quad (3).$$

In the exact same way, for Λ, from the harmonic quadrilaterals ΓΔΕΙ, ΗΘΙΓ, ΑΒΓΗ, ΕΖΗΑ, respectively, we also find that:

$$\frac{υ_ε}{ε}=\frac{υ_ι}{ι},\ \frac{υ_ι}{ι}=\frac{υ_γ}{γ},\ \frac{υ_γ}{γ}=\frac{υ_η}{η},\ \frac{υ_η}{η}=\frac{υ_α}{α}. \quad (4).$$

Thus, from relations (3), (4), relation (1) occurs, which is true for point Λ too.

This means that the desired point Λ' is Λ or Λ'≡Λ.

Hence the desired point, is the point where the diagonals of the Asteroid Decagon coincide, from which the pentagon is created.

(b). Converse.

I.e., if for the inscribed in circle (o) Asteroid Pentagon ΑΕΙΓΗΑ a point Λ exists, for which relation (1) is true, then we will prove that this pentagon is Harmonic.

In fact, if for Λ relation (1) is true and ΑΛ, ΕΛ, ΙΛ, ΓΛ, ΗΛ, re-intersect circle (o) at points Ζ, Κ, Δ, Θ, Β, respectively, then because of (1), it will be for example: $$\frac{υ_α}{α}=\frac{υ_ε}{ε}. \quad (3).$$

This means that in triangle ΕΑΙ, ΕΚ is symmedian (Theorem in § 11-14.2 in Geometry [6]), thus ΕΚ, ΙΑ will be conjugate with respect to circle (o) [Auxiliary Proposition 6 (direct) in Part E] and thus quadrilateral ΙΚΑΕ is Harmonic.

In the exact similar way, if we take into consideration relation (1), we also find that quadrilaterals: ΑΒΓΗ, ΓΔΕΙ, ΕΖΗΑ, ΗΘΙΓ, are Harmonic.

Therefore, since for example quadrilateral ΙΚΑΕ is harmonic and it is ΑΖ∩ΔΙ∩ΕΚ≡Λ, quadrilateral ΔΕΖΚ will also be harmonic (Criterion 34).

In a similar way we find that the following quadrilaterals are harmonic too:

ΖΗΘΒ, ΘΙΚΔ, ΚΑΒΖ, ΒΓΔΘ.

Thus all ten quadrilaterals, in relation (2) are harmonic. This means that the convex decagon ΑΒΓΔΕΖΗΘΙΚ is also harmonic, thus the Asteroid Decagon ΑΔΗΚΖΙΒΕΘΑ is also Harmonic (Proposition 108).

This shows that, since points Ζ, Κ, Δ, Θ, Β exist and this Asteroid Decagon is harmonic, then the Asteroid Pentagon ΑΕΙΓΗΑ will be harmonic too (definition § 16.1 e).

Remarks.

(a). Proposition 118 (criterion), we believe to be new and firstly appear here.

(b). Equivalent Propositions (Criteria) we have proved to be true for Harmonic Hexagons (Criterion 112), Octagons (Criterion 113), etc.

(c). As we already saw, equivalent Criterion is true for the Asteroid non regular Pentagon ΑΓΕΗΙ [Proposition 117 (§ 19.26)].

(d). Relevant Propositions are 34, 106-108, 112-117.

Comment.

After Criterion 118 of Asteroid Harmonic Pentagons, we also notice that:

(a). Point Lemoine in a triangle, has the property, according to which, this point is far from the sides of the triangle, distances analogous to its respective sides (the triangle's).

(b). As it is known and easily proven {Criterion 8ι(87) in our book [31]} , the intersection of the diagonals on each harmonic quadrilateral, has the same, as above, property. Hence the distance of this intersection from the sides of the quadrilateral, are analogous, to the respective sides of the quadrilateral. This is why the intersection of the diagonals of the harmonic quadrilateral, was characterized as the Lemoine point of the harmonic quadrilateral.

(c). According to Criterion 112 above in the Harmonic Hexagon, the intersection of its diagonals (main), has the same, as above, property, and this is why we named it the Lemoine point of the Harmonic Hexagon.
(d). According to Criterion 113 above for every Harmonic Octagon as we saw, the intersection of its diagonals (main), has the same, property.
(e). According to Criterion 114 above for every Harmonic Decagon, the intersection of its diagonals (main), has the same, property.
(f). According to Criteria 115, 116 and 117 above for every Asteroid Harmonic Octagon, every Asteroid Harmonic Decagon , as well as for every Harmonic Pentagon, respectively, the intersection of its diagonals (main), has the same, as above, property.
(g). According to Criterion 118 above for every convex Asteroid Harmonic Pentagon, the intersection of the diagonals (main) of the decagon from which this is created, has the same, as above, property, thus we reach the following conclusion:

Conclusion.

After the aforementioned and in equivalence to these, it occurs that:

The point of convergence of the diagonals of every Harmonic Decagon, is a LEMOINE POINT for its twin Asteroid Harmonic Pentagons as well.

Another formation of Criterion 118.

After the above, Criterion 118 (Criterion of Harmonic being of Asteroid Inscribed Pentagons), can be stated in the following way as well:

Every asteroid pentagon inscribed in circle, is harmonic, if and only if, it has a LEMOINE POINT. (This coincides with the convergence point of the diagonals of the Harmonic Decagon, from which it is created).

Alternative Definition of the Harmonic Asteroid Pentagon.

Therefore, after the above, from now on we can also use for Asteroid Harmonic Pentagons, the following alternative Definition:

«We call Asteroid Harmonic Pentagon each cyclic asteroid pentagon, if and only if, it has a LEMOINE POINT».

19.28. General Conclusion of Harmonic Decagons and their Twin Pentagons.

We have proved in Criterion 114, that the convex Harmonic Decagon has a Lemoine point which coincides with the intersection of its diagonals (main). Additionally we have proved in Criterion 116, that the respective Asteroid Harmonic Decagon of this Harmonic Decagon, has a Lemoine point that also coincides with tha above intersection of the diagonals (main) of the respective convex Harmonic Decagon (they obviously have main diagonals). Moreover we have proven in Criteria 117 and 118 above, that both the two convex Harmonic Pentagons, (twin), as well as their respective Asteroid Harmonic Pentagons (twin), that its apexes are the one by one apexes of a convex Harmonic Decagon, have Lemoine points that also coincide with the above intersection, of the diagonals (main) of the Harmonic Decagon, from which they are created.

Hence, each convex Harmonic Decagon, its twin convex Harmonic Pentagons, its Asteroid Harmonic Decagon and its twin Asteroid Harmonic Pentagons, that their sides are diagonals (not main) of this Harmonic Decagon, have a common Lemoine point that coincides, for all these, with the intersection of the diagonals (main) of the convex Harmonic Decagon.

19.29, General Conclusion concerning Multilaterals.

If we take into consideration Criteria 112-118 and if we continue for a Harmonic Dodecagon, etc, we will realize that there are true equivalent conclusions to the ones in Criteria 112-118 as well, thus we reach the following General Conclusion:

The convergence point of the diagonals of every Harmonic Multilateral, is a LEMOINE POINT for this Harmonic Multilateral.

19.30. General Formation of Criteria 112-118 for Multilaterals as well.

If we take into consideration the above, as well as Criteria 112-118, we easily realize that the following Criterion about Multilaterals is true as well:

Each convex or Asteroid Multilateral inscribed in circle, with concurrent diagonals (main), is Harmonic, if and only if, is has a LEMOINE POINT, which coincides with the convergence point of its diagonals as well, or of the diagonals of the convex Multilateral to which this one belongs (As for example the Pentagon, Heptagon, etc, that don't have main diagonals, but the convex Decagon, Tetradecagon, etc, respectively, do have, to which they belong as twins).

19. 31. Alternative Definition of the Harmonic Multilateral.

Consequently, due to the above, from now on we can use for Harmonic Multilaterals as well, the alternative Definition below:

«We call Harmonic Multilateral each convex or asteroid cyclic Multilateral, if and only if, this has a LEMOINE POINT.

19.32. Significant Property of the Simple Harmonic Hexagon.

Extension to Proposition 84, from a line to a circle.

Proposition 119 (Figure 64).

For every Simple Harmonic Hexagon ΑΒΓΔΕΖ, it is': $\frac{AB}{B\Gamma}\cdot\frac{\Gamma\Delta}{\Delta E}\cdot\frac{EZ}{ZA}=1.$ (1).

Proof.

1st Way (Based on Auxiliary Proposition 5 in Part E).

Since, according to Proposition 104 (§ 19.11), every Harmonic Hexagon has concurrent diagonals (main), according to the direct of Auxiliary Proposition 5 (§ 2.5) in Part E, relation (1) above will in fact be true.

2nd Way (Based on the relevant Definition).

Since this hexagon is harmonic, its quadrilaterals ΑΒΓΕ, ΓΔΕΑ, ΕΖΑΓ will be harmonic too (definition § 16.1b), for which it will be respectively:

$$\frac{AB}{B\Gamma}=\frac{AE}{E\Gamma},\ \frac{\Gamma\Delta}{\Delta E}=\frac{\Gamma A}{AE},\ \frac{EZ}{ZA}=\frac{E\Gamma}{\Gamma A}. \qquad (2).$$

Thus, from relations (2), if we multiply the respective parts and after the proper reductions, we get: $\frac{AB}{B\Gamma}\cdot\frac{\Gamma\Delta}{\Delta E}\cdot\frac{EZ}{ZA}=\frac{AE}{E\Gamma}\cdot\frac{\Gamma A}{AE}\cdot\frac{E\Gamma}{\Gamma A}=1 \Rightarrow (1).$

Remarks.

(a). Proposition 119, we believe to be new and firstly appear here.

(b). As we already saw in paragraph 16.1 a, equivalent Proposition, is true for the Harmonic Quadrilateral ΑΒΓΔ as well (For which it is $\frac{AB}{B\Gamma}\cdot\frac{\Gamma\Delta}{\Delta A}=1$), while Equivalent Propositions we will prove to be true for Harmonic Octagons, Decagons, etc (Propositions 120, 121, etc, respectively).

(c). This Proposition is not a Criterion of Harmonic being, as its converse is not true.

(d). This Property is true for Harmonic Collinear Groups of Six Points as well (See Proposition 84).

(e). Relevant Propositions are 84, 104, 120, 121.

19.33. Significant Property of the Simple Harmonic Octagon.

Extension to Proposition 85, from a line to a circle.

Proposition 120 (Figure 65).

For every Simple Harmonic Octagon ΑΒΓΔΕΖΗΘ and its respective Asteroid Octagon ΑΖΓΘΕΒΗΔΑ, it is:

$$\frac{AB}{B\Gamma}\cdot\frac{\Gamma\Delta}{\Delta E}\cdot\frac{EZ}{ZH}\cdot\frac{H\Theta}{\Theta A}=\frac{AZ}{Z\Gamma}\cdot\frac{\Gamma\Theta}{\Theta E}\cdot\frac{EB}{BH}\cdot\frac{H\Delta}{\Delta A}=1. \qquad (1).$$

Proof.

Since, according to the relevant Definition (§ 16.1c), for the Harmonic Octagon ΑΒΓΔΕΖΗΘ, for example quadrialterals ΑΒΓΖ, ΕΖΗΒ, are Harmonic, the pair of points Β, Ζ, harmonically divides the two pairs of points Α, Γ and Ε, Η. Thus, according to Proposition 38 (§ 8.75) and since it is ΑΒ/ΒΓ=ΑΖ/ΖΓ, it will also be:

$$\frac{AB^2}{B\Gamma^2}=\frac{AZ^2}{Z\Gamma^2}=\frac{AH}{H\Gamma}\cdot\frac{AE}{E\Gamma}. \qquad (2).$$

Moreover, since quadrilaterals ΓΔΕΘ, ΗΘΑΔ, are Harmonic too (definition) and since it is ΓΔ/ΔΕ=ΓΘ/ΘΕ, in the exact same way we find that it is:

$$\frac{\Gamma\Delta^2}{\Delta E^2}=\frac{\Gamma\Theta^2}{\Theta E^2}=\frac{\Gamma A}{AE}\cdot\frac{\Gamma H}{HE}. \qquad (3),$$

If we continue as above, we also find that:

$$\frac{EZ^2}{ZH^2}=\frac{EB^2}{BH^2}=\frac{E\Gamma}{\Gamma H}\cdot\frac{EA}{AH}, \qquad \frac{H\Theta^2}{\Theta A^2}=\frac{H\Delta^2}{\Delta A^2}=\frac{HE}{EA}\cdot\frac{H\Gamma}{\Gamma A}. \qquad (4).$$

Thus, from relations (2)-(4), if we multiply their respective parts and after the proper reductions we have relation (1).

Remarks.

(a). This Proposition is not a Criterion of Harmonic being, as its converse is not true.

(b). We notice that equivalent Propositions, are true for the Harmonic Quadrilateral ΑΒΓΔ as well (For which it is $\frac{AB}{B\Gamma}\cdot\frac{\Gamma\Delta}{\Delta A}$=1) and for Harmenic HExagon ΑΒΓΔΕΖ too (Proposition 119).

(c). This Property is true for Harmonic Collinear Groups of Eight Points as well [See Proposition 85 (§ 15.28)].

(d). The second part in (1), shows us that Proposition 120, is true for the Asteroid Harmonic Octagon ΑΖΓΘΕΒΗΔΑ.

(e).This Proposition we believe to be new and firstly appear here.

(f). Relevant Propositions are 38, 85, 119, 121.

19.34. Significant Property of the Simple Harmonic Decagon..

Extension to Proposition 86, from a line to circle.

Proposition 121 (Figure 81).

For every Simple Harmonic Decagon ΑΒΓΔΕΖΗΘΙΚ and its matching Asteroid Decagon ΑΘΕΒΙΖΓΚΗΔΑ, it is:

$$\frac{AB}{B\Gamma}\cdot\frac{\Gamma\Delta}{\Delta E}\cdot\frac{EZ}{ZH}\cdot\frac{H\Theta}{\Theta I}\cdot\frac{IK}{KA}=\frac{A\Theta}{\Theta E}\cdot\frac{EB}{BI}\cdot\frac{IZ}{Z\Gamma}\cdot\frac{\Gamma K}{KH}\cdot\frac{H\Delta}{\Delta A}=1. \qquad (1).$$

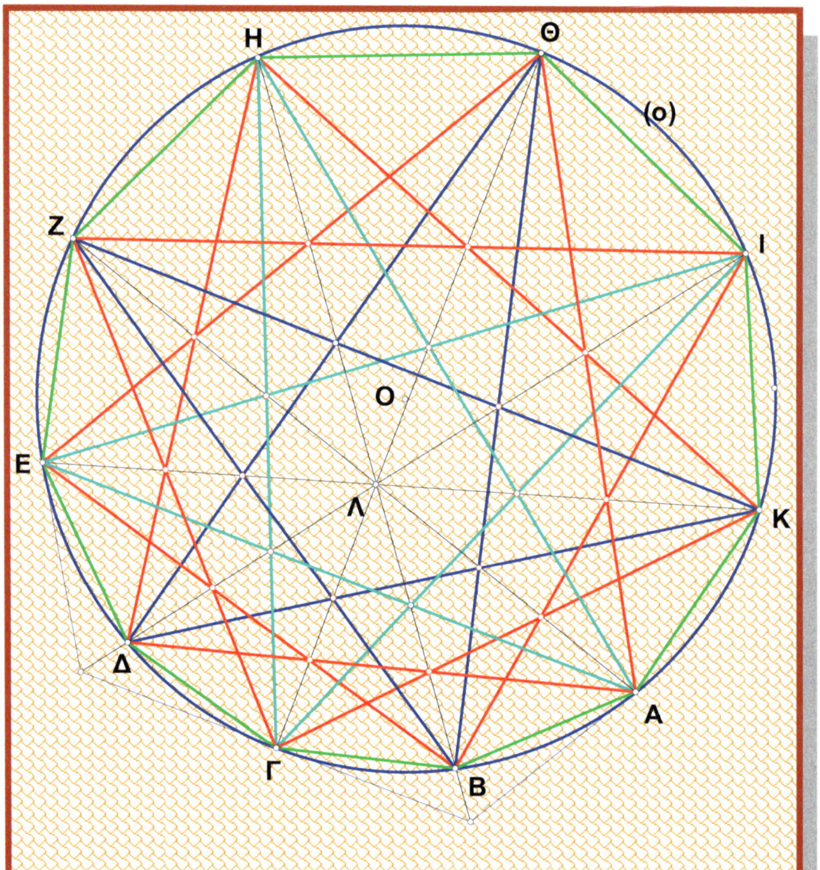

Σχήμα 81.

Proof.

1st Way (Based on Proposition 38).

Since, according to the relevant definition (§ 16.1d), for the Harmonic Decagon ΑΒΓΔΕΖΗΘΙΚ, for example quadrilaterals ΑΒΓΗ and ΖΗΘΒ, are Harmonic, the pair of points Β, Η harmonically divides the two pairs of points Α, Γ and Ζ, Θ. Thus, according to Proposition 38 (§ 8.75) and since it is ΑΒ/ΒΓ=ΑΗ/ΗΓ, it will also be:

$$\frac{AB^2}{B\Gamma^2}=\frac{AH^2}{H\Gamma^2}=\frac{A\Theta}{\Theta\Gamma}\cdot\frac{AZ}{Z\Gamma}. \qquad (2).$$

Additionally, since quadrilaterals ΓΔΕΙ and ΘΙΚΔ, are Harmonic as well (definition) and since it is ΓΔ/ΔΕ=ΓΙ/ΙΕ, exactly similarly, we also find that it is:

$$\frac{\Gamma\Delta^2}{\Delta E^2}=\frac{\Gamma I^2}{IE^2}=\frac{\Gamma K}{KE}\cdot\frac{\Gamma\Theta}{\Theta E}. \qquad (3),$$

If we continue as above, we also find that:

$$\frac{EZ^2}{ZH^2}=\frac{EA^2}{AH^2}=\frac{EB}{BH}\cdot\frac{EK}{KH},$$

$$\frac{H\Theta^2}{\Theta I^2}=\frac{H\Gamma^2}{\Gamma I^2}=\frac{H\Delta}{\Delta I}\cdot\frac{HB}{BI},$$

$$\frac{IK^2}{KA^2}=\frac{IE^2}{EA^2}=\frac{IZ}{ZA}\cdot\frac{I\Delta}{\Delta A}. \qquad (4).$$

Hence, if we multiply, the respective parts in relations: (2), (3), (4), we get:

$$\frac{AB^2}{B\Gamma^2}\cdot\frac{\Gamma\Delta^2}{\Delta E^2}\cdot\frac{EZ^2}{ZH^2}\cdot\frac{H\Theta^2}{\Theta I^2}\cdot\frac{IK^2}{KA^2}=\frac{A\Theta}{\Theta E}\cdot\frac{EB}{BI}\cdot\frac{IZ}{Z\Gamma}\cdot\frac{\Gamma K}{KH}\cdot\frac{H\Delta}{\Delta A}=$$

$$\frac{AH^2}{H\Gamma^2}\cdot\frac{\Gamma I^2}{IE^2}\cdot\frac{EA^2}{AH^2}\cdot\frac{H\Gamma^2}{\Gamma I^2}\cdot\frac{IE^2}{EA^2}=1, \qquad (5).$$

as the third part in (5) is reducted and becomes equal to 1.

Hence, from (5), we easily get relation (1).

2nd Way (Based on Proposition 37).

Since, according to the relevant definition (§ 16.1d) for the Harmonic Decagon ΑΒΓΔΕΖΗΘΙΚ, for example quadrilaterals ΑΒΓΗ and ΖΗΘΒ, are Harmonic, the pair of points Β, Η, harmonically divides both pairs of points

A, Γ and Z, Θ. Thus, according to Proposition 37 (§ 8.74) and since, $\frac{AB}{B\Gamma}=\frac{AH}{H\Gamma}$, it will be :

$\frac{AB}{B\Gamma}\cdot\frac{\Gamma Z}{ZB}\cdot\frac{B\Theta}{\Theta A}=1$ (6) and $\frac{AH}{H\Gamma}\cdot\frac{\Gamma Z}{ZB}\cdot\frac{B\Theta}{\Theta A}=1.$ (7).

Moreover, since quadrilaterals ΓΔΕΙ andı ΘΙΚΔ, are also Harmonic (definition) and since it is $\frac{\Gamma\Delta}{\Delta E}=\frac{\Gamma I}{IE}$, exactly similarly we find that:

$\frac{\Gamma\Delta}{\Delta E}\cdot\frac{E\Theta}{\Theta\Delta}\cdot\frac{\Delta K}{K\Gamma}=1$ (8) and $\frac{\Gamma I}{IE}\cdot\frac{E\Theta}{\Theta\Delta}\cdot\frac{\Delta K}{K\Gamma}=1.$ (9).

If we continue as above, we also find that:

$\frac{EZ}{ZH}\cdot\frac{HK}{KZ}\cdot\frac{ZB}{BE}=1,$ (10) and $\frac{EA}{AH}\cdot\frac{HK}{KZ}\cdot\frac{ZB}{BE}=1,$ (11).

$\frac{H\Theta}{\Theta I}\cdot\frac{IB}{B\Theta}\cdot\frac{\Theta\Delta}{\Delta H}=1,$ (12) and $\frac{H\Gamma}{\Gamma I}\cdot\frac{IB}{B\Theta}\cdot\frac{\Theta\Delta}{\Delta H}=1,$ (13).

$\frac{IK}{KA}\cdot\frac{A\Delta}{\Delta K}\cdot\frac{KZ}{ZI}=1,$ (14) and $\frac{IE}{EA}\cdot\frac{A\Delta}{\Delta K}\cdot\frac{KZ}{ZI}=1.$ (15).

Thus, if we multiply the respective parts, of relations: (6), (8), (10), (12), (14), we get:

$$\frac{AB}{B\Gamma}\cdot\frac{\Gamma\Delta}{\Delta E}\cdot\frac{EZ}{ZH}\cdot\frac{H\Theta}{\Theta I}\cdot\frac{IK}{KA}=\frac{A\Theta}{\Theta E}\cdot\frac{EB}{BI}\cdot\frac{IZ}{Z\Gamma}\cdot\frac{\Gamma K}{KH}\cdot\frac{H\Delta}{\Delta A}. \quad (16).$$

Similarly from (7), (9), (11), (13), (15), it easily occurs that:

$$\frac{A\Theta}{\Theta E}\cdot\frac{EB}{BI}\cdot\frac{IZ}{Z\Gamma}\cdot\frac{\Gamma K}{KH}\cdot\frac{H\Delta}{\Delta A}=\frac{AH}{H\Gamma}\cdot\frac{\Gamma I}{IE}\cdot\frac{EA}{AH}\cdot\frac{H\Gamma}{\Gamma I}\cdot\frac{IE}{EA}=1, \quad (17).$$

As the second part in (17) is reducted and becomes equal to 1.

Therefore, from (16) and (17), we easily get relation (1).

3rd Way (Based on the combination of Propositions 38 and 37).

Working similarly to way 1 above, from the first and third part in (2), (3), (4), we get:

$$\frac{AB^2}{B\Gamma^2}\cdot\frac{\Gamma\Delta^2}{\Delta E^2}\cdot\frac{EZ^2}{ZH^2}\cdot\frac{H\Theta^2}{\Theta I^2}\cdot\frac{IK^2}{KA^2}=\frac{A\Theta}{\Theta E}\cdot\frac{EB}{BI}\cdot\frac{IZ}{Z\Gamma}\cdot\frac{\Gamma K}{KH}\cdot\frac{H\Delta}{\Delta A}. \quad (18).$$

Consequently, working as in way 2 above and only from (6), (8), (10), (12), (14), we get (16).

Hence from (16) and (18), we easily get:

$$\frac{AB^2}{B\Gamma^2}\cdot\frac{\Gamma\Delta^2}{\Delta E^2}\cdot\frac{EZ^2}{ZH^2}\cdot\frac{H\Theta^2}{\Theta I^2}\cdot\frac{IK^2}{KA^2}=\frac{AB}{B\Gamma}\cdot\frac{\Gamma\Delta}{\Delta E}\cdot\frac{EZ}{ZH}\cdot\frac{H\Theta}{\Theta I}\cdot\frac{IK}{KA}\Rightarrow$$

$\frac{AB}{B\Gamma}\cdot\frac{\Gamma\Delta}{\Delta E}\cdot\frac{EZ}{ZH}\cdot\frac{H\Theta}{\Theta I}\cdot\frac{IK}{KA}=1$, thus from this and (16), desired relation (1) occurs.

Remarks.

(a). Proposition 121 is not a Criterion of Harmonic being, as its converse is not true.

(b). We notice that equivalent Propositions, are true for the Harmonic Quadrilateral ΑΒΓΔ as well (For which it is $\frac{AB}{B\Gamma}\cdot\frac{\Gamma\Delta}{\Delta A}=1$), for the Harmonic Hexagon ΑΒΓΔΕΖ and for Octagon ΑΒΓΔΕΖΗΘ (Propositions 119 and 120, respectively).

(c). This Property is true for Harmonic Collinear Groups of Ten Points as well (Proposition 86).

(d). The second part in (1), shows us that Proposition 121, is true for the Asteroid Harmonic Decagon ΑΘΕΒΙΖΓΚΗΔΑ.

(e).This Proposition we believe to be new and firstly appear here.

(f). Relevant Propositions are 37, 38, 86, 119, 120.

19.35. General Conclusion.

Due to Propositions 119-121 above and if we work in the same way as in their proofs, we reach the following conclusion:

Equivalent Proposition, to Propositions 119-121, can be made and proven to be true for any other Harmonic Multilateral with an even number of sides.

19.36. Thirteenth Criterion of Harmonic being of an Inscribed in Circle Quadrilateral.

Notable Property of the Lemoine Point of a Simple Harmonic Quadrilateral.

Criterion 122 (Figure 82).

Every convex Simple Quadrilateral ΑΒΓΔ inscribed in circle (o), is harmonic, if and only if, it is: $\frac{\alpha'}{\beta'}.\frac{\gamma'}{\delta'}=1$, **(1).**

where α', β', γ', δ' is the distance of the convergence point K of the diagonals of this quadrilateral, from its side ΑΒ=α, ΒΓ=β, ΓΔ=γ, ΔΑ=δ, respectively.

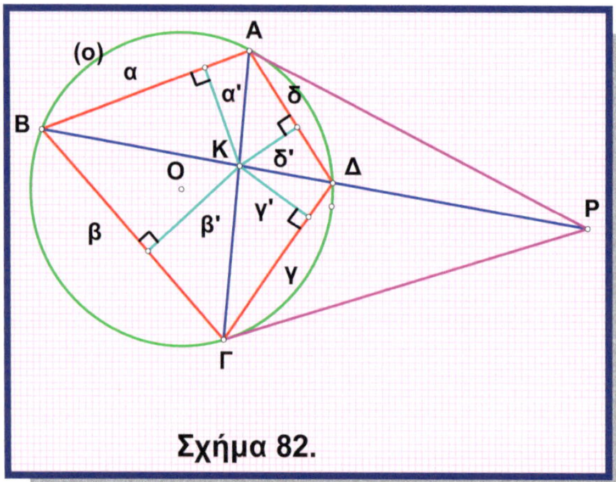

Σχήμα 82.

Proof.

(a). Direct.

If quadrilateral Α-ΒΓΔ is harmonic, we will prove that relation (1) is true.

1st Way.

Since quadrilateral ΑΒΓΔ is harmonic, as it is mentioned in the remarks of Propositions 112-118, the intersection of its diagonals, is a Lemoine Point of this quadrilateral, for which, according to known Proposition {Γεωμετρίες [31] § 8ι(87)(Criterion) and [32] IV page 54}, it will be:

$$\frac{\alpha}{\alpha'}=\frac{\beta}{\beta'}=\frac{\gamma}{\gamma'}=\frac{\delta}{\delta'}=\lambda \quad \text{or} \quad \alpha=\lambda.\alpha',\ \beta=\lambda.\beta',\ \gamma=\lambda.\gamma',\ \delta=\lambda.\delta'. \qquad (2).$$

After all, according to the known Criterion {Paragraph 2β(21) της Γεωμετρίας [31]}, it is: $\frac{\alpha}{\beta}.\frac{\gamma}{\delta}=1$, which due to (2), becomes: $\frac{\lambda\alpha'}{\lambda\beta'}.\frac{\lambda\gamma'}{\lambda\delta'}=1$ and from which we get the proven relation (1).

2nd Way.

Since for example it is ∠AKB=∠ΓKΔ and ∠KAB=∠KΔΓ triangles KAB, KΔΓ are similar, thus according to what we know, it will be: $\frac{\alpha}{\gamma}=\frac{\alpha'}{\gamma'}$. (3).

In the exact same way, we also find that: $\frac{\beta}{\delta}=\frac{\beta'}{\delta'}$. (4).

Thus, from relations (3), (4), if we multiply their respective parts, we get:

$$\frac{\alpha}{\beta}\cdot\frac{\gamma}{\delta}=\frac{\alpha'}{\beta'}\cdot\frac{\gamma'}{\delta'}. \quad (5).$$

Thus, if we take into consideration the known Criterion 2β(21) in book [31], then (5) becomes: $\frac{\alpha}{\beta}\cdot\frac{\gamma}{\delta}=\frac{\alpha'}{\beta'}\cdot\frac{\gamma'}{\delta'}=1$ or $\frac{\alpha'}{\beta'}\cdot\frac{\gamma'}{\delta'}=1$, which is the proven relation (1).

(b). Converse.

I.e., if (1) is true, we will prove that quadrilateral ΑΒΓΔ, is harmonic.

We find, as in the direct of way 2 above, that relations (3) and (4) are true, from which we get (5).

Hence, since (1) is true as well, then (5), becomes:

$\frac{\alpha}{\beta}\cdot\frac{\gamma}{\delta}=\frac{\alpha'}{\beta'}\cdot\frac{\gamma'}{\delta'}=1$, or $\frac{\alpha}{\beta}\cdot\frac{\gamma}{\delta}=1$. This, according to the converse of the known Criterion 2β(21), means that quadrilateral ΑΒΓΔ, is actually harmonic.

Remarks.

(a). Proposition 122 is the <u>thirteenth criterion of harmonic being of a quadrilateral.</u> We have come up with the first ten criteria of harmonic being of the harmonic quadrilateral, earlier and they are included in book [31]. Criteria 11, 12, 14, are the Propositions 33, 35, 133, respectively.

(b). As we will see below, equivalent Propositions, are true for Hexagon ΑΒΓΔΕΖ, (For which it is: $\frac{\alpha'}{\beta'}\cdot\frac{\gamma'}{\delta'}\cdot\frac{\varepsilon'}{\zeta'}=1$), for Octagon, for Decagon, etc, that are not Criteria of Harmonic being though.

(c). This Proposition we believe to be a new Criterion and it firstly appears here.

(d). Relevant Propositions are 123-127.

19.37. Significant Property of the Lemoine Point of a Simple Harmonic Hexagon.

Extension to Proposition 122 to a Harmonic Hexagon.

Proposition 123 (Figure 74).

For every Simple Harmonic Hexagon ΑΒΓΔΕΖ, it is:

$$\frac{\upsilon_\alpha}{\upsilon_\beta}.\frac{\upsilon_\gamma}{\upsilon_\delta}.\frac{\upsilon_\varepsilon}{\upsilon_\zeta}=1, \qquad (1).$$

where υ_α, υ_β, υ_γ,..., υ_ζ, is the distance of the convergence point K of the diagonals (main) of this hexagon, from its side ΑΒ=α, ΒΓ=β, ΓΔ=γ, ...,ΖΑ=ζ, respectively.

Proof.

1st Way (Based on Propositions 104, 112 and 119).

Since hexagon ΑΒΓΔΕΖ is harmonic, according to Proposition 104 (§ 19.11), it will have concurrent diagonals at point K (Lemoine Point of the Hexagon) for which, according to Criterion 112 (§ 19.21), it will be:

$$\frac{\upsilon_\alpha}{\alpha}=\frac{\upsilon_\beta}{\beta}=\frac{\upsilon_\gamma}{\gamma}=\frac{\upsilon_\delta}{\delta}=\frac{\upsilon_\varepsilon}{\varepsilon}=\frac{\upsilon_\zeta}{\zeta}=\frac{1}{\lambda} \qquad \text{or}$$

$$\alpha=\lambda.\upsilon_\alpha,\ \beta=\lambda.\upsilon_\beta,\ \gamma=\lambda.\upsilon_\gamma,\ \delta=\lambda.\upsilon_\delta,\ \varepsilon=\lambda.\upsilon_\varepsilon,\ \zeta=\lambda.\upsilon_\zeta. \qquad (2).$$

After all, according to Proposition 119 (§ 19.32), it is: $\frac{\alpha}{\beta}.\frac{\gamma}{\delta}.\frac{\varepsilon}{\zeta}=1$, which due to (2), becomes: $\frac{\lambda.\upsilon_\alpha}{\lambda.\upsilon_\beta}.\frac{\lambda.\upsilon_\gamma}{\lambda.\upsilon_\delta}.\frac{\lambda.\upsilon_\varepsilon}{\lambda.\upsilon_\zeta}=1$ and from which we get the proven relation (1).

2nd Way (Based on Propositions 104 and 119.).

Since Hexagon ΑΒΓΔΕΖ is Harmonic, according to Proposition 104, it will have concurrent diagonals at point K (Lemoine Point of the Hexagon).

Since for example it is ∠AKB=∠ΔKE and ∠KAB=∠KEΔ triangles KAB, KEΔ are similar, thus according to what we know, it will be:

$$\frac{\alpha}{\delta}=\frac{\upsilon_\alpha}{\upsilon_\delta}. \qquad (3).$$

In the exact same way, we also find that:

$$\frac{\varepsilon}{\beta}=\frac{\upsilon_\varepsilon}{\upsilon_\beta}, \quad \frac{\gamma}{\zeta}=\frac{\upsilon_\gamma}{\upsilon_\zeta}. \qquad (4).$$

Hence, from relations (3), (4), if we multiply their respective parts, we get:

$$\frac{\alpha}{\beta}\cdot\frac{\gamma}{\delta}\cdot\frac{\varepsilon}{\zeta}=\frac{\upsilon_\alpha}{\upsilon_\beta}\cdot\frac{\upsilon_\gamma}{\upsilon_\delta}\cdot\frac{\upsilon_\varepsilon}{\upsilon_\zeta}. \qquad (5).$$

Thus, if we take into consideration Proposition 119 (§ 19.32), according to which it is $\frac{\alpha}{\beta}\cdot\frac{\gamma}{\delta}\cdot\frac{\varepsilon}{\zeta}=1$, then from (5) we get: $\frac{\upsilon_\alpha}{\upsilon_\beta}\cdot\frac{\upsilon_\gamma}{\upsilon_\delta}\cdot\frac{\upsilon_\varepsilon}{\upsilon_\zeta}=1$, which is the proven relation (1).

3rd Way (Based on Propositions 104 and 119).

Since Hexagon ΑΒΓΔΕΖ is Harmonic, according to Proposition 104, it will have concurrent diagonals at point K (Lemoine Point of the Hexagon).

Since for example it is ∠AKB=∠ΔKE, according to known Theorem, for triangles KAB, KΔE, it will be:

$$\frac{KA.KB}{K\Delta.KE}=\frac{(KAB)}{(K\Delta E)}=\frac{\alpha.\upsilon_\alpha}{\delta.\upsilon_\delta} \quad \text{or} \quad \frac{KA.KB}{K\Delta.KE}=\frac{\alpha.\upsilon_\alpha}{\delta.\upsilon_\delta}. \qquad (6).$$

In the exact same way, we also find that:

$$\frac{KE.KZ}{KB.K\Gamma}=\frac{(KEZ)}{(KB\Gamma)}=\frac{\varepsilon.\upsilon_\varepsilon}{\beta.\upsilon_\beta} \quad \text{or} \quad \frac{KE.KZ}{KB.K\Gamma}=\frac{\varepsilon.\upsilon_\varepsilon}{\beta.\upsilon_\beta}. \qquad (7).$$

Similarly we find that:

$$\frac{K\Gamma.K\Delta}{KZ.KA}=\frac{\gamma.\upsilon_\gamma}{\zeta.\upsilon_\zeta}. \qquad (8).$$

Thus, from relations (6), (7), (8) if we multiply the respective parts and after the proper reductions, we get:

$$1=\frac{\alpha}{\beta}\cdot\frac{\gamma}{\delta}\cdot\frac{\varepsilon}{\zeta}\cdot\frac{\upsilon_\alpha}{\upsilon_\beta}\cdot\frac{\upsilon_\gamma}{\upsilon_\delta}\cdot\frac{\upsilon_\varepsilon}{\upsilon_\zeta}. \qquad (9).$$

Thus, if we take into consideration Proposition 119 (§ 19.32) according to which it is $\frac{\alpha}{\beta}\cdot\frac{\gamma}{\delta}\cdot\frac{\varepsilon}{\zeta}=1$, then (9), becomes: $1=1.\frac{u_\alpha}{u_\beta}\cdot\frac{u_\gamma}{u_\delta}\cdot\frac{u_\varepsilon}{u_\zeta}$ or

$\frac{u_\alpha}{u_\beta}\cdot\frac{u_\gamma}{u_\delta}\cdot\frac{u_\varepsilon}{u_\zeta}=1$, which is the proven relation (1).

Remarks.

(a). Proposition 123 is not a Criterion of Harmonic being, as its converse is not true.

(b). We notice that equivalent Propositions, are also true for the Harmonic Quadrilateral ΑΒΓΔ (For which, according to Proposition 122, it is $\frac{u_\alpha}{u_\beta}\cdot\frac{u_\gamma}{u_\delta}=1$), for the Harmonic Octagon, for the Harmonic Decagon, etc.

(c). Proposition 123 is also true for the Asteroid Harmonic Octagon, Decagon, etc.

(e). This Proposition we believe is a new Proposition and firstly appears here.

(f). Relevant Propositions are 104, 112, 119, 122, 124-127.

19.38. Significant Property of the Lemoine Point of a Simple Harmonic Octagon.

Extension to Proposition 123 to the Harmonic Octagon.

Proposition 124 (Figure 75).

For every convex Simple Harmonic Octagon ΑΒΓΔΕΖΗΘ, it is:

$$\frac{u_\alpha}{u_\beta}\cdot\frac{u_\gamma}{u_\delta}\cdot\frac{u_\varepsilon}{u_\zeta}\cdot\frac{u_\eta}{u_\theta}=1, \quad (1).$$

where u_α, u_β, u_γ,..., u_θ, is the distance of the convergence point K of the diagonals (main) of this octagon, from its side ΑΒ=α, ΒΓ=β, ΓΔ=γ, ...,ΘΑ=θ, respectively.

Proof.

Since octagon ΑΒΓΔΕΖΗΘ is Harmonic, according to Proposition 105 (§ 19.12), it will have concurrent diagonals at point K (Lemoine point of the octagon) for which, according to Criterion 113 (§ 19.22), it will be:

$$\frac{\upsilon_\alpha}{\alpha}=\frac{\upsilon_\beta}{\beta}=\frac{\upsilon_\gamma}{\gamma}=\frac{\upsilon_\delta}{\delta}=\frac{\upsilon_\varepsilon}{\varepsilon}=\frac{\upsilon_\zeta}{\zeta}=\frac{\upsilon_\eta}{\eta}=\frac{\upsilon_\theta}{\theta}=\frac{1}{\lambda} \text{ or}$$

$\alpha=\lambda.\upsilon_\alpha$, $\beta=\lambda.\upsilon_\beta$, $\gamma=\lambda.\upsilon_\gamma$, $\delta=\lambda.\upsilon_\delta$, $\varepsilon=\lambda.\upsilon_\varepsilon$, $\zeta=\lambda.\upsilon_\zeta$, $\eta=\lambda.\upsilon_\eta$, $\theta=\lambda.\upsilon_\theta$. (2).

After all, according to Proposition 120 (§ 19.33), it is: $\frac{\alpha}{\beta}.\frac{\gamma}{\delta}.\frac{\varepsilon}{\zeta}.\frac{\eta}{\theta}=1$, which due to (2), becomes: $\frac{\lambda.\upsilon_\alpha}{\lambda.\upsilon_\beta}.\frac{\lambda.\upsilon_\gamma}{\lambda.\upsilon_\delta}.\frac{\lambda.\upsilon_\varepsilon}{\lambda.\upsilon_\zeta}.\frac{\lambda.\upsilon_\eta}{\lambda.\upsilon_\theta}=1$ and from which we directly get proven relation (1).

Remarks.

(a). Proposition 124 is not a Criterion of Harmonic being, as its converse is not true.

(b). We notice that equivalent Propositions, are also true for the Harmonic Quadrilateral ΑΒΓΔ (For which, according to Proposition 122, it is $\frac{\upsilon_\alpha}{\upsilon_\beta}.\frac{\upsilon_\gamma}{\upsilon_\delta}=1$), for the Harmonic Hexagon, for the Harmonic Decagon, etc.

(c). Proposition 124 is also true for the Asteroid Harmonic Octagon, Decagon, etc.

(e). This Proposition we believe is a new Proposition and firstly appears here.

(f). Relevant Propositions are 105, 113, 120, 122, 123, 125-127, etc.

19.39. Significant Property of the Lemoine Point of a Harmonic Decagon. Extension to Proposition 124 to the Harmonic Decagon.

Proposition 125 (Figure 76).

For every convex Simple Harmonic Decagon ΑΒΓΔΕΖΗΘΙΚ, it is:

$$\frac{\alpha'}{\beta'}.\frac{\gamma'}{\delta'}.\frac{\varepsilon'}{\zeta'}.\frac{\eta'}{\theta'}.\frac{\iota'}{\kappa'}=1, \qquad (1).$$

where α', β', γ',..., κ', is the distance of the convergence point Λ of the diagonals (main) of this decagon, from its side ΑΒ=α, ΒΓ=β, ΓΔ=γ, ...,ΚΑ=κ, respectively.

Proof.

1st Way (Based on Propositions 106, 114 and 121).

Since decagon ΑΒΓΔΡΖΗΘΙΚ is harmonic, according to Proposition 106 (§ 19.13), it will have concurrent diagonals at point Λ (Lemoine Point of the decagon) for which, according to Criterion 114 (§ 19.23), it will be:

$\frac{\alpha}{\alpha'}=\frac{\beta}{\beta'}=\frac{\gamma}{\gamma'}=\frac{\delta}{\delta'}=\frac{\varepsilon}{\varepsilon'}=\frac{\zeta}{\zeta'}=\frac{\eta}{\eta'}=\frac{\theta}{\theta'}=\frac{\iota}{\iota'}=\frac{\kappa}{\kappa'}=\lambda$ or $\alpha=\lambda.\alpha'$, $\beta=\lambda.\beta'$, $\gamma=\lambda.\gamma'$, $\delta=\lambda.\delta'$, $\varepsilon=\lambda.\varepsilon'$, $\zeta=\lambda.\zeta'$, $\eta=\lambda.\eta'$, $\theta=\lambda.\theta'$, $\iota=\lambda.\iota'$, $\kappa=\lambda.\kappa'$. (2).

After all, according to Proposition 121 (§ 19.34), it is: $\frac{\alpha}{\beta}.\frac{\gamma}{\delta}.\frac{\varepsilon}{\zeta}.\frac{\eta}{\theta}.\frac{\iota}{\kappa}=1$, which due to (2), becomes: $\frac{\lambda\alpha'}{\lambda\beta'}.\frac{\lambda\gamma'}{\lambda\delta'}.\frac{\lambda\varepsilon'}{\lambda\zeta'}.\frac{\lambda\eta'}{\lambda\theta'}.\frac{\lambda\iota'}{\lambda\kappa'}=1$ and from which we get the proven relation (1).

2nd Way (Based on Propositions 106 and 121).

Since Decagon ΑΒΓΔΡΖΗΘΙΚ is Harmonic, according to Proposition 106 (§ 19.13), it will have concurrent diagonals at point Λ (Lemoine Point of the decagon).

Since for example it is ∠ΑΛΒ=∠ΗΛΖ and ∠ΛΑΒ=∠ΛΗΖ triangle ΛΑΒ, ΛΗΖ are similar, thus according to what we know, it will be: $\frac{\alpha}{\zeta}=\frac{\alpha'}{\zeta'}$. (3).

In the exact same way, we also find that:

$$\frac{\eta}{\beta}=\frac{\eta'}{\beta'},\quad \frac{\gamma}{\theta}=\frac{\gamma'}{\theta'},\quad \frac{\iota}{\delta}=\frac{\iota'}{\delta'},\quad \frac{\varepsilon}{\kappa}=\frac{\varepsilon'}{\kappa'}. \qquad (4).$$

Hence, from relations (3), (4), if we multiply their respective parts, we get:

$$\frac{\alpha}{\beta}.\frac{\gamma}{\delta}.\frac{\varepsilon}{\zeta}.\frac{\eta}{\theta}.\frac{\iota}{\kappa}=\frac{\alpha'}{\beta'}.\frac{\gamma'}{\delta'}.\frac{\varepsilon'}{\zeta'}.\frac{\eta'}{\theta'}.\frac{\iota'}{\kappa'}. \qquad (5).$$

Thus, if we take into consideration Proposition 121 (§ 19.34), according to which it is $\frac{\alpha}{\beta}.\frac{\gamma}{\delta}.\frac{\varepsilon}{\zeta}.\frac{\eta}{\theta}.\frac{\iota}{\kappa}=1$, then (5) becomes: $\frac{\alpha'}{\beta'}.\frac{\gamma'}{\delta'}.\frac{\varepsilon'}{\zeta'}.\frac{\eta'}{\theta'}.\frac{\iota'}{\kappa'}=1$, which is the proven relation (1).

3rd Way (Based on Propositions 106 and 121).

Since Decagon ΑΒΓΔΡΖΗΘΙΚ is Harmonic, according to Proposition 106 (§ 19.13), it will have concurrent diagonals at point Λ (Lemoine Point of the decagon).

Since for example it is ∠ΑΛΒ=∠ΗΛΖ, according to known Theorem, for triangles ΛΑΒ, ΛΗΖ, it will be: $\frac{ΛΑ.ΛΒ}{ΛΖ.ΛΗ}=\frac{(ΛΑΒ)}{(ΛΖΗ)}=\frac{α.α'}{ζ.ζ'}$. (6).

In the exact same way, we also find that:

$$\frac{ΛΒ.ΛΓ}{ΛΗ.ΛΘ}=\frac{(ΛΒΓ)}{(ΛΗΘ)}=\frac{β.β'}{η.η'} \text{ ή } \frac{ΛΗ.ΛΘ}{ΛΒ.ΛΓ}=\frac{η.η'}{β.β'}. \quad (7).$$

Similarly we find that:

$$\frac{ΛΓ.ΛΔ}{ΛΘ.ΛΙ}=\frac{γ.γ'}{θ.θ'},\ \frac{ΛΙ.ΛΚ}{ΛΔ.ΛΕ}=\frac{ι..ι'}{δ.δ'},\ \frac{ΛΕ.ΛΖ}{ΛΚ.ΛΑ}=\frac{ε.ε'}{κ.κ'}. \quad (8).$$

Thus, from relations (6), (7), (8) if we multiply the respective parts and after the proper reductions, we get:

$$1=\frac{α}{β}.\frac{γ}{δ}.\frac{ε}{ζ}.\frac{η}{θ}.\frac{ι}{κ}.\frac{α'}{β'}.\frac{γ'}{δ'}.\frac{ε'}{ζ'}.\frac{η'}{θ'}.\frac{ι'}{κ'}. \quad (9).$$

Thus, if we take into consideration Proposition 121 (§ 19.34) according to which it is $\frac{α}{β}.\frac{γ}{δ}.\frac{ε}{ζ}.\frac{η}{θ}.\frac{ι}{κ}=1$, then (9), becomes: $1=1.\frac{α'}{β'}.\frac{γ'}{δ'}.\frac{ε'}{ζ'}.\frac{η'}{θ'}.\frac{ι'}{κ'}$ or

$\frac{α'}{β'}.\frac{γ'}{δ'}.\frac{ε'}{ζ'}.\frac{η'}{θ'}.\frac{ι'}{κ'}=1$, which is the proven relation (1).

Remarks.

(a). Proposition 125 is not a Criterion of Harmonic being, as its converse is not true.

(b). We notice that equivalent Propositions, are also true for the Harmonic Quadrilateral ΑΒΓΔ (For which, according to Proposition 122, it is $\frac{α'}{β'}.\frac{γ'}{δ'}=1$), for the Harmonic Hexagon ΑΒΓΔΕΖ (For which, according to

Proposition 123, it is: $\frac{\alpha'}{\beta'}.\frac{\gamma'}{\delta'}.\frac{\varepsilon'}{\zeta'}$=1), for the Harmonic Octagon (Proposition 124), etc.

(c). Equivalent Proposition to Proposition 125, is also true for the Asteroid Harmonic Decagon ΑΔΗΚΓΖΙΒΕΘΑ (Proposition 127).

(d). This Proposition we believe is a new Proposition and firstly appears here.

(f). Relevant Propositions are 106, 114, 121-124, 126, 127.

19.40. Significant Property of the Lemoine Point of an Asteroid Harmonic Octagon.

Extension to Proposition 124 to the Asteroid Harmonic Octagon.

Proposition 126 (Figure 77).

For every Asteroid Harmonic Octagon ΑΔΗΒΕΘΓΖΑ, it is:

$$\frac{u_\alpha}{u_\beta}.\frac{u_\gamma}{u_\delta}.\frac{u_\varepsilon}{u_\zeta}.\frac{u_\eta}{u_\theta}=1, \quad (1).$$

where u_α, u_β, u_γ,..., u_θ, is the distance of the convergence point K of the diagonals (main) of this octagon, from its side ΑΔ=α, ΔΗ=β, ΗΒ=γ, ...,ΖΑ=θ, respectively.

Proof.

Since octagon ΑΔΗΒΕΘΓΖΑ is Harmonic, according to Proposition 105 (§ 19.12), it will have concurrent diagonals at point K (Lemoine point of the octagon) for which, according to Criterion 115 (§ 19.24), it will be:

$$\frac{u_\alpha}{\alpha}=\frac{u_\beta}{\beta}=\frac{u_\gamma}{\gamma}=\frac{u_\delta}{\delta}=\frac{u_\varepsilon}{\varepsilon}=\frac{u_\zeta}{\zeta}=\frac{u_\eta}{\eta}=\frac{u_\theta}{\theta}=\frac{1}{\lambda} \text{ or}$$

$\alpha=\lambda.u_\alpha$, $\beta=\lambda.u_\beta$, $\gamma=\lambda.u_\gamma$, $\delta=\lambda.u_\delta$, $\varepsilon=\lambda.u_\varepsilon$, $\zeta=\lambda.u_\zeta$, $\eta=\lambda.u_\eta$, $\theta=\lambda.u_\theta$. (2).

After all, according to Proposition 120 (§ 19.33), it is: $\frac{\alpha}{\beta}.\frac{\gamma}{\delta}.\frac{\varepsilon}{\zeta}.\frac{\eta}{\theta}$=1, which due to (2), becomes: $\frac{\lambda.u_\alpha}{\lambda.u_\beta}.\frac{\lambda.u_\gamma}{\lambda.u_\delta}.\frac{\lambda.u_\varepsilon}{\lambda.u_\zeta}.\frac{\lambda.u_\eta}{\lambda.u_\theta}$=1 and from which we directly get proven relation (1).

Remarks.

(a). Proposition 126 is not a Criterion of Harmonic being, as its converse is not true.

(b). We notice that equivalent Propositions, are also true for the Harmonic Quadrilateral ΑΒΓΔ, for which, according to Proposition 122, it is $\frac{υ_α}{υ_β}.\frac{υ_γ}{υ_δ}=1$, for the Harmonic Hexagon, for the Harmonic Decagon, etc.

(c). Proposition 126, is also true for the Asteroid Harmonic Decagon (Proposition 127), etc.

(d). This Proposition we believe is a new Proposition and firstly appears here.

(f). Relevant Propositions are 105, 115, 120, 122–125, 127, etc.

19.41. Significant Property of the Lemoine Point of an Asteroid Harmonic Decagon.

Extension to Proposition 125 to the Asteroid Harmonic Decagon.

Proposition 127 (Figure 78).

For every Asteroid Harmonic Decagon ΑΔΗΚΓΖΙΒΕΘΑ, it is:

$$\frac{υ_α}{υ_β}.\frac{υ_γ}{υ_δ}.\frac{υ_ε}{υ_ζ}.\frac{υ_η}{υ_θ}.\frac{υ_ι}{υ_κ}=1, \qquad (1).$$

where $υ_α$, $υ_β$, $υ_γ$,..., $υ_κ$, is the distance of the convergence point Λ of the diagonals (main) of this decagon, from its side ΑΔ=α, ΔΗ=β, ΗΚ=γ, ...,ΕΘ=ι, ΘΑ=κ, respectively.

Proof.

1st Way (Based on Propositions 106, 116 and 121).

Since decagon ΑΔΗΚΓΖΙΒΕΘΑ is harmonic, according to Proposition 106 (§ 19.13), it will have concurrent diagonals at point Λ (Lemoine Point of the decagon) for which, according to Criterion 116 (§ 19.25), it will be:

$$\frac{υ_α}{α}=\frac{υ_β}{β}=\frac{υ_γ}{γ}=\frac{υ_δ}{δ}=\frac{υ_ε}{ε}=\frac{υ_ζ}{ζ}=\frac{υ_η}{η}=\frac{υ_θ}{θ}=\frac{υ_ι}{ι}=\frac{υ_κ}{κ}=\frac{1}{λ} \quad \text{or}$$

$α=λ.υ_α$, $β=λ.υ_β$, $γ=λ.υ_γ$, $δ=λ.υ_δ$, $ε=λ.υ_ε$, $ζ=λ.υ_ζ$, $η=λ.υ_η$, $θ=λ.υ_θ$, $ι=λ.υ_ι$, $κ=λ.υ_κ$.(2).

After all, according to Proposition 121 (§ 19.34), it is: $\frac{\alpha}{\beta}.\frac{\gamma}{\delta}.\frac{\varepsilon}{\zeta}.\frac{\eta}{\theta}.\frac{\iota}{\kappa}=1$, which due to (2), becomes: $\frac{\lambda.\upsilon_\alpha}{\lambda.\upsilon_\beta}.\frac{\lambda.\upsilon_\gamma}{\lambda.\upsilon_\delta}.\frac{\lambda.\upsilon_\varepsilon}{\lambda.\upsilon_\zeta}.\frac{\lambda.\upsilon_\eta}{\lambda.\upsilon_\theta}.\frac{\lambda.\upsilon_\iota}{\lambda.\upsilon_\kappa}=1$ and from which we get the proven relation (1).

2nd Way (Based on Propositions 106 and 121).

Since Decagon ΑΔΗΚΓΖΙΒΕΘΑ is Harmonic, according to Proposition 106 (§ 19.13), it will have concurrent diagonals at point Λ (Lemoine Point of the decagon).

Since for example it is ∠ΑΛΔ=∠ΙΛΖ and ∠ΛΑΔ=∠ΛΙΖ triangles ΛΑΔ, ΛΙΖ are similar, thus according to what we know, it will be: $\frac{\alpha}{\zeta}=\frac{\upsilon_\alpha}{\upsilon_\zeta}$. (3).

In the exact same way, we also find that:

$$\frac{\eta}{\beta}=\frac{\upsilon_\eta}{\upsilon_\beta},\ \frac{\gamma}{\theta}=\frac{\upsilon_\gamma}{\upsilon_\theta},\ \frac{\iota}{\delta}=\frac{\upsilon_\iota}{\upsilon_\delta},\ \frac{\varepsilon}{\kappa}=\frac{\upsilon_\varepsilon}{\upsilon_\kappa}. \quad (4).$$

Hence, from relations (3), (4), if we multiply their respective parts, we get:

$$\frac{\alpha}{\beta}.\frac{\gamma}{\delta}.\frac{\varepsilon}{\zeta}.\frac{\eta}{\theta}.\frac{\iota}{\kappa}=\frac{\upsilon_\alpha}{\upsilon_\beta}.\frac{\upsilon_\gamma}{\upsilon_\delta}.\frac{\upsilon_\varepsilon}{\upsilon_\zeta}.\frac{\upsilon_\eta}{\upsilon_\theta}.\frac{\upsilon_\iota}{\upsilon_\kappa}. \quad (5).$$

Thus, if we take into consideration Proposition 121 (§ 19.34), according to which it is $\frac{\alpha}{\beta}.\frac{\gamma}{\delta}.\frac{\varepsilon}{\zeta}.\frac{\eta}{\theta}.\frac{\iota}{\kappa}=1$, then (5) becomes: $\frac{\upsilon_\alpha}{\upsilon_\beta}.\frac{\upsilon_\gamma}{\upsilon_\delta}.\frac{\upsilon_\varepsilon}{\upsilon_\zeta}.\frac{\upsilon_\eta}{\upsilon_\theta}.\frac{\upsilon_\iota}{\upsilon_\kappa}=1$, which is the proven relation (1).

3rd Way (Based on Propositions 106 and 121).

Since Decagon ΑΔΗΚΓΖΙΒΕΘΑ is Harmonic, according to Proposition 106 (§ 19.13), it will have concurrent diagonals at point Λ (Lemoine Point of the decagon).

Since for example it is ∠ΑΛΔ=∠ΙΛΖ, according to known Theorem, for triangles ΛΑΔ, ΛΙΖ, it will be: $\frac{ΛΑ.ΛΔ}{ΛΖ.ΛΙ}=\frac{(ΛΑΔ)}{(ΛΖΙ)}=\frac{\alpha.\upsilon_\alpha}{\zeta.\upsilon_\zeta}$. (6).

In the exact same way, we also find that:

$$\frac{ΛΔ.ΛΗ}{ΛΙ.ΛΒ}=\frac{(ΛΔΗ)}{(ΛΙΒ)}=\frac{\beta.\upsilon_\beta}{\eta.\upsilon_\eta} \text{ ή } \frac{ΛΙ.ΛΒ}{ΛΔ.ΛΗ}=\frac{\eta.\upsilon_\eta}{\beta.\upsilon_\beta}. \quad (7).$$

Similarly we find that:

$$\frac{ΛΗ.ΛΚ}{ΛΒ.ΛΕ}=\frac{γ.υ_γ}{θ.υ_θ},\ \frac{ΛΕ.ΛΘ}{ΛΚ.ΛΓ}=\frac{ι.υ_ι}{δ.υ_δ},\ \frac{ΛΓ.ΛΖ}{ΛΘ.ΛΑ}=\frac{ε.υ_ε}{κ.υ_κ}. \qquad (8).$$

Thus, from relations (6), (7), (8) if we multiply the respective parts and after the proper reductions, we get:

$$1=\frac{α}{β}.\frac{γ}{δ}.\frac{ε}{ζ}.\frac{η}{θ}.\frac{ι}{κ}.\frac{υ_α}{υ_β}.\frac{υ_γ}{υ_δ}.\frac{υ_ε}{υ_ζ}.\frac{υ_η}{υ_θ}.\frac{υ_ι}{υ_κ}. \qquad (9).$$

Thus, if we take into consideration Proposition 121 (§ 19.34) according to which it is $\frac{α}{β}.\frac{γ}{δ}.\frac{ε}{ζ}.\frac{η}{θ}.\frac{ι}{κ}=1$, then (9), becomes:

$1=1.\frac{υ_α}{υ_β}.\frac{υ_γ}{υ_δ}.\frac{υ_ε}{υ_ζ}.\frac{υ_η}{υ_θ}.\frac{υ_ι}{υ_κ}$ or $\frac{υ_α}{υ_β}.\frac{υ_γ}{υ_δ}.\frac{υ_ε}{υ_ζ}.\frac{υ_η}{υ_θ}.\frac{υ_ι}{υ_κ}=1$, which is the proven relation (1).

Remarks.

(a). Proposition 127 is not a Criterion of Harmonic being, as its converse is not true.

(b). We notice that equivalent Propositions, are also true for the Harmonic Quadrilateral ΑΒΓΔ, for which, according to Proposition 122, it is $\frac{α'}{β'}.\frac{γ'}{δ'}=1$, for the Harmonic Hexagon ΑΒΓΔΕΖ, for which, according to Proposition 123, it is: $\frac{α'}{β'}.\frac{γ'}{δ'}.\frac{ε'}{ζ'}=1$, for the Harmonic Octagon, etc.

(c). Equivalent Proposition to Proposition 127, is also true for the Asteroid Harmonic Octagon (Proposition 126).

(d). This Proposition we believe is a new Proposition and firstly appears here.

(f). Relevant Propositions are 106, 114-116, 121-126.

19.42. General Conclusion.

Due to Propositions 122-127 above and if we work in the same way as in their proofs, we reach the following result:

Equivalent Proposition, to Propositions 119-127, can be made and proven to be true for any other Harmonic Multilateral convex or asteroid, with an even number of sides.

19.43. Criterion of Harmonic being of a Simple Inscribed Pentagon.

Extension to Criterion 117. (Theorem analogous to Gergonne Theorem for the Harmonic 5-gon).

Criterion 128 (Figure 83).

We consider the inscribed in circle (o), convex pentagon ΑΓΕΗΙ and its respective (its polar converse), circumscribed (tangential) pentagon αγεηι (where α the intersection of the tangents at points Ε, Η, where γ the intersection of the tangents at points Η, Ι, and so on and so forth).

Pentagon ΑΓΕΗΙ is Harmonic, if and only if, Αα, Γγ, Εε, Ηη, Ιι, coincide at point let's say Λ.

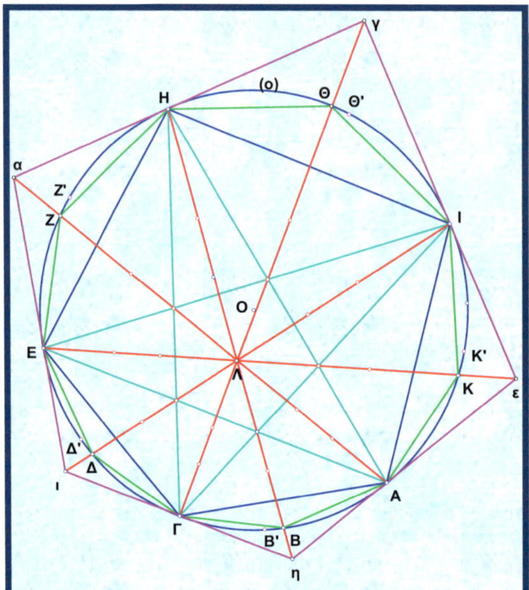

Σχήμα 83.

Proof.

(a). Direct.

I.e., if it is Aα∩Γγ∩Eε∩Hη∩Iι ≡Λ , (1).

We will prove that AΓEHI is harmonic.

According to the relevant definition, for AΓEHI to be Harmonic (§ 16.1e), on circle (o), we need to have points B, Δ, Z, Θ, K, such that the convex Decagon ABΓΔEZHΘIK, would be Harmonic.

In fact, if Aα, Γγ, Eε, Hη, Iι, intersect circle (o), at points Z, Θ, K, B, Δ, respectively, then since for example AZ runs through pole α of EH, then quadrilateral EZHA would be Harmonic.

Similarly we find that the rest four quadrilaterals of the pentagon are Harmonic, i.e. HΘIΓ, IKAE, ABΓH, ΓΔEI, are Harmonic, then, according to Proposition 111 (§19.19), Decagon ABΓΔEZHΘIK is Harmonic, since, from the implication, relation (1) is also true.

Thus, since for the intersections B, Δ, Z, Θ, K, Decagon ABΓΔEZHΘIK is harmonic, then according to the relevant definition (§ 16.1e) pentagon AΓEHI, is Harmonic as well.

(b). Converse.

I.e., if pentagon AΓEHI is Harmonic, we will prove that Aα, Γγ, Eε, Hη, Iι, coincide (Theorem analogous to Theorem Gergonne for the Harmonic 5-gon).

Since pentagon AΓEHI is Harmonic, according to the relevant definition (§ 16.1 e), in circle (o), we will have points B', Δ', Z', Θ', K', for which the convex decagon AB'ΓΔ'EZ'HΘ'IK', is Harmonic.

Therefore, since AB'ΓΔ'EZ'HΘ'IK', would be Harmonic, then its ten quadrilaterals will be harmonic too: AB'ΓH, B'ΓΔ'Θ', ΓΔ'EI, Δ'EZ'K', EZ'HA, Z'HΘ'B', HΘ'IΓ, Θ'IK'Δ', IK'AE, K'AB' Z'. (2).

Thus, since for example quadrilateral EZ'HA would be Harmonic, then AZ', will be conjugate to EH, hence AZ' will run through pole α of EH. This means that then Z' is the intersection of Aα with circle (o). But the intersection of Aα with circle (o) is called Z. This means that Z'≡Z.

In the same way we also find that: B'≡B, Δ'≡Δ, Θ'≡Θ, K'≡K.

Therefore, since B', Δ', Z', Θ', K', were received in such way that decagon

ΑΒ'ΓΔ'ΕΖ'ΗΘ'ΙΚ', would be Harmonic, then ΑΒΓΔΕΖΗΘΙΚ would also be harmonic, hence its diagonals ΑΖ, ΓΘ, ΕΚ, ΗΒ, ΙΔ, or Αα, Γγ, Εε, Ηη, Ιι, will coincide at point let's say Λ (Proposition 106, § 19.13).

Conclusion.

Due to the above Criterion of Harmonic Pentagons (Criterion of Harmonic being of convex inscribed pentagons), we reach the following conclusion:

Each convex pentagon ΑΓΕΗΙ inscribed in circle (o), is harmonic, if and only if, for it and for its respective (its polar converse), circumscribed (tangential) pentagon αγεηι (where α is the intersection of the tangents at points Ε, Η, where γ is the intersection of the tangents at points Η, Ι, and so on and so forth), diagonals Αα, Γγ, Εε, Ηη, Ιι, of the decagon ΑηΓιΕαΗγΙε, are concurrent.

Therefore, from now on we can use for Harmonic pentagons, the following alternative Definition:

«Harmonic Pentagon is called, each convex pentagon ΑΓΕΗΙ inscribed in circle (o), if for it and its respective (its polar converse), circumscribed (tangential) pentagon αγεηι (where α is the intersection of tangents at points Ε, Η, where γ is the intersection of tangents at points Η, Ι, and so on and so forth), diagonals Αα, Γγ, Εε, Ηη, Ιι, of the decagon ΑηΓιΕαΗγΙε, are concurrent».

Remarks.

(a). The above Criterion of Harmonic being of Pentagons, is rather significant, as it allows us to have an alternativer simpler DEFINITION, from our initial definition, about Harmonic Pentagons, as well as because it has many applications, as we will realize from now on.

(b). This Criterion we believe is a new Proposition and firstly appears here.

(c). It is obvious that the converse of Criterion 128, is a Theorem analogous to Gergonne Theorem for the Harmonic 5-gon.

(d). Relevant Propositions are 109, 110, 111, 129, 130.

19.44. Criterion of Harmonic being of a Simple Inscribed Pentagon.

Extension to Criteria 117 and 128.

Criterion 129 (Figure 83).

We consider the inscribed in circle (o) convex pentagon ΑΓΕΗΙ.
If on circle (o) we have points Β, Δ, Ζ, Θ, Κ, for which the five quadrilaterals:

ΑΒΓΗ, ΓΔΕΙ, ΕΖΗΑ, ΗΘΙΓ, ΙΚΑΕ, (1).

are Harmonic and Decagon ΑΒΓΔΕΖΗΘΙΚ is convex with concurrent diagonals (main), then and only then pentagon ΑΓΕΗΙ is Harmonic.

Proof.

(a). Direct.

I.e., if for the convex decagon ΑΒΓΔΕΖΗΘΙΚ, it is:

$$ΑΖ \cap ΓΘ \cap ΕΚ \cap ΗΒ \cap ΙΔ \equiv Λ, \qquad (2)$$

And if the quadrilaterals in relation (1), are Harmonic, we will prove that pentagon ΑΓΕΗΙ is Harmonic too.

1st Way (Based on Proposition 128).

If the 5-gon αγεηι is the respective circumscribed one of the 5-gon ΑΓΕΗΙ (where α is the intersection of the tangents at points Ε, Η, where γ is the intersection of the tangents at points Η, Ι, and so on and so forth), since, for example quadrilateral ΕΖΗΑ, is Harmonic from the hypothesis, then ΑΖ, will run through pole α of ΕΗ. Similarly we also find that ΓΘ, ΕΚ, ΗΒ, ΙΔ, will run through poles γ, ε, η, ι, of ΗΙ, ΙΑ, ΑΓ, ΓΕ, respectively, hence relation (2) becomes:

$$Αα \cap Γγ \cap Εε \cap Ηη \cap Ιι \equiv Λ. \qquad (3).$$

Hence, since (3) is true, according to the direct of Criterion 128 (§ 19.43), pentagon ΑΓΕΗ, is Harmonic.

2nd Way (Based on Proposition 111).

Since Decagon ΑΒΓΔΕΖΗΘΙΚ, has concurrent diagonals and relation (1) is true, it will be harmonic (Direct of Criterion 111).

Hence, we see that for Pentagon ΑΓΕΗΙ there are points Β, Δ, Ζ, Θ, Κ, for which decagon ΑΒΓΔΕΖΗΘΙΚ, is harmonic.

This, according to the relevant definition (§ 16.1 e), means that pentagon ΑΓΕΗΙ is harmonic too.

(b). Converse.

I.e., if Pentagon ΑΓΕΗΙ is Harmonic, we will prove that on circle (o), there are points Β, Δ, Ζ, Θ, Κ, for which on decagon ΑΒΓΔΕΖΗΘΙΚ, ΑΖ, ΓΘ, ΕΚ, ΗΒ, ΙΔ, coincide and the five quadrilaterals in relation (1), are Harmonic.

1st Way (Based on Proposition 128).

Since pentagon ΑΓΕΗΙ is harmonic, according to the converse of Criterion 128, Αα, Γγ, Εε, Ηη, Ιι, will be concurrent (where α α is the intersection of the tangents at points Ε, Η, where γ is the intersection of the tangents at points Η, Ι, and so on and so forth).

Hence, if Ζ, Θ, Κ, Β, Δ, are the intersections Αα, Γγ, Εε, Ηη, Ιι, with the circle (o), then for Ζ, Θ, Κ, Β, Δ, decagon ΑΒΓΔΕΖΗΘΙΚ will in fact have ΑΖ, ΓΘ, ΕΚ, ΗΒ, ΙΔ concurrent and the five quadrilaterals in (1) are harmonic, as ΑΖ, ΓΘ, ΕΚ, ΗΒ, ΙΔ, run through poles α, γ, ε, η, ι, of ΕΗ, ΗΙ, ΙΑ, ΑΓ, ΓΕ respectively.

2nd Way (Based on Proposition 106).

Actually, since pentagon ΑΓΕΗΙ, is Harmonic, according to the relevant definition (§ 16.1e), there will be points Β, Δ, Ζ, Θ, Κ, for which decagon ΑΒΓΔΕΖΗΘΙΚ, will be Harmonic, thus, according to Proposition 106 (§ 19.13), its diagonals will be concurrent. Hence it will be ΑΖ∩ΓΘ∩ΕΚ∩ΗΒ∩ΙΔ ≡Λ.

Moreover, since for points Β, Δ, Ζ, Θ, Κ, decagon ΑΒΓΔΕΖΗΘΙΚ, is Harmonic, according to the relevant definition of the Harmonic Decagon (§ 16.1d), it will have Harmonic all its ten quadrilaterals, among which there are also the five quadrilaterals in relation (1). Hence the five quadrilaterals in relation (1), are in fact Harmonic.

Conclusion.

Due to the above Criterion of Harmonic Pentagons (Criterion of Harmonic being of convex inscribed pentagons), this conclusion is reached:

Every convex pentagon ΑΓΕΗΙ inscribed in circle (o), is harmonic, if and only if, on circle (o) there are points Β, Δ, Ζ, Θ, Κ, for which the following five quadrilaterals ΑΒΓΗ, ΓΔΕΙ, ΕΖΗΑ, ΗΘΙΓ, ΙΚΑΕ, are Harmonic and decagon ΑΒΓΔΕΖΗΘΙΚ is convex with concurrent diagonals (main).

Alternative Definition of the Harmonic Pentagon.

Consequently, from now on we can use for Harmonic pentagons as well, the alternative Definition below:

«Harmonic Pentagon is called, every convex pentagon ΑΓΕΗΙ inscribed in circle (o), if and only if, on circle (o) there are points Β, Δ, Ζ, Θ, Κ, for which the five following quadrilaterals ΑΒΓΗ, ΓΔΕΙ, ΕΖΗΑ, ΗΘΙΓ, ΙΚΑΕ, are Harmonic and decagon ΑΒΓΔΕΖΗΘΙΚ is convex with concurrent diagonals (main)».

Remarks.

(a). The above Criterion of Harmonic being of Pentagons, is rather significant, as it allows us to have an alternative simpler DEFINITION, from our initial definition, for Harmonic Pentagons, as well as because it finds many applications, as we will later realize.

(b). This Criterion we believe to be new and firstly appear here.

(c). Relevant Propositions are 106, 111, 117, 128, 130, 131.

19.45. Criterion of Harmonic being of the Inscribed Pentagon.

Extension to Criteria 117, 128 and 129.

Does the Harmonic Pentagon have an Analogous Gergonne Point?

Criterion 130 (Figure 83).

Every Harmonic Pentagon and its respective tangential or its polar converse pentagon, are homologous, or in another formation:

The tangential or polar converse pentagon, of every Harmonic Pentagon, has an analogous Gergonne points, and conversely:

If the tangential or polar converse pentagon, inscribed in circle of a pentagon, has an analogous Gergonne point, then the second pentagon is Harmonic.

Proof.

(a). Direct.

I.e. if ΑΓΕΗΙ is a Harmonic Pentagon inscribed in circle (o) and αγεηι (where α is the intersection of the tangents at E, H, where γ is the intersection of the tangents at H, I, etc), is its tangential or polar converse pentagon, we will prove that it is: $A\alpha \cap \Gamma\gamma \cap E\varepsilon \cap H\eta \cap I\iota \equiv \Lambda$, (1)

(Λ is analogous to the Gergonne point of the triangle).

Since pentagon ΑΓΕΗΙ is Harmonic, according to the converse of Criterion 128 (§ 19.43), relation (1) will in fact be true. [In which relation Λ is also the convergence point of the diagonals of the Harmonic Decagon ΑΒΓΔΕΖΗΘΙΚ, which is also the Lemoine point of the Harmonic Decagon, as well of the Harmonic Pentagon ΑΓΕΗΙ (Criteria 114 and 117)].

(b). Converse.

I.e., if for the convex inscribed in circle (o) pentagon ΑΓΕΗΙ and for its tangential Pentagon αγεηι, relation (1) is true, we will then prove that ΑΓΕΗΙ is Harmonic.

Since from the hypothesis relation (1) is true, according to the direct of Criterion 128 (§ 19.43), pentagon ΑΓΕΗΙ is in fact Harmonic.

Conclusion.

Due to the above and since the convergence point of diagonals Λ of the respective Harmonic Decagon is also the Lemoine point of this Harmonic Decagon [Criterion 114 (§ 19.23)], as well as of its twin respective Harmonic Pentagons [Criterion 117 (§ 19.26)], the following conclusion is reached:

«The tangential or the Polar converse pentagon of ever Harmonic Pentagon, has a point analogous to the Gergonne point of a triangle, which coincides with the common Lemoine Point of the Harmonic Pentagon, as well as its respective Harmonic Decagon, which is a convergence point of the diagonals of this Harmonic Decagon».

Remarks.

(a). We believe that equivalent Propositions, are true for Triangle, etc.

(b). This Proposition, we believe to be a new Proposition and firstly appear here.

(c). Actually Propositions 128 and the above 130 are the same, but in another formation.

(d). Relevant Propositions are 114, 117, 128, 129, 131.

19.46. Significant Property of the Lemoine Point of a Simple Harmonic Pentagon.

Extension to Propositions 117, 128-130.

Proposition 131 (Figure 84).

We consider a Harmonic Pentagon ΑΓΕΗΙ inscribed in circle (o) and the Lemoine point Λ of it.

If ΑΛ, ΓΛ, ΕΛ, ΗΛ, ΙΛ re-intersect circle (o), at points Ζ, Θ, Κ, Β, Δ, respectively, then pentagon ΒΔΖΘΚ will be Harmonic too and it shares its Lemoine point Λ with ΑΓΕΗΙ.

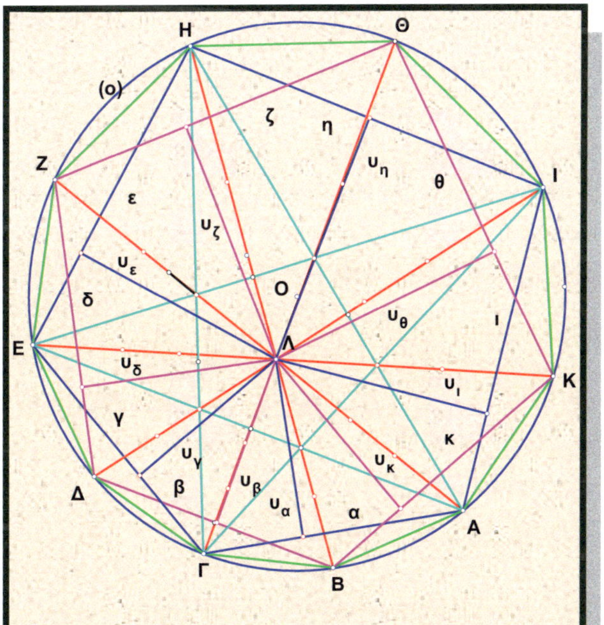

Σχήμα 84.

Proof.

Since pentagon ΑΓΕΗΙ is Harmonic, according to Criterion 117 (§ 19.26), it will in fact have a Lemoine point Λ.

If $υ_α$, $υ_γ$, $υ_ε$, $υ_η$, $υ_ι$, is the distance of Λ from side ΑΓ=α, ΓΕ=γ, ΕΗ=ε, ΗΙ=η, ΙΑ=ι, respectively, according to Criterion 117 (direct) (§ 19.26), it will be:

$$\frac{υ_α}{α}=\frac{υ_γ}{γ}=\frac{υ_ε}{ε}=\frac{υ_η}{η}=\frac{υ_ι}{ι}, \qquad (1).$$

Thus it will also be for example: $$\frac{υ_ε}{ε}=\frac{υ_η}{η}. \qquad (2).$$

Therefore in triangle ΕΗΙ, ΗΛ or ΗΒ is the symmedian (converse of Theorem 11-14.2 in Geometry [6]).

Hence, according to the direct of the Auxiliary Proposition 6 (§ 2.6) in Part E, ΕΙ, ΗΒ, will be conjugate, thus quadrilateral ΒΕΗΙ will be Harmonic.

Consequently, since quadrilateral ΒΕΗΙ is Harmonic and from the implication it is also $ΚΕ \cap ΒΗ \cap ΔΙ \equiv Λ$, according to Proposition 34 (§ 8.71) quadrilateral ΚΒΔΗ, will be Harmonic too.

Thus, since quadrilateral ΚΒΔΗ is Harmonic, then ΔΚ and ΒΗ will be conjugate, hence ΒΗ will be the symmedian of ΔΚ in triangle ΚΒΔ [Proposition 6 (§ 2.6) in Part E] (converse)].

In this way, all the points on ΒΗ, hence Λ as well, will be far from sides ΒΚ=κ, ΒΔ=β distance, $υ_κ$, $υ_β$ analogous to these sides respectively (direct of Theorem 11-14.2 in Γεωμετρία [6]).

Hence, it will be: $$\frac{υ_κ}{κ}=\frac{υ_β}{β}. \qquad (3).$$

In the exact same way we prove that the distance of Λ from the pairs of sides ΔΒ-ΔΖ, ΖΔ-ΖΘ, ΘΖ-ΘΚ, ΚΘ-ΚΒ, are analogous to these sides.

Thus the following relations are true:

$$\frac{υ_β}{β}=\frac{υ_δ}{δ},\ \frac{υ_δ}{δ}=\frac{υ_ζ}{ζ},\ \frac{υ_ζ}{ζ}=\frac{υ_θ}{θ},\ \frac{υ_θ}{θ}=\frac{υ_κ}{κ}. \qquad (4).$$

From relations (3), (4) it occurs that relation (1) is true, i.e. that the distance of Λ from each side of the pentagon ΒΔΖΘΚ is analogous to its respective sides. This means that pentagon ΒΔΖΘΚ is Harmonic and Λ is a Lemoine point for this pentagon as well [converse of Proposition 117 (§ 19.26) and the comment after its conclusion].

Remarks.

(a). Equivalent Propositions can be formed for hexagons as well (Corollary 3 of Criterion 139), etc. This time as well, the proofs are carried through with in similar ways, as for pentagons ΑΓΕΗΙ and ΒΔΖΘΚ.

(b). Proposition 131, we believe to be new and firstly appear here.

(c). Relevant Propositions are 34, 114, 117, 128, 129, 130, 139.

19.47. Significant Property of the Simple Harmonic Decagon.

Proposition 132 (Figure 84).

The twin pentagons, of every Harmonic Decagon, are harmonic.

Proof.

I.e., if the inscribed in circle (o) decagon ΑΒΓΔΕΖΗΘΙΚ, is harmonic, we will prove that its twin pentagons ΑΓΕΗΙ and ΒΔΖΘΚ, are harmonic as well.

In fact, pentagon ΑΓΕΗΙ is harmonic, as, according to the relevant definition (§ 16.1 e), there on circle (o) points Β, Δ, Ζ, Θ, Κ, for which decagon Α-ΒΓΔΕΖΗΘΙΚ, is harmonic.

In the exact same way we prove that pentagon ΒΔΖΘΚ, is harmonic as well.

Remarks.

(a). Proposition 132, we believe to be new and firstly appear here.

(b). Relevant Propositions are 117, 128, 129, 130, 131, 139.

19.48. The Fourteenth Criterion of Harmonic being of the Inscribed Quadrilateral. {The first Ten Criteria are already included in our book [31], while the rest three are 33, 35 and 122}.

Criterion 133 (Figure 28).

If ΑΓΒΔ is a convex quadrilateral inscribed in circle (O,ρ) or (κ) and M is the midpoint of its diagonals ΓΔ, then this quadrilateral is harmonic, if and only if, MΓ is the bisector of angle AMB.

Proof.

(a). Direct.

I.e., if ∠AMΓ=∠ΓMB, we will prove that quadrilateral ΑΓΒΔ is harmonic.

We draw line OM which we suppose intersects AB at point P. In triangle MAB, MΓ is from the hypothesis the bisector of ∠AMB and it is ΓM⊥OP. Thus, according to known Theorem of the bisector of a triangle (triangle MAB), the intersection α of AB and ΓΔ and P will harmonically divide segment AB, thus α will be a point on the polar of P and since ΓM⊥OP then M will be a second point on this polar, αM or ΓΔ will be the polar of P. We then see that diagonal AB, runs through pole P of ΓΔ. This means that diagonals AB, ΓΔ are conjugate, with respect to circle (κ), hence according to the relevant definition (Γεωμετρία [6] paragraph 8-4.10), quadrilateral ΑΓΒΔ is harmonic.

(b). Converse.

I.e., if quadrilateral ΑΓΒΔ is harmonic, we will then prove that ∠AMΓ=∠ΓMB.

In fact, since quadrilateral ΑΓΒΔ is harmonic then its diagonal AB, will run through pole P of its diagonal ΓΔ (definition of the Harmonic Quadrilateral). Moreover, through P as it is known, OM will also run, the perpendicular to ΓΔ. This means that pole P of ΓΔ, is the intersection of AB and OM.

Hence, α, P will harmonically divide points A, B and since MΓ⊥OP, according to the known Theorem about the bisectors of a triangle (converse), in triangle AMB, ΓM or ΓΔ will be the bisector of angle AMB.

(d). Investigation.

1/. In the special case where ΓΔ runs through the midpoint N of AB or N≡α, then for it to be ∠AMΓ=∠ΓMB or ∠AMα=∠αMB, triangle AMB should be isosceles. But then Mα⊥AB and since MΓ⊥OP, it will be AB//OM. Hence in this case, for AΓBΔ to be harmonic it needs to be AB⊥ΓΔ (or ΓΔ to be the perpendicular bisector of AB) and therefore P will be in infinity.

2/. It is obvious that for the same Criterion 133, it is also true for the mid-point N of the other diagonal AB of the quadrilateral AΓBΔ. Hence NB is the bisector of angle ΓNΔ.

Remarks.

(a). Proposition 133 is the fourteenth criterion of harmonic being of a quadrilateral. We have come up with the first ten criteria of harmonic being of a harmonic quadrilateral, earlier and they are included in book [31]. Criteria 11, 12, 13, are the Propositions 33, 35, 122, respectively.

(b). Proposition 133, we believe to be new and firstly appear here.

(c). Relevant Propositions are 33, 35, 122.

19.49. Construction of a Harmonic Quadrilateral based on a Harmonic Group of Four.

Easy Way of Construction of a Harmonic Quadrilateral.

Extension to known Construction, from a Line to a Circle.

Construction 134 (Figure 85).

On circle (κ) we consider the pair of points A, B. On the same circle (κ) set a second pair of points Γ, Δ, which harmonically divide points A, B at such parts, that their ratio is μ/ν (where μ, ν are given segments). I.e. that it is:

$$\frac{A\Gamma}{\Gamma B}=\frac{A\Delta}{\Delta B}=\frac{\mu}{\nu}, \qquad (1).$$

Or in a second formation:

In circle (κ) inscribe a harmonic quadrilateral AΓBΔ whose apexes A, B are given and whose apexes Γ, Δ, harmonically divide its apexes A, B at parts that their ratio is μ/ν. Hence relation (1) should be true.

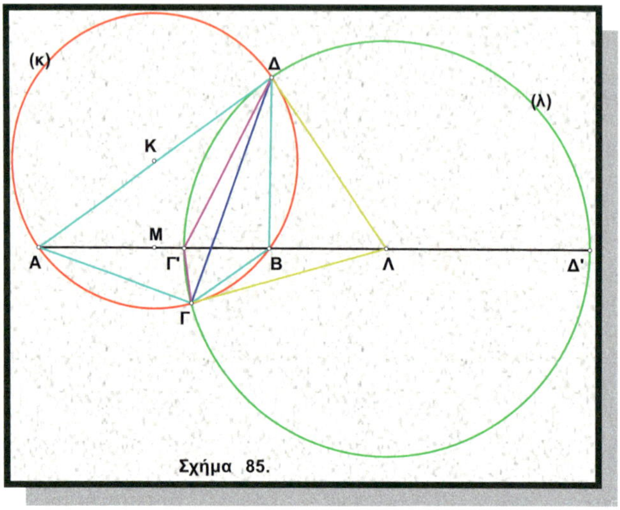

Σχήμα 85.

Solution.

(a). Analysis.

We suppose that quadrilateral ΑΓΒΔ is the desired one. Then it will be inscribed in circle (κ) and it will be harmonic, thus according to criterion 2 of harmonic being of a quadrilateral [Proposition 2β(21) in our book [31] or in Γεωμετρία [6] paragraph 8-4.10], it will have:

$$ΑΓ.ΒΔ=ΑΔ.ΒΓ \text{ or } \frac{ΑΓ}{ΓΒ}=\frac{ΑΔ}{ΔΒ}. \qquad (2).$$

Moreover, from the hypothesis, it is as well: $\frac{ΑΓ}{ΓΒ}=\frac{ΑΔ}{ΔΒ}=\frac{μ}{ν}$. (3).

Hence, if Γ' is the intersection of the bisector of angle Γ with ΑΒ, it will be:

$$\frac{ΑΓ}{ΓΒ}=\frac{ΑΓ'}{Γ'Β}=\frac{μ}{ν}. \qquad (4).$$

Thus, from (3), (4) ⇒ $\frac{ΑΔ}{ΔΒ}=\frac{ΑΓ'}{Γ'Β}=\frac{μ}{ν}$. (5).

Relation (5), shows that ΔΓ' is a bisector of angle Δ as well.

{The proof that the bisectors of angles Γ, Δ intersect ΑΒ at the same point Γ', can also be based on criterion 3 of harmonic being of the quadrilateral, paragraph 2β(22) of our book [31] above}.

If Δ' is the conjugate harmonic of Γ' with respect to Α, Β and we draw the Apollonian circle with diameter Γ'Δ', then this, as it is known, will also run through Γ, Δ. Hence Γ, Δ are the intersections of circle (κ) and the Apollonian circle with diameter Γ'Δ'.

Thus since Γ', Δ', are constructed in the known way (Construction 237 page 234 in book Γεωμετρία [21]), we are led to the construction below.

(b). Construction.

With points Γ', Δ', we harmonically divide segment AB with ratio μ/ν in the known way (Problem 237 page 234 in Γεωμετρία [21]).

With segment Γ'Δ' as diameter, we draw the Apollonian circle (λ), which intersects the given circle (κ), at points Γ, Δ. We say that quadrilateral ΑΓΒΔ is the desired harmonic one.

(c). Proof.

This quadrilateral is inscribed in circle (κ), due to its construction.

Moreover, since circle (λ) is Apollonian for triangle ΓΑΒ, ΓΓ' will be the bisector of angle Γ. Similarly ΔΓ' will be the bisector of angle Δ, in triangle ΔΑΒ.

Thus, it will be: $\frac{A\Gamma'}{\Gamma'B}=\frac{\mu}{\nu}=\frac{A\Gamma}{\Gamma B}=\frac{A\Delta}{\Delta B}$, or $\frac{A\Gamma}{\Gamma B}=\frac{A\Delta}{\Delta B}$ or $A\Gamma.B\Delta=A\Delta.B\Gamma$. This, according to the converse of criterion 2 {Proposition 2β(21) in our book [31]} and which is easily proven, this quadrilateral will be harmonic.

(c). Investigation.

Since Γ' and Δ' can always be constructed and since Γ' will lie inside AB and thus in circle (κ), while Δ' outside of it, thus then circle (λ) will be always constructible, therefore these circles will intersect with each other and hence we will always have points Γ, Δ.

In the case where ν=μ, then we easily realize that Γ' coincides with the midpoint of AB and thus Δ' will be in infinity. Additionally, then Γ, Δ will coincide with the midpoints of the two arcs of AB, then triangles ΓΑΒ, ΔΑΒ will be isosceles and ΓΔ will be the perpendicular bisector of AB and hence it will be the diameter of circle (κ) and will run through point Γ'≡M (midpoint of AB).

Remarks.

(a). Construction 134 we believe to be new and firstly appear here.

(b). It is obvious that Construction 134, is an extension to a known Problem {Construction 237, page 234 in book [21]}, from a line to a circle.

(c). Relevant Constructions are 99, 135.

19.50. Two Easy Ways of Construction of a Harmonic Quadrialteral.

Extension to known Construction, from a Line to a Circle.

Construction 135 (Figure 86).

On circle (O,ρ) or (κ), we are given three points Α, Γ, Β. On the same circle set a point Δ, which with Γ harmonically divides the pair of points Α, Β, or the quadrilateral ΑΓΒΔ would be harmonic.

Solution.

Thoughts.

Since the pair of points Γ, Δ, should harmonically divide, the pair of points Α, Β. Hence it should be: $\frac{A\Gamma}{\Gamma B}=\frac{A\Delta}{\Delta B}$ or ΑΓ.ΒΔ=ΑΔ.ΒΓ, or quadrilateral ΑΓΒΔ should be harmonic {Known Criterion 2 of harmonic being of the quadrilateral, Proposition 2β(21) in our book [31], or Γεωμετρία [6] paragraph 8-4.10}.

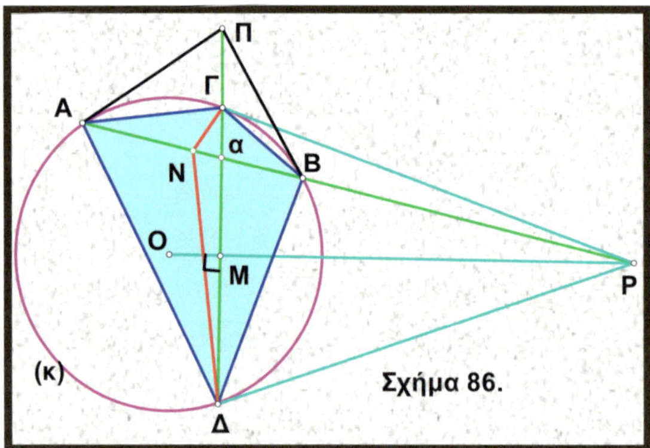

Σχήμα 86.

In other words we are asked to construct a harmonic quadrilateral, whose three apexes Α, Γ, Β we are given and we are asked to set the fourth one Δ, which obviously should lie on circle (κ).

To achieve this, we have to be based on one of the fourteen Criteria of the harmonic quadrilateral or its definition.

1st Way (Based on its definition).

(a). Analysis.

According to the definition of the Harmonic Quadrilateral (Γεωμετρία [6] paragraph 8-4.10), its diagonals should be conjugate with respect to the circle, thus for example its diagonal ΓΔ should run through pole Π of AB. Hence, it will run through two of its known points Γ, Π, thus we are led to the construction below.

(b). Construction.

We find, in the known way pole Π of AB, with respect to the circle. We draw ΠΓ which intersects (κ) at Δ. We say that the desired point is Δ.

(c). Proof.

Since ΓΔ, from its construction, runs through pole Π of AB, the diagonals of quadrialteral ΑΓΒΔ will be conjugate, thus according to the definition, of the Harmonic Quadrilateral, quadrilateral ΑΓΒΔ will in fact be harmonic and thus Γ, Δ will harmonically divide points A, B.

Therefore, Δ is the desired point.

2nd Way (Based on criterion 14).

(a). Analysis.

A rather easy way, as we will realize, is if we depend on Criterion 14 of the Harmonic being of a Quadrilateral, that we gave above, with Proposition 133 (§ 19.48) and that then we are led to the construction below.

(b). Construction.

We define midpoint N of AB, we draw ΓN which we suppose forms with AB an angle φ.

Using N as an apex and from the side of AB that ΓN is not lying, we draw a line that forms with AB angle φ, which intersects circle (κ) at Δ. We say that the desired point is Δ.

(c). Proof.

According to the direct of Criterion 14 (Proposition 133), since ∠ΓNB = ∠BNΔ=φ, quadrilateral ΑΓΒΔ will be harmonic, hence Γ, Δ, will also harmonically divide points A, B.

Remarks.

(a). Another easy way is also this construction to be made based on criterion 11, which is given above, in Proposition 33 (§ 8.5), according to which it is AN^2 = NΓ.NΔ, thus we find in the known way that NΔ=α, since we know AN and NΓ. In this way, Δ will belong to circle (κ) as well as to circle (N, α).

(b). It is obvious that, we can find at least 14, such ways of construction of a harmonic quadrilateral, i.e. as many as our so far known criteria of a Harmonic Quadrilateral.

(c). In this Problem it is possible, instead of point Γ, to be given the ratio ΑΓ/ΓΒ=μ/ν [(Construction 134 (§ 19.49)].

(d). It is obvious that Proposition 135, is an extension to known Proposition (Συμπλήρωμα Γεωμετρίας [2] § 6.6, application 1, page 46), from a line to a circle.

(e). Relevant Propositions are 61, 99, 134

19.51. Criterion of Harmonic being of an Inscribed Trapezium.

Special Trapezium.

Criterion 136 (Figure 87).

Each isosceles trapezium is harmonic, if and only if its equal sides are equal to the side of a square whose diagonal is equal to the diagonal of this trapezium, or to state it otherwise:

Each Harmonic Trapezium, is isosceles and its sides are equal to the side of a square whose diagonal is equal to the diagonal of this trapezium, and conversely,

each isosceles trapezium, whose equal sides are equal to the side of a square whose diagonal is equal to the diagonal of this trapezium, is harmonic.

Solution

(a). Direct.

We consider the Harmonic Trapezium ΑΒΓΔ, με ΑΒ//ΓΔ.

Since the given trapezium is a harmonic quadrilateral, according to the relevant definition of the Harmonic Quadrilateral (§ 16.1a), it will be cyclic to a circle let's say (Ο, ρ) or (ο) and hence it will be isosceles with let's say ΑΔ=ΒΓ=β, ΑΓ=ΒΔ=λ and with ΑΒ=α, ΓΔ=γ. We will prove that it is: $\lambda=\beta\sqrt{2}$.

In fact, according to Criterion 2β(21) in our Γεωμετρία [31], or to Theorem 8-4.10-in Γεωμετρία [6], it will be: $\beta\beta=\alpha\gamma$ or $\beta^2=\alpha\gamma$. (1).

Moreover, since this quadrilateral is cyclic, according to the 1st Ptolemy's Theorem and if we take into consideration relation (1), it will be:

$\lambda\lambda=\alpha\gamma+\beta\beta$ or $\lambda^2=\beta^2+\beta^2$ or $\lambda^2=2\beta^2$ or $\lambda=\beta\sqrt{2}$. (2).

(Additionally it is as well $\lambda^2=\alpha\gamma+\alpha\gamma=2\alpha\gamma$ or $\lambda^2=2\alpha\gamma$ or

$\lambda=\sqrt{2\alpha\gamma}$). (3).

This means that, in fact side β of the trapezium, according to relation (2), is equal to the side of a square whose diagonal is equal to diagonal λ of this trapezium.

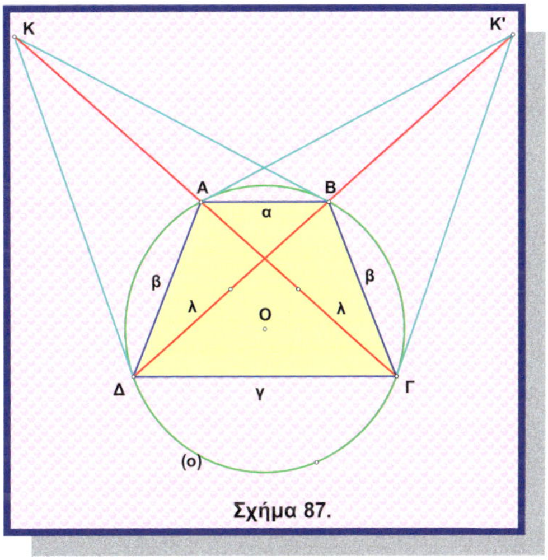

Σχήμα 87.

(b). Converse.

I.e., if an isosceles trapezium ΑΒΓΔ (ΑΔ=ΒΓ), has its equal sides ΑΔ=ΒΓ=β, equal to the side of a square whose diagonal is equal to the diagonal of the given trapezium, or if it is $\lambda=\beta\sqrt{2}$, we will show that it is harmonic.

In fact then, it will be: $\lambda^2=2\beta^2$ or $\lambda^2=\beta^2+\beta^2$. (4).

Additionally since the given trapezium is isosceles, it will be cyclic for a circle (O, ρ) or(o), hence the 1st Ptolemy's Theorem will be true. Thus, it will also be: $\lambda\lambda=\alpha\gamma+\beta\beta$ or $\lambda^2=\beta^2+\alpha\gamma$. (5).

Hence, from relations (4), (5) $\Rightarrow \beta^2+\beta^2=\beta^2+\alpha\gamma$ or $\beta^2=\alpha\gamma$ ή $\beta\beta=\alpha\gamma$. (6).

Therefore since (6) is true, according to Criterion 2β(21) (converse) above, trapezium ΑΒΓΔ, will be harmonic.

(d). Investigation.

--<u>In the special case</u> where the diagonals of the isosceles trapezium are equal to the diameter of the circle, then (2) becomes:

$\lambda=2\rho=\beta\sqrt{2}$ or $\beta=\rho\sqrt{2}$.

This shows that the non-parallel sides β of the trapezium, are equal to the side of the inscribed in circle square, then obviously this trapezium will be square, as then the trapezium should be also rectangle and thus $\alpha=\gamma$, $\alpha\gamma=\beta^2=\alpha^2=\gamma^2$ ή $\alpha=\beta=\gamma$.

--<u>In the special case</u> the large base of the isosceles trapezium is equal to the diameter of the circumscribed circle ($\gamma=2\rho$), then:

From the right triangle ΑΓΔ it occurs that $\beta^2+\lambda^2=(2\rho)^2=4\rho^2$, which due to relation (2), becomes:

$$\beta^2+2\beta^2=4\rho^2 \quad \text{or} \quad 3\beta^2=4\rho^2 \qquad \text{or} \qquad \beta=\frac{2}{3}.\rho\sqrt{3}. \qquad (7).$$

Relation (7) means that then side β is equal to the 2/3 of the side of an equilateral triangle inscribed in the same circle (Ο,ρ).

Evem then, (7) due to (1) becomes:

$$\sqrt{\alpha\gamma}=\frac{2}{3}.\rho\sqrt{3} \quad \text{or} \quad \sqrt{\alpha 2\rho}=\frac{2}{3}.\rho\sqrt{3} \qquad \text{or} \qquad \alpha=\frac{2}{3}.\rho=\frac{\gamma}{3}. \qquad (8).$$

Conclusion.

From the above we conclude that:

«When the large base γ of a Harmonic Trapezium, becomes the biggest possible equal to 2ρ, then the small base α, becomes the smallest possible equal to $\frac{2}{3}.\rho$ or $=\frac{\gamma}{3}$.»

Corollary.

Every isosceles trapezium, that its large base is equal to the diameter of its circumscribed circle and its small base is equal to the one third of the big one, then it is harmonic.

Proof.

In fact then, if we take into consideration (7) and the first in (8), it will be: $\alpha\gamma=\frac{4}{9}.\rho^2.3=(\frac{2}{3}.\rho.\sqrt{3})^2=\beta\beta$ ή $\alpha\gamma=\beta\beta$. This means that this trapezium, is harmonic.

Remarks.

(a). From Criterion 136, we conclude that this Harmonic Trapezium is a «Special Trapezium».

(b). For the Harmonic Trapezium ΑΒΓΔ, bases ΑΒ, ΓΔ with diagonals ΑΓ, ΒΔ, are connected through the following relation $\lambda^2=2\alpha\gamma$ [relation (3)].

(c). Relevant Propositions are 137, 138.

19.52. Three Ways of Construction of a Harmonic Trapezium.

Extension to Construction 90, from a Line to a Circle.

Construction 137 (Figure 88).

In circle (O, R) inscribe a Harmonic Trapezium ΑΓΒΔ whose we are given one diagonal ΑΒ=2τ and such that ΑΒ=ΓΔ=2τ ⩽2R (where τ given segment), or in another formation:

On circle (O, R) we consider the pair of points Α, Β. On the same circle set a second pair of points Γ, Δ which harmonically divide points Α, Β and such that ΑΒ=ΓΔ=2τ ⩽2R (where τ given segment).

Solution 1st (Using poles and polars).

(a). Analysis.

If ΑΓΒΔ is the desired Harmonic Trapezium, then according to what we know ΓΔ should run through pole N of our known string ΑΒ.

Moreover it should also be:

ΑΒ=ΓΔ=2τ ⩽2R.

Thus, since N is constructible and circle (O, R) should cut from line NΔ a segment ΓΔ=ΑΒ= 2τ, we are led to the construction below.

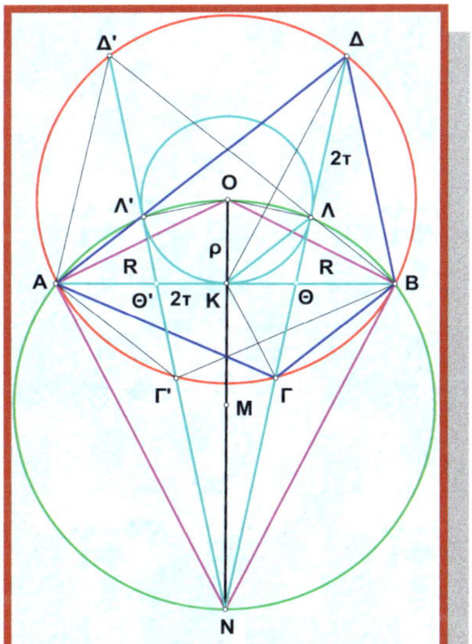

Σχήμα 88.

(b). Construction.

We put, in the known way, on circle (O, R)

a string AB=2τ, we set the pole N of chord AB and from N we draw, in the known way, a tangent NΛ to the circle (O, ρ) homocentric to circle (O, R), whose radius is ρ equal to the distance OK of O from AB, which intersects circle (O, R) at points Γ, Δ (Γ lies between N, Δ). We say that quadrilateral AΓBΔ is the desired Harmonic Trapezium.

(c). Proof.

Since ΓΔ runs through pole N of AB, strings AB, ΓΔ are conjugate, hence quadrilateral AΓBΔ is harmonic (Definition of Harmonic being of an inscribed quadrilateral, Γεωμετρία [6] paragraph 8-4,10).

Additionally, since ρ=OΛ=OK, then ΓΔ is equal to the chord AB=2τ, thus also AΓ=BΔ (isosceles trapezium).

(d). Investigation.

We easily realize that this problem has two solutions, symmetrical with respect to ON, as tangents from N to circle (O, ρ), can always be drawn two (NΓΔ and NΓ'Δ').

(See relevantly the investigation in Criterion 136).

Solution 2nd (Metrical).

(a). Analysis.

If AΓBΔ is the desired Harmonic Trapezium, then since it is harmonic, due to what we know, it will be convex, cyclic and it will have:

$$A\Gamma.B\Delta=A\Delta.B\Gamma. \quad (1).$$

Moreover, from the implication it is: $AB=\Gamma\Delta=2\tau$. (2).

In this way, we easily realize that this inscribed quadrilateral will be an isosceles trapezium, thus we suppose that it has:

$$A\Delta//B\Gamma \text{ and } A\Gamma=B\Delta. \quad (3).$$

Therefore (1) due to (3) becomes: $A\Gamma^2=B\Delta^2=A\Delta.B\Gamma$. (4).

If we take into consideration (2), (3) and if we apply Ptolemy's Theorem, on this trapezium, we will have:

$$AB.\Gamma\Delta= AB^2=\Gamma\Delta^2=A\Gamma.B\Delta+A\Delta.B\Gamma=A\Gamma^2+A\Delta.B\Gamma=4\tau^2. \quad (5).$$

Relation (5), due to (1), (2) (4), becomes:

$$AB^2= \Gamma\Delta^2=2A\Gamma^2 =2A\Delta.B\Gamma = 4\tau^2 \text{ or } A\Gamma^2= 2\tau^2 \text{ or } A\Gamma=B\Delta= \tau\sqrt{2}. \quad (6).$$

We thus see that AΓ is constructed as the diagonal of a square with side τ.

Thus, we are led to the construction below.

(b). Construction.

On given circle (O,R) we put, in the known way, a string AB=2τ.

We construct the equal sides AΓ=BΔ= $τ\sqrt{2}$, as we mentioned above and draw circles (A,AΓ), (B,BΔ), that intersect circle (O, R) at pairs of points Γ, Δ' and Γ', Δ, respectively.

We say that each one from quadrilaterals AΓBΔ, AΓ'BΔ' are the desired trapezium.

(c). Proof.

AΓBΔ is in fact the desired one, as:

It is inscribed on circle (O, R), it has AB=2τ and AΓ=BΔ= $τ\sqrt{2}$, thus this quadrilateral will be isosceles and hence it will have equal diagonals

AB=ΓΔ=2τ.

Besides, from the above and according to Ptolemy's Theorem it will be: $AB^2= ΓΔ^2=AΓ^2 +AΔ.BΓ = 4τ^2$ or $(τ\sqrt{2})^2 +AΔ.BΓ =2τ^2 +AΔ.BΓ= 4τ^2$ or since $AΓ= τ\sqrt{2}$ or $AΓ^2=2τ^2$, it will also be: $AΔ.BΓ=2τ^2 =AΓ^2=AΓ.AΓ=AΓ.BΔ$ or AΔ.BΓ=AΓ.BΔ. This means that quadrialteral AΓBΔ is harmonic, as it is also inscribed. Thus it is the desired one. Similarly we also find that quadrilateral AΓ'BΔ' is the solution to the problem.

Solution 3rd (With the application of Criterion 136).

This proof is carried through with as mentioned in way 2 above, but this time applying Criterion 136, where needed, after we find that AΓ=BΔ=$τ\sqrt{2}$, through relation 2τ=AΓ$\sqrt{2}$, that in Criterion 136 (relation 2), we have proved to be true.

(c). Investigation.

From the above we realize that the problem has always two solutions.

Remarks.

(a). According to the above solution 2, it is AΓ=BΔ= $τ\sqrt{2}$.

Hence, each one from the non parallel side of every harmonic trapezium, is equal to the diagonal of a square with side equal to half the diagonal of this trapezium.

(b). A third formation of Proposition 137, is the following:
«In circle (O, R), inscribe a harmonic trapezium ΑΓΒΔ, whose we are given the diagonal $AB=2\tau$».
(c). Proposition 137 we believe to be new and firstly appear here, while it is an extension to Construction 90, from a line to a circle.
(d). Relevant Propositions are 89, 90, 136, 138.

19.53. Construction of a Harmonic Quadrilateral.
Extension to Construction 89, from a Line to a Circle.
Generalization of Construction 137.

Construction 138 (Figure 89).

In given circle (O, R) inscribe a quadrilateral ΑΓΒΔ, which is harmonic and whose diagonals $AB=2\tau \leq 2R$, $\Gamma\Delta=2\tau' \leq 2R$ we are given (where τ, τ' given segments),

Or in another formation:

On given circle (O, R) we consider a pair of points A, B. On the same circle set a second pair of points Γ, Δ, that harmonically divide pair of points A, B and such that $AB=2\tau \leq 2R$, $\Gamma\Delta=2\tau' \leq 2R$ (where τ, τ' given segments).

Solution .

We think and work here in the exact same way as in the first solution of Construction 137, with the difference that ΝΓΔ, is drawn tangent from pole N to a circle homocentric to (O, R) and whose radius is ρ, equal to the perpendicular side of a right triangle, with hypotenuse R and its other perpendicular side, equal to τ'. I.e. $\rho^2=R^2-\tau'^2$.

(c). Investigation.

We easily realize that this problem always has two solutions, as long as it is: $2\tau \leq 2R$ and $2\tau' \leq 2R$.

Remarks.

(a). Construction 138 we believe to be new and firstly appear here.

(b). It is obvious, that Construction 138, is a generalization to Construction 137, while this is also an extension to Construction 89, from a line to a circle.

(c). Relevant Propositions are 89, 90, 136, 137.

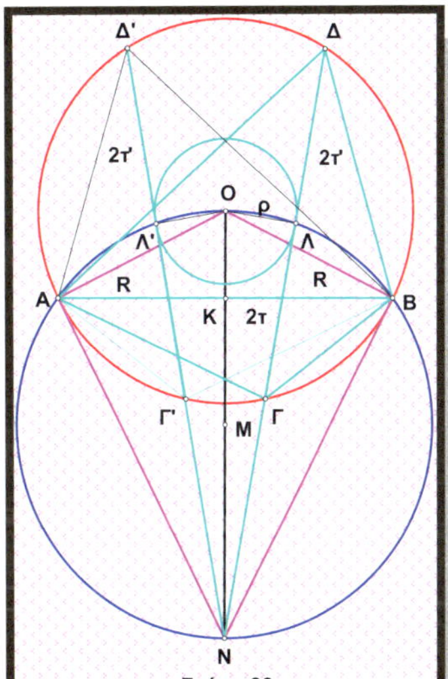

Σχήμα 89.

19.54. Fourth Criterion of Harmonic being of an Inscribed Hexagon.

Criterion 139 (Figure 64).

If the three diagonals (main) of a convex inscribed in circle hexagon, are respectively the carriers of the symmedians of at least one of the twin triangles of this hexagon, then and only then this hexagon is harmonic.

Proof.

(a). Direct.

We consider a harmonic hexagon ΑΒΓΔΕΖ inscribed in circle (o), with ΑΔ∩ΒΕ∩ΓΖ≡Κ (Proposition 104). We will show that ΑΔ, ΒΕ, ΓΖ, are the carriers of the symmedians of the twin triangles ΑΓΕ and ΒΔΖ.

In fact, according to the definition of the Harmonic Hexagons (§ 16.1b), for example quadrilateral ΓΔΕΑ will also be harmonic, hence, according to the definition of Harmonic Quadrilaterals (§ 16.1a), diagonals ΑΔ, ΓΕ of this quadrilateral, will be conjugate, with respect to circle (o).

Thus we see that ΑΔ is conjugate to side ΓΕ of the triangle ΑΓΕ, thus ΑΔ will be the carrier of the symmedian of the same triangle (converse of the Auxiliary Proposition 6, § 2.6, Part E).

Similarly we find that for the harmonic quadrilateral ZABΔ, AΔ is conjugate to the side BZ of triangle ΔBZ, hence AΔ will be the carrier of the symmedian of the triangle ΔBZ.

Working in a similar way, we find that the rest of what is asked is true.

Therefore, we see that AΔ, BE, ΓZ are symmedians of both triangles AΓE and BΔZ (carriers).

(b). Converse.

I.e., if the given hexagon ABΓΔEZ is convex, inscribed in circle (O, R) and its diagonals AΔ, BE, ΓZ are symmedians (carriers) for example of the twin triangle AΓE or BΔZ, then this hexagon is harmonic.

In fact, if AΔ, BE, ΓZ are symmedians (carriers) for example of the twin triangle AΓE, then, according to Auxiliary Proposition 6, § 2.6, in Part E (direct), pairs AΔ-ΓE, BE-AΓ, ΓZ-EA will be conjugate. Hence, according to the definition of Harmonic Quadrilaterals (§ 16.1a), quadrilaterals ΓΔEA, EZAΓ, ABΓE will be Harmonic, thus according to Criterion 109 (§ 19.17, direct) of Harmonic Hexagons, hexagon ABΓΔEZ is harmonic.

We would reach the same result if we are given that AΔ, BE, ΓZ are symmedians (carriers) of the twin triangle BΔZ.

Corollary 1.

The three apexes of each triangle and the three intersections of its symmedian with its circumscribed circle, are apexes to a Harmonic Hexagon.

Proof.

We consider triangle AΓE. If its symmedians intersect its circumscribed circle at points Δ, Z, B respectively, then hexagon ABΓΔEZ is harmonic.

The proof of this obviously, is a direct consequence of the converse of the above Proposition 139 of the Harmonic Hexagons.

Corollary 2.

If on a convex inscribed in circle hexagon, with concurrent diagonals, the point of convergence of its diagonals is the Lemoine point of at least one of its twin triangles, then and only then this hexagon is harmonic.

Proof.

(a). Direct.

We consider a harmonic hexagon ΑΒΓΔΕΖ. According to Proposition 104, It will have ΑΔ∩ΒΕ∩ΓΖ≡Κ. We will show Κ is the Lemoine point of triangles ΑΓΕ, ΒΔΖ.

According to the direct of the above Criterion 139, ΑΔ, ΒΕ, ΓΖ, are the carriers of the symmedians of triangles ΑΓΕ, ΒΔΖ, hence their convergence point Κ, is the Lemoine point of both triangles ΑΓΕ, ΒΔΖ.

(b). Converse.

I.e., if on given convex hexagon ΑΒΓΔΕΖ, inscribed in circle (O, R), it is ΑΔ∩ΒΕ∩ΓΖ≡Κ and Κ is the Lemoine point of triangle ΑΓΕ or of ΒΔΖ, we will show that this hexagon is harmonic.

If for example Κ is the Lemoine point of triangle ΑΓΕ, then ΑΔ, ΒΕ, ΓΖ, that run through Κ, will also be symmedians to triangle ΑΓΕ, thus according to the converse of Criterion 139 above, hexagon ΑΒΓΔΕΖ is harmonic.

Remark.

It is obvious that the proof of Corollary 2, can be based on Auxiliary Proposition 6, § 2.6, in Part E.

Corollary 3.

We are given a triangle ΑΓΕ, the Lemoine Point Κ of this triangle and (o) the circumscribed circle (o) of the same triangle. If ΑΚ, ΓΚ, ΕΚ, re-intersect circle (o) at points Δ, Ζ, Β, respectively, prove that Κ is a Lemoine point of triangle ΒΔΖ as well.

Proof.

1st Way (Based on Corollary 1).

According to Corollary 1 above, since K is a Lemoine point for triangle AΓE, apexes A, Γ, E, and the intersections Δ, Z, B, of AK, ΓK, EK, with circle (o), will be apexes of the Harmonic Hexagon ABΓΔEZ.

Hence, since hexagon ABΓΔEZ, is harmonic, according to the relevant definition (§ 16.1b), it will have its following quadrialterals harmonic ΔEZB, ZABΔ, BΓΔZ.

Thus, since for example quadrilateral ΔEZB is harmonic then BE and ΔZ will be conjugate with respect to circle (o), hence for triangle BΔZ, BE will be its symmedian (Converse of Auxiliary Proposition 6, § 2.6, in Part E).

Similarly, we also find that AΔ and ΓZ are symmedians of triangle BΔZ and since AΔ, BE, ΓZ, are concurrent from the implication at K, it will be the Lemoine point for triangle BΔZ too.

2nd Way (Analytical).

Since, for example AΔ will be the symmedian of triangle AΓE then AΔ, ΓE, will be conjugate with respect to circle (o) (Direct of the Auxiliary Proposition 6, § 2.6, in Part E).

Thus, quadrilateral ΓΔEA will be harmonic.

In the exact same way we find that quadrilaterals EZAΓ, ABΓE, are harmonic as well.

Hence, since for example quadrilateral ΓΔEA is harmonic and AΔ∩BE∩ΓZ≡K, quadrilateral ZABΔ will also be harmonic (Direct of Proposition 34, § 8.71). This means that AΔ and BZ, are conjugate with respect to circle (o).

Therefore, AΔ is the symmedian of triangle BΔZ (Converse of the Auxiliary Proposition 6, § 2.6, in Part E).

In a similar way we find that symmedians of triangle BΔZ are BE, ΓZ too, and since these run through K, this is a Lemoine point for triangle BΔZ.

Remark.

Equivalent Proposition to Corollary 3, is also true for the Harmonic Decagon (Proposition 131).

General Remarks.

(a). Criterion 139 we believe to be new and firstly appear here.

(b). The 1st, 2nd, 3rd, 4th, 5th, Criteria about the Harmonic Hexagon, are given in Propositions 96, 109, 112, 139, 140, respectively.

(c). Relevant Propositions are 96, 104, 109, 112, 131, 140.

19.55. Fifth Criterion of Harmonic being of an Inscribed Hexagon.

Extension to Criterion 139.

Criterion 140 (Figure 90).

If we have a convex inscribed in circle hexagon, and each one of its three diagonals (main) runs through one of three non consecutive apexes of the hexagon respectively, which hexagon has as apexes the six intersections of the one by one sides of the respective circumscribed (tangential) hexagon of the given inscribed hexagon, then and only then this hexagon is harmonic.

Proof.

(a). Direct.

We consider a Harmonic Hexagon ΑΒΓΔΕΖ, with ΑΔ∩ΒΕ∩ΓΖ≡Κ (Proposition 104) and if Α'Β'Γ'Δ'Ε'Ζ' is its respective circumscribed hexagon (Polar converse), with (O, R) the respective (directing) circle, where Α∈Ζ'Α', Β∈Α'Β', ..., $Α'_1$≡Ζ'Α'∩Β'Γ', $Β'_1$≡Α'Β'∩Γ'Δ',... $Ζ'_1$≡Ε'Ζ'∩Α'Β', we will prove that:

$Α'_1$∈ΒΕ, $Γ'_1$∈ΑΔ, $Ε'_1$∈ΓΖ and $Β'_1$∈ΓΖ, $Δ'_1$∈ΒΕ, $Ζ'_1$∈ΑΔ. (1).

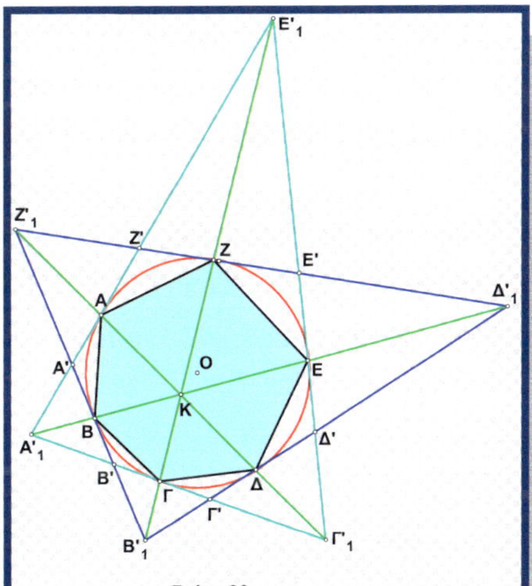

Σχήμα 90.

In fact, since hexagon ABΓΔEZ is harmonic, then for example quadrilateral ABΓE will also be harmonic (initial definition of the Harmonic Hexagon, § 16.1b), thus, according to the relevant definition of Harmonic Quadrilaterals § 16.1a, its diagonals BE, AΓ will be conjugate and thus BE will run through pole A'_1 of AΓ or $A'_1 \in BE$.

Similarly we find that the rest of what is asked in true.

Hence that it will also be: $\Gamma'_1 \in A\Delta$, $E'_1 \in \Gamma Z$, $B'_1 \in \Gamma Z$, $\Delta'_1 \in BE$, $Z'_1 \in A\Delta$.

Conclusions.

--From the above it becomes obvious that for the harmonic hexagon A-BΓΔEZ, the following $A'_1BKE\Delta'_1$, $B'_1\Gamma KZE'_1$, $\Gamma'_1\Delta KAZ'_1$, are lines.

--Additionally we easily realize that for every harmonic hexagon, the Gergonne points of the two pairs of the respective (Polar converse) triangles $A\Gamma E$-$A'_1\Gamma'_1E'_1$ and $B\Delta Z$-$B'_1\Delta'_1Z'_1$, coincide with the common convergence point of the diagonals of the harmonic hexagon.

(b). Converse.

I.e., if for the given hexagon ABΓΔEZ which is convex and inscribed in circle (O, R), it is: $A'_1 \in BE$, $\Gamma'_1 \in A\Delta$, $E'_1 \in \Gamma Z$, or $B'_1 \in \Gamma Z$, $\Delta'_1 \in BE$, $Z'_1 \in A\Delta$, we will prove that ABΓΔEZ is a Harmonic Hexagon.

In fact, if for example it is: $A'_1 \in BE$, $\Gamma'_1 \in A\Delta$, $E'_1 \in \Gamma Z$, then since A'_1, Γ'_1, E'_1, are obviously the poles of AΓ, ΓE, EA respectively, then pairs BE-AΓ, AΔ-ΓE, ΓZ-AE will be conjugate, with respect to circle (O,R), hence according to the relevant definition § 16.1a, quadrilaterals ABΓE, ΓΔEA, EZAΓ, will be harmonic.

Consequently, according to the direct of Criterion 109 about Harmonic Hexagons, then hexagon ABΓΔEZ is harmonic.

Similarly we prove that hexagon ABΓΔEZ is also harmonic when $B'_1 \in \Gamma Z$, $\Delta'_1 \in BE$, $Z'_1 \in A\Delta$.

Corollary.

The three apexes, of each triangle and the three intersections of the three Gergonne cevians (terminology § 2) of its respective (tangential) triangle, with their respective circle, are apexes to a Harmonic Hexagon.

Proof.

We consider triangle ΑΓΕ and its respective (tangential) triangle $A'_1\Gamma'_1E'_1$ (where $A\in A'_1E'_1$, $\Gamma\in A'_1\Gamma'_1$, $E\in \Gamma'_1E'_1$) with respective circle (O, R). If $B\equiv EA'_1\cap(O,R)$, $\Delta\equiv A\Gamma'_1\cap(O, R)$, $Z\equiv\Gamma E'_1\cap(O, R)$, then hexagon ΑΒΓΔΕΖ is harmonic.

The proof of this is obviously, a direct consequence of the converse of Criterion 140 above about Harmonic Hexagons, as then it is as well: $A'_1\in BE$, $\Gamma'_1\in A\Delta$, $E'_1\in\Gamma Z$.

Remarks.

(a). Criterion 140 we believe to be new and firstly appear here.

(b). The 1st, 2nd, 3rd, 4th, 6th, 7th, Criteria of the Harmonic Hexagon, are mentioned in Propositions 96, 109, 112, 139, 141, 142, respectively.

(c). Relevant Propositions are 96, 109, 112, 139, 141, 142.

19.56. Sixth Criterion of Harmonic being of an Inscribed Hexagon.

Extension to Criteria 67, 109, 140.

Criterion 141 (Figure 64).

If a convex inscribed in circle (O, R), hexagon ΑΒΓΔΕΖ, has its diagonals (main) concurrent and at least two out of its six quadrilaterals ΑΒΓΕ, ΒΓΔΖ, ΓΔΕΑ, ΔΕΖΒ, ΕΖΑΓ, ΖΑΒΔ, are harmonic, except for those pairs whose two quadrilaterals have as shared diagonal one of the diagonals of the given hexagon (I.e., the following pairs of quadrilaterals are excluded: ΑΒΓΕ-ΔΕΖΒ, ΓΔΕΑ-ΖΑΒΔ, ΕΖΑΓ-ΒΓΔΖ), then and only then this hexagon is harmonic.

Proof.

We discern two cases:

1st Case.

The two harmonic quadrilaterals belong to only one of the following two groups of three quadrilaterals:

ΑΒΓΕ-ΓΔΕΑ-ΕΖΑΓ, or ΔΕΖΒ-ΖΑΒΔ-ΒΓΔΖ. (1).

(a). Direct.

We consider a Harmonic Hexagon ΑΒΓΔΕΖ. We will prove that

AΔ∩BE∩ΓZ≡K and that all the pairs of quadrilaterals that occur from (1), are harmonic.

In fact, since hexagon ABΓΔEZ is harmonic it will be AΔ∩BE∩ΓZ≡K (Proposition 104 of Harmonic Hexagons).

Additionally, according to the initial definition of Harmonic Hexagons (§ 16.1b), all the quadrilaterals in (1), are harmonic.

Therefore, all the pairs of quadrilaterals in the 1st Case will also be harmonic.

(b). Converse.

I.e., if the given hexagon ABΓΔEZ which is convex, inscribed in circle (O, R), has AΔ∩BE∩ΓZ≡K and whose for example quadrilaterals EZAΓ and ABΓE are harmonic. We will prove that ABΓΔEZ is a harmonic hexagon.

We only need to prove that then its quadrilateral ΓΔEA is harmonic as well [Based on the direct of Criterion 109 about Harmonic Hexagons (Second Criterion)].

1st Way.

In fact since quadrilaterals EZAΓ and ABΓE are harmonic, the pairs of ΓZ-AE and BE-AΓ will be conjugate with respect to circle (O, R) and therefore ΓZ and EB are symmedians of the triangle AΓE (converse of Lemma 6 in Part E) and hence their intersection ΓZ∩EB≡K coincides with the Lemoine point of triangle AΓE.

Besides, from the implication AΔ runs through K, too. This shows that AΔ is the third symmedian of triangle AΓE. Thus AΔ will be conjugate to the third side ΓE of this triangle, as AΔ is symmedian of triangle AΓE (direct of Lemma 6 in Part E). This means that quadrilateral ΓΔEA is harmonic as well, thus hexagon ABΓΔEZ will be harmonic [Direct of Criterion 109 about Harmonic Hexagons (Criterion second)].

2nd Way.

Since, from the implication it is AΔ∩BE∩ΓZ≡K, according to Lemma 5 in Part E, it will be:

$$\frac{AB}{B\Gamma}\cdot\frac{\Gamma\Delta}{\Delta E}\cdot\frac{EZ}{ZA}=1. \qquad (2).$$

Additionally, since from the implication quadrilaterals ΑΒΓΕ and ΕΖΑΓ are harmonic, it will be: $\frac{AB}{B\Gamma}=\frac{AE}{E\Gamma}$, $\frac{EZ}{ZA}=\frac{E\Gamma}{\Gamma A}$. (3).

Thus, relation (2) due to (3), becomes: $\frac{AE}{E\Gamma}\cdot\frac{\Gamma\Delta}{\Delta E}\cdot\frac{E\Gamma}{\Gamma A}=1$ or ΑΕ.ΓΔ=ΔΕ.ΓΑ.

The latter relation shows that quadrilateral ΓΔΕΑ is harmonic too, hence hexagon ΑΒΓΔΕΖ will be harmonic too [Direct of Proposition 109 about Harmonic Hexagons (Criterion second)].

2nd Case.

Each of the two harmonic quadrilaterals belongs to each from the two following groups of three quadrilaterals:

ΑΒΓΕ-ΓΔΕΑ-ΕΖΑΓ, ΔΕΖΒ-ΖΑΒΔ-ΒΓΔΖ. (4).

(a). Direct.

We consider Harmonic Hexagon ΑΒΓΔΕΖ. We will prove that ΑΔ∩ΒΕ∩ΓΖ≡Κ and all the pairs of quadrilaterals that occur from (4), are harmonic.

In fact, since hexagon ΑΒΓΔΕΖ is harmonic it will be: ΑΔ∩ΒΕ∩ΓΖ≡Κ (Proposition 104 about Harmonic Hexagons).

Additionally, according to the initial definition of Harmonic Hexagons (§ 16.1b), all the quadrilaterals in (4), are harmonic.

Consequently, the pairs of quadrilaterals in the 2nd Case will be harmonic too [relation (4)].

(b). Converse.

I.e., instead of the quadrilaterals ΕΖΑΓ and ΑΒΓΕ that we received as harmonic in the 1st Case, here we are given for example that quadrilaterals Ε-ΖΑΓ and ΔΕΖΒ are harmonic as well.

Since, from the implication it is ΑΔ∩ΒΕ∩ΓΖ≡Κ, according to Criterion (Lemma) 34, or 207, then for example quadrilateral ΑΒΓΕ will also be harmonic, as ΔΕΖΒ is harmonic (implication).

Thus, we notice that hexagon ΑΒΓΔΕΖ will have ΑΔ∩ΒΕ∩ΓΖ≡Κ and its quadrilaterals ΕΖΑΓ and ΑΒΓΕ harmonic, hence the 2nd Case, traces back to the 1st one above (Case).

Remarks.

(a). Criterion 141 we believe to be new and firstly appear here.

(b). The 1st, 2nd, 3rd, 4th, 5th, 7th, Criteria of the Harmonic Hexagon, are given in Propositions 96, 109, 112, 139, 140, 142, respectively.

(c). Relevant Propositions are 34, 96, 109, 112, 139, 140, 142, 207.

19.57. Seventh Criterion of Harmonic being of an Inscribed Hexagon.

Extension to Criterion 141.

Criterion 142 (Figure 90).

If convex inscribed in circle (O, R) hexagon ΑΒΓΔΕΖ, has concurrent diagonals (main) and at least two of these each one runs through one of the apexes of the hexagon $A'_1B'_1\Gamma'_1\Delta'_1E'_1Z'_1$ that its apexes are six one by one intersections of the sides of the respective circumscribed (tangential) hexagon A'B'Γ'Δ'E'Z' of the given hexagon ΑΒΓΔΕΖ, apart from the three pairs of opposite apexes of hexagon $A'_1B'_1\Gamma'_1\Delta'_1E'_1Z'_1$, (I.e., the following pairs of apexes are excluded; A'_1 - Δ'_1, B'_1-E'_1, Γ'_1- Z'_1), then and only then this hexagon is harmonic.

Proof.

(a). Direct.

We consider a Harmonic Hexagon ΑΒΓΔΕΖ, with $A\Delta \cap BE \cap \Gamma Z \equiv K$ (Proposition 104), inscribed in circle (O, R) and its respective circumscribed hexagon (its polar converse) A'B'Γ'Δ'E'Z' ($A \in Z'A'$, $B \in A'B'$,.....).

If $A'_1 \equiv Z'A' \cap B'\Gamma'$, $B'_1 \equiv A'B' \cap \Gamma'\Delta'$,... $Z'_1 \equiv E'Z' \cap A'B'$, we will prove that: A'_1 - $\Delta'_1 \in BE$, B'_1-$E'_1 \in \Gamma Z$, Γ'_1- $Z'_1 \in A\Delta$, or $BE \equiv A'_1\Delta'_1$, $\Gamma Z \equiv B'_1E'_1$, $A\Delta \equiv \Gamma'_1Z'_1$ (their carriers).

Then, since quadrilaterals ΑΒΓΕ, ΒΓΔΖ, ΓΔΕΑ, ΔΕΖΒ, ΕΖΑΓ, ΖΑΒΔ, will be harmonic (definition of Harmonic Hexagons, § 16.1b), for example BE, ΑΓ will also be conjugate, with respect to circle (O, R), hence BE will run through pole A'_1 of ΑΓ, as quadrilateral ΑΒΓΕ is harmonic. Similarly we also find that $B'_1 \in \Gamma Z$, $\Gamma'_1 \in A\Delta$,...[For the proof of this, obviously, we can depend on the direct of Proposition 140 about Harmonic Hexagons (5th Criterion)].

(b). Converse.

We discern two cases:

1st Case.

The two diagonals of hexagon ΑΒΓΔΕΖ run through two one by one apexes of the hexagon $A'_1B'_1\Gamma'_1\Delta'_1E'_1Z'_1$.

I.e., hexagon ΑΒΓΔΕΖ is harmonic, if $A'_1 \in BE$ and $\Gamma'_1 \in A\Delta$, or $A'_1 \in BE$ and $E'_1 \in \Gamma Z$, or $\Gamma'_1 \in A\Delta$ and $E'_1 \in \Gamma Z$, or $B'_1 \in \Gamma Z$ and $\Delta'_1 \in BE$, or $B'_1 \in \Gamma Z$ and $Z'_1 \in A\Delta$, or $\Delta'_1 \in BE$ and $Z'_1 \in A\Delta$, while it is as well: $A\Delta \cap BE \cap \Gamma Z \equiv K$.

In fact, if for example it is $A'_1 \in BE$ and $\Gamma'_1 \in A\Delta$, then obviously BE will be conjugate to ΑΓ, as BE runs through pole A'_1 of ΑΓ (implication). Therefore quadrilateral ΑΒΓΕ is harmonic (definition of the Harmonic Quadrilateral, § 16.1a).

Similarly we also find that quadrilateral ΓΔΕΑ is harmonic too, since $\Gamma'_1 \in A\Delta$. Hence, hexagon ΑΒΓΔΕΖ is harmonic [Converse of Criterion 141, about Harmonic Hexagons (Case 1)].

2nd Case.

The two diagonals of ΑΒΓΔΕΖ run through two consecutive apexes of hexagon $A'_1B'_1\Gamma'_1\Delta'_1E'_1Z'_1$.

I.e., hexagon ΑΒΓΔΕΖ is harmonic, if $A\Delta \cap BE \cap \Gamma Z \equiv K$ and $A'_1 \in BE$ and $B'_1 \in \Gamma Z$, or $B'_1 \in \Gamma Z$ and $\Gamma'_1 \in A\Delta$,...

Hence, if for example it is $A'_1 \in BE$ and $B'_1 \in \Gamma Z$, then we find as above (1st Case), that quadrilaterals ΑΒΓΕ and ΒΓΔΖ are harmonic, hence hexagon ΑΒΓΔΕΖ is harmonic too [Converse of Proposition 141 about Harmonic Hexagons (Case 2)], as here it is as well:. $A\Delta \cap BE \cap \Gamma Z \equiv K$.

Remarks.

(a). Criterion 142 we believe to be new and firstly appear here.

(b). The 1st, 2nd, 3rd, 4th, 5th, 6th, 8th, 9th, Criteria of the Harmonic Hexagon, are given in Propositions 96, 109, 112, 139, 140, 141, 143, 147, respectively.

(c). Relevant Propositions are 96, 109, 112, 139, 140, 141, 143, 147.

19.58. Eighth Criterion of Harmonic being of an Inscribed Hexagon.

Extension to Criterion 142.

Criterion 143 (Figure 91).

If a convex inscribed in circle (O, R) hexagon ΑΒΓΔΕΖ, has concurrent diagonals (main) and at least two of its diagonals run one by one through an intersection of its opposite sides, then and only then this hexagon is harmonic.

Proof.

(a). Direct.

We consider a Harmonic Hexagon ΑΒΓΔΕΖ, inscribed in circle (O, R).

If $A_2 \equiv ZA \cap \Gamma\Delta$, $B_2 \equiv AB \cap \Delta E$, $\Gamma_2 \equiv B\Gamma \cap EZ$, prove that:

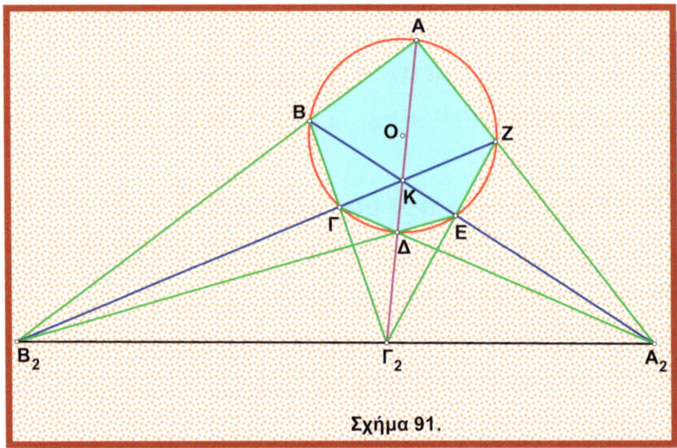

Σχήμα 91.

$$A_2 \in BE,\ B_2 \in \Gamma Z,\ \Gamma_2 \in A\Delta \text{ and } A\Delta \cap BE \cap \Gamma Z \equiv K.$$

In fact, since hexagon ΑΒΓΔΕΖ is harmonic (implication), it will have for example the following quadrilaterals harmonic: ΑΒΓΕ and ΔΕΖΒ (definition §16.1b), hence, according to the converse of Lemma 51 (§ 9.12), or 208, it will be: $A\Delta \cap BE \cap \Gamma Z \equiv K$ and $A_2 \in BE$.

Similarly we also find that: $B_2 \in \Gamma Z$, $\Gamma_2 \in A\Delta$ {It is known [26], that $A_2B_2\Gamma_2$ is a line (Pascal Line)}.

(b). Converse.

I.e., if the given convex inscribed in circle (O, R) hexagon ΑΒΓΔΕΖ, has $A\Delta \cap BE \cap \Gamma Z \equiv K$ and $A_2 \in BE$, $B_2 \in \Gamma Z$, or $A_2 \in BE$, $\Gamma_2 \in A\Delta$ or $B_2 \in \Gamma Z$, $\Gamma_2 \in A\Delta$, then this hexagon is harmonic.

1st Way.

In fact, if for example $A_2 \in BE$ and $B_2 \in \Gamma Z$, then according to a Proposition that has been published with one paper of ours in magazine [Z] (issue 3-4/1978, or mentioned in [31] § 2 α (54), it will also be $\Gamma_2 \in A\Delta$.

This is easily proven:

In fact, since $A_2 \in BE$ and $B_2 \in \Gamma Z$, then BEA_2 and $Z\Gamma B_2$, will be lines, thus, according to Pappus' Theorem, the intersections:

$B\Gamma \cap EZ \equiv \Gamma_2$, $EB_2 \cap \Gamma A_2 \equiv \Delta$, $BB_2 \cap ZA_2 \equiv A$, will be collinear or $A\Delta\Gamma_2$ will be a line, or $\Gamma_2 \in A\Delta$.

Hence, since $A\Delta \cap BE \cap \Gamma Z \equiv K$ and it is for example $A_2 \in BE$, according to the direct of Lemma 51 (§ 9.12), quadrilaterals ΑΒΓΕ and ΔΕΖΒ will also be harmonic.

Similarly we find that pairs of quadrilaterals: ΒΓΔΖ, ΕΖΑΓ and ΓΔΕΑ, ΖΑΒΔ, are harmonic too.

Consequently, according to the initial definition (§ 16.1b) of Harmonic Hexagons, hexagon ΑΒΓΔΕΖ will be Harmonic too.

2nd Way.

If for example $A_2 \in BE$ and $B_2 \in \Gamma Z$, since it also is $A\Delta \cap BE \cap \Gamma Z \equiv K$ (implication). According to the direct of Lemma 51 (§ 9.12), quadrilaterals ΑΒΓΕ, ΔΕΖΒ and ΒΓΔΖ, ΕΖΑΓ, will be harmonic.

Thus, according to Proposition 141 of Harmonic Hexagons (Criterion 6) (converse, Case 1 or 2), hexagon ΑΒΓΔΕΖ will be harmonic too.

Remarks.

(a). Criterion 143 we believe to be new and firstly appear here.

(b). The 1st, 2nd, 3rd, 4th, 5th, 6th, 7th ,9th, Criteria about the Harmonic Hexagon, are given in Propositions 96, 109, 112, 139, 140, 141, 142, 147, respectively.

(c). Relevant Propositions are 51, 96, 109, 112, 139, 140, 141, 142, 147.

19.59. Property of Harmonic Hexagons.

Extension to Proposition 88, from a line to a circle.

Proposition 144 (Figure 64).

For every Harmonic Hexagon ΑΒΓΔΕΖ, it is:

$$AB.\Gamma\Delta.EZ = B\Gamma.\Delta E.ZA = \frac{A\Delta.BE.\Gamma Z}{8}. \quad (1).$$

Proof.

According to the initial definition of Harmonic Hexagons (§ 16.1b), quadrilaterals ΑΒΓΕ, ΒΓΔΖ, ΓΔΕΑ, ΔΕΖΒ, ΕΖΑΓ, ΖΑΒΔ, will all be harmonic.

Hence, according to known Theorem § 285 in Γεωμετρία [26] (Remark 2), or [6] § 8-4.10, of harmonic quadrilaterals. for example for quadrilaterals ΑΒΓΕ, ΓΔΕΑ, ΕΖΑΓ, it will be respectively:

$$AB.\Gamma E=B\Gamma.EA=\frac{BE.A\Gamma}{2},\ \Gamma\Delta.EA=\Delta E.A\Gamma=\frac{A\Delta.\Gamma E}{2},\ EZ.A\Gamma=ZA.\Gamma E=\frac{Z\Gamma.EA}{2}.\quad (2).$$

From relations (2) if we multiply their respective parts and after the proper reductions, we easily get relation (1).

Remarks.

(a). Proposition 144, we believe to be new and firstly appear here, while we have already included it in our book [1] (volume 3, page 1326, Proposition 40H).

(b). This Proposition is an extension to Proposition 88, from a line to a circle.

(c). Equivalent Proposition is true for the Harmonic Quadrilateral as well (Γεωμετρία [6] , § 8-4.10}.

(d). Relevant Propositions are 87, 88.

19.60. Construction of a Simple Harmonic Hexagon.
(Construction of a Harmonic Cocyclic Group of Six).
Extension to Construction 62, from a Line to a Circle.

Construction 145 (Figure 92).

On given circle (o), set six ordered points Α, Β, Γ, Δ, Ε, Ζ, which are the apexes to an arbitrary Simple Harmonic Hexagon, or constitute a Harmonic Cocyclic Group of Six Points,

or to state it in a simpler way:

Construct an arbitrary Simple Harmonic Hexagon.

Solution.

Case 1, Based on three consecutive points.

(a). Analysis.

On circle (o) we put, if we are not given, for example the three first consecutive points Α, Β, Γ, of the desired Simple Harmonic Hexagon.

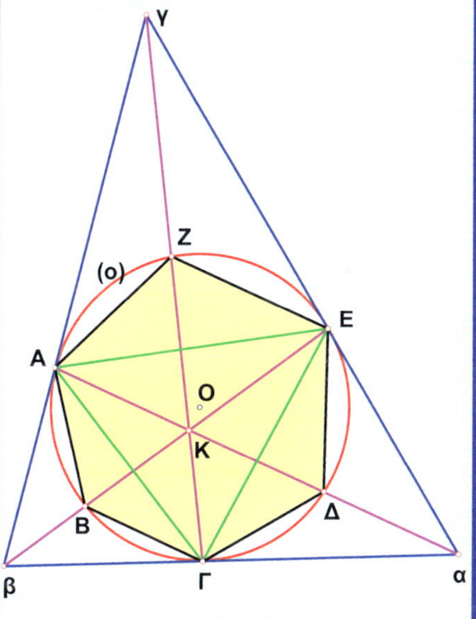

Σχήμα 92.

Hence, if the group of six points Α, Β, Γ, Δ, Ε, Ζ, is the desired one, then this, according to the relevant definition (§ 16.1b), should have all its six quadrilaterals below harmonic: ΑΒΓΕ - ΒΓΔΖ – ΓΔΕΑ - ΔΕΖΒ – ΕΖΑΓ - ΖΑΒΔ. (1).

We notice that based on the three points Α, Β, Γ, we can define Ε, if we use the first harmonic group of four points in (1), in such way that Ε with Β, harmonically divide Α, Γ. This is achieved in the way we mentioned above, in Construction 135 (§ 19.50).

Similarly, consequently, from the third harmonic group of four points in (1), we can define Δ, in such way that Δ and Α , harmonically divide Γ, Ε, as Ε is already defined.

Now we are left to construct Ζ.

From the fifth harmonic group of four points in (1), we can define Ζ, in such way that Ζ and Γ, harmonically divide Α, Ε.

Due to the above we are led to the construction below.

(b). Construction.

Based on points Α, Β, Γ, we define as mentioned above points Ε, Δ, Ζ as well and we say that the group of six points Α, Β, Γ, Δ, Ε, Ζ are the apexes of the desired Simple Harmonic Hexagon.

(c). Proof.

For the above group of six points to be the desired one, all the quadrialterals in (1), need to be harmonic (definition § 16.1b).

From their construction the first, third and fifth of the quadrilaterals in (1), are harmonic.

Now we are left to prove that also the second, fourth and sixth of the quadrilaterals in (1), are harmonic.

Since quadrilaterals ΑΒΓΕ, ΓΔΕΑ, ΕΖΑΓ, are harmonic, it will be:

$$\frac{AB}{B\Gamma}\cdot\frac{\Gamma E}{EA}=1,\quad \frac{\Gamma\Delta}{\Delta E}\cdot\frac{EA}{A\Gamma}=1,\quad \frac{EZ}{ZA}\cdot\frac{A\Gamma}{\Gamma E}=1. \qquad (2).$$

Hence, from (2), it occurs that:

$$\frac{AB}{B\Gamma}\cdot\frac{\Gamma\Delta}{\Delta E}\cdot\frac{EZ}{ZA}=1. \qquad (3).$$

Relation (3), according to Auxiliary Proposition 5 in Part E, means that:

$$A\Delta\cap BE\cap \Gamma Z'\equiv K. \qquad (4).$$

Let us note that relation (4) can also be proven in the following way:

If α, β, γ, are the poles of ΓΕ, ΑΓ, ΕΑ, respectively, then $\alpha\in A\Delta$, $\beta\in BE$, $\gamma\in\Gamma Z$, (Criterion 140, direct) and since, as we easily realize, triangle αβγ is tangential (polar converse) to triangle ΑΕΓ, with respect to circle (o), according to Gergonne Theorem, it will be:

$$A\alpha\cap E\beta\cap\Gamma\gamma\equiv K, \text{ ή } A\Delta\cap BE\cap\Gamma Z'\equiv K. \qquad (4).$$

Hence, since it is $A\Delta\cap BE\cap\Gamma Z'\equiv K$, and since quadrilateral ΑΒΓΕ is harmonic, according to Proposition 34 (§ 8.71), quadrilateral ΔΕΣΒ will be harmonic too.

In a similar way we find that quadrilaterals ΖΑΒΔ and ΒΓΔΖ are harmonic as well.

Therefore, in fact hexagon ΑΒΓΔΕΖ, is harmonic.

Remark.

It is obvious that we would reach the same result if we receive three other consecutive points, instead of Α, Β, Γ.

Case 2, Based on three one by one points.

1st Way.

In this case on circle (o) we put or we are given, let there be three one by one points Α, Γ, Ε.

Case 2, traces back to case 1 §(a), from the point where in case 1 we set a point Ε, so we continue, as mentioned there, not only for Δ and Ζ, but also for B that we don't have.

It is obvious that we would reach the same result, if in case 2 we receive three points B, Δ, Z, instead of A, Γ, E.

2nd Way.

(a). Analysis.

If the desired Harmonic Hexagon is ΑΒΓΔΕΖ, whose we are given for example its apexes A, Γ, E then for its construction, we just need to define the rest apexes B, Δ, Z.

If we draw the circumscribed circle (O, R) for triangle AΓE, then apexes B, Δ, Z will belong to the circle (O, R) (definition § 16.1b).

Additionally, if we draw the respective (tangential) triangle αβγ of triangle AΓE, then, according to the direct of Proposition 140 about Harmonic Hexagons (Criterion 5), then AΔ, BE, ΓZ will respectively run through the fixed points α, β, γ, so we trace to the construction below.

(b). Construction.

On triangle AΓE we circumscribe circle (O, R) and we construct the respective (tangential) triangle of triangle AΓE.

We draw Aα, Γγ, Eβ, that intersect (O, R) at points Δ, Z, B, respectively, and we say that the desired Harmonic Hexagon is ΑΒΓΔΕΖ.

(c). Proof.

In fact, hexagon ΑΒΓΔΕΖ is the desired one, since from its Construction and according to the Corollary of Criterion 140, it is harmonic.

(It is possible to have another way of construction of hexagon ΑΒΓΔΕΖ, if we draw circle AΓE, we draw the symmedians of triangle AΓE and depend on Criterion 139 and mainly on Corollary 1 that follows. Specifically, if for triangle AΓE we draw its symmedians or we define its Lemoine Point K and AK, ΓK, EK intersect the circumscribed circle of triangle AΓE at points Δ, Z, B respectively, then hexagon ΑΒΓΔΕΖ, is obviously the desired one, as apart from everything else, according to Corollary 1 of Criterion 139, it will be harmonic).

(d). Investigation.

It is obvious that the Problem can always reach a solution.

Remarks.

(a). Construction 145, is obviously an extension to Construction 62, from a Group of Six Collinear Harmonic Points to a Cocyclic one.

(b). The Problem above we believe to be new and firstly appear here.

(c). Relevant Constructions are 34, 62, 100, 135, 139, 140.

19.61. Property of the Harmonic Hexagons.

The Analogous Brianchon Point coincides with the Point of Convergence of the Diagonals at Harmonic Hexagons.

Extension of the Newton's Theorem to Harmonic Hexagons.

Criterion 146 (Figure 93).

The point of convergence of the diagonals of each Harmonic Hexagon, coincides with the Analogous Brianchon Point {[6](§ 11-11)}, of its respective circumscribed hexagon (tangential or polar converse).

Proof.

I.e., if ΑΒΓΔΕΖ is a Harmonic Hexagon, Α'Β'Γ'Δ'Ε'Ζ' its respective circumscribed (tangential) hexagon (where Α∈Α'Ζ', Β∈Α'Β', ...), (O, R) its respective (directing) circle, ΑΔ∩ΒΕ∩ΓΖ≡Κ the point of convergence of the diagonals (Proposition 104 about the Harmonic Hexagons) and

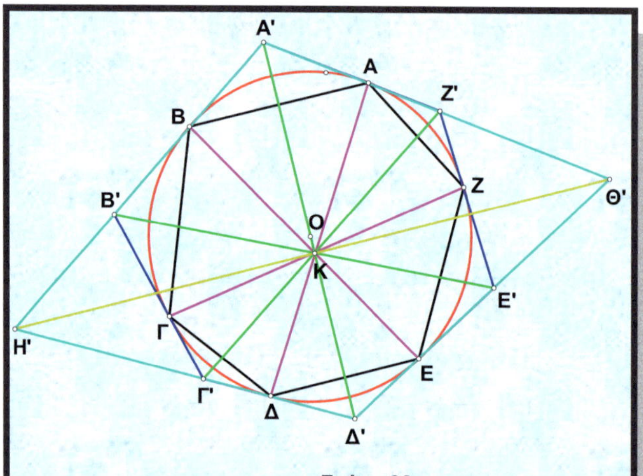

Σχήμα 93.

A'Δ'∩B'E'∩Γ'Z'≡K' the Brianchon point {[6](§ 11-11), we will prove that K≡K'.

The proof of this is obviously just the application of Lemma 8 in Part E.

In fact, since the diagonals of the Harmonic Hexagon ΑΒΓΔΕΖ are concurrent at K (Proposition 104 about the Harmonic Hexagons), according to Lemma 8 in Part E, this will coincide with Brianchon Point K' {[6](§ 11-11)}, of its respective circumscribed hexagon A'B'Γ'Δ'E'Z'.

Remarks.

(a). An equivalent proposition is true for all Harmonic Multilaterals as well and its proof is based on Proposition 1δ(1) of the book [31}.

(b). Generalization to Proposition 8, for 2n-laterals, is given in § 1δ(1) of the book [31]. Both these Propositions are obviously a generalization to the relevant Newton's Theorem, which is true, as known, for quadrialterals.

(c). The above Lemmas 8 in Part E and 1δ(1) in book [31] were to proven to be rather valuable for the proof of Propositions concerning Harmonic Multilaterals, with an even number of sides (See also equivalent Proposition 206, for a Harmonic Decagon).

(d). Relevant Proposition 8 in Part E.

(e). Extension to Proposition 146 to a Harmonic Decagon, in Proposition 206 (§ 29.7).

19.62. Amazing Harmony.

Ninth Criterion of Harmonic being of an Inscribed Hexagon.

Pair of Triangle Quadruple Homological (unprecedented).

Extension to Criterion 143 (Another discussion on Criterion 143).

Criterion 147 (Figure 94).

If a convex inscribed in circle (O, R) hexagon ΑΒΓΔΕΖ, has its twin triangles ΑΓΕ and ΒΔΖ quadruple homological {[7] § 1249} then and only then this hexagon is harmonic.

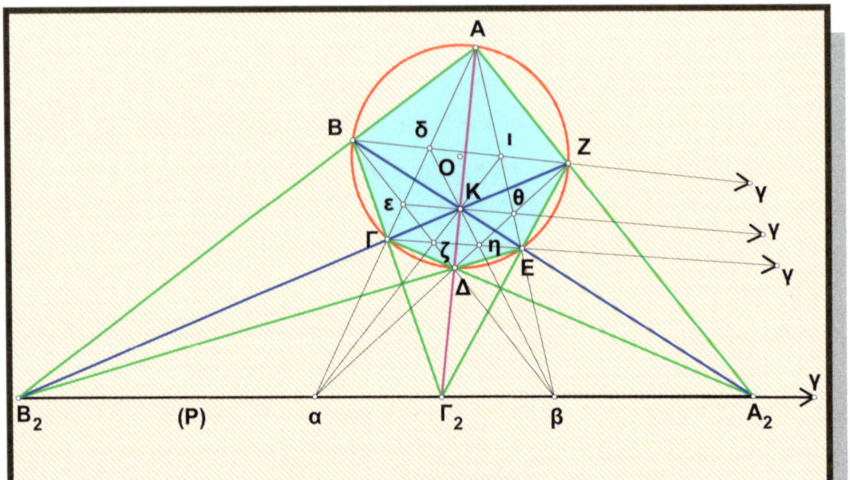

Σχήμα 94.

Proof.

(a). Direct.

We consider a harmonic hexagon ΑΒΓΔΕΖ, inscribed in circle (O, R).

We will prove that, for its twin triangles ΑΓΕ and ΒΔΖ, it is ΑΔ∩ΒΕ∩ΓΖ≡Κ, ΖΑ∩ΓΔ∩ΒΕ≡A_2, ΑΒ∩ΔΕ∩ΓΖ≡B_2, ΒΓ∩ΕΖ∩ΔΑ≡ $Γ_2$. I.e., these triangles are quadruple homological {terminology in [7] § 1249}.

This is in fact true, according to the direct of Criterion 143 (§ 19.58).

(b). Converse.

I.e., if the given convex inscribed in circle (O, R) hexagon ΑΒΓΔΕΖ, has its triangles ΑΓΕ and ΒΔΖ, quadruple homological, hence if it is ΑΔ∩ΒΕ∩ΓΖ≡Κ, ΖΑ∩ΓΔ∩ΒΕ≡A_2, ΑΒ∩ΔΕ∩ΓΖ≡B_2, ΒΓ∩ΕΖ∩ΔΑ≡ $Γ_2$, then we will prove that this hexagon is harmonic.

In fact, according to the converse of Criterion 143, this is true.

Amazing Harmony.

<u>In figure 94, the amazing facts below are happening:</u>

It is known [26], that $A_2B_2Γ_2$ is a line (Pascal Line), as well as polar of point K, while A_2, B_2, $Γ_2$, are the three homology centers in triangles ΑΒΕ and ΒΔΖ, while the fourth homology center is obviously K.

Moreover, intersections $A\Gamma \cap \Delta Z \equiv \alpha$, $B\Delta \cap AE \equiv \beta$, $BZ \cap \Gamma E \equiv \gamma$, are collinear and they constitute the one axis of homology of these triangles, which coincides with line $A_2B_2\Gamma_2$, as points α, β, γ, belong to line $A_2B_2\Gamma_2$, given that they are conjugate to K, with respect to circle (O, R).
Consequently, all points B_2, α, B_2, β, Γ_2, γ, are collinear.
The other three homology axes of triangles ΑΓΕ, ΒΔΖ, are the following αζι, βηδ, γθε (where $A\Gamma \cap BZ \equiv \delta$, $A\Gamma \cap B\Delta \equiv \varepsilon$, $B\Delta \cap \Gamma E \equiv \zeta$, $\Gamma E \cap \Delta Z \equiv \eta$, $EA \cap \Delta Z \equiv \theta$, $ZB \cap EA \equiv \iota$), while αζι, βηδ, γθε, run through K, as ζΚι, ηΚδ, θΚε, are obviously Pascal lines, hence αζΚι, βηΚδ, γθΚε, are the polars of A_2, B_2, Γ_2, respectively, etc.

Remarks.

(a). Criterion 147 we believe to be new and firstly appear here.

(β). This Criterion is rather significant and unprecedented, as it is known to us until now that, that two triangles are thrice homological, though not quadruple homological.

(c). The 1st, 2nd, 3rd, 4th, 5th, 6th, 7th, 8th, Criteria of the Harmonic Hexagon, are given in Propositions 96, 109, 112, 139, 140, 141, 142, 143, respectively.

(d). Equivalent Criterion is true as well for the twin triangles $A'\Gamma'E'$, $B'\Delta'Z'$ of the tangential hexagon $A'B'\Gamma'\Delta'E'Z'$ of the Harmonic Hexagon ΑΒΓΔΕΖ (Figure 95), as well as for the twin triangles $A'_1\Gamma'_1E'_1$, $B'_1\Delta'_1Z'_1$, of the same figure 95, that follows (Construction 148).

(e). Criterion 147, is obviously another discussion on Criterion 143 and its proof can be based on Criterion 51.

(f). Relevant Propositions are 96, 109, 112, 139, 140, 141, 142, 143, 148.

19.63. Construction of a Pair of Triangles Quadruple Homological.

Construction 148 (Figures 94 and 95).

Construct a pair of coplanar triangles quadruple homological, if we are given one of them.

Solution.

1st Solution (Figure 94).

(a). Analysis.

If we take into consideration Criterion 147 above (§ 19.62, converse), we see that if two triangles ΑΓΕ, ΒΔΖ are inscribed in circle and they are quadruple homological {terminology in [7], § 1249}, then hexagon ΑΒΓΔΕΖ is harmonic. If, the twin triangles of a Harmonic Hexagon are a solution to the given problem, then we are led to the construction below.

(b). Construction.

Based on the given arbitrary triangle ΑΓΕ, we construct the Simple Harmonic Hexagon ΑΒΓΔΕΖ, in the exact same way as mentioned above in Construction 145 (§ 19.60) (Construction of a Simple Harmonic Hexagon, case 2).

We say that the twin triangles ΑΓΕ, ΒΔΖ of the Harmonic Hexagon ΑΒΓΔΕΖ, are a solution to the Problem.

(c). Proof.

In fact, according to Criterion 147 (direct) above, as hexagon ΑΒΓΔΕΖ is harmonic, it will be:

$ΑΔ \cap ΒΕ \cap ΓΖ \equiv K$, $ΑΖ \cap ΒΕ \cap ΓΔ \equiv A_2$, $ΑΒ \cap ΖΓ \cap ΕΔ \equiv B_2$, $ΒΓ \cap ΑΔ \cap ΕΖ \equiv Γ_2$. (1).

Thus in fact twin triangles ΑΓΕ, ΒΔΖ are quadruple homological, with homology centers K, A_2, B_2, $Γ_2$, respectively.

(d). Investigation.

Since the Simple Harmonic Hexagon ΑΒΓΔΕΖ, can always be constructed (Construction 145), here as well we can always reach a solution to this Problem.

Conclusion.

So we see that the twin triangles of each Simple Harmonic Hexagon, are a solution to the Problem above, as these are quadruple homological.

Comment.

Due to the above the following question is raised. Is there a solution for

which hexagon ΑΒΓΔΕΖ is not Harmonic? The answer will be given to us in solution 2 below.

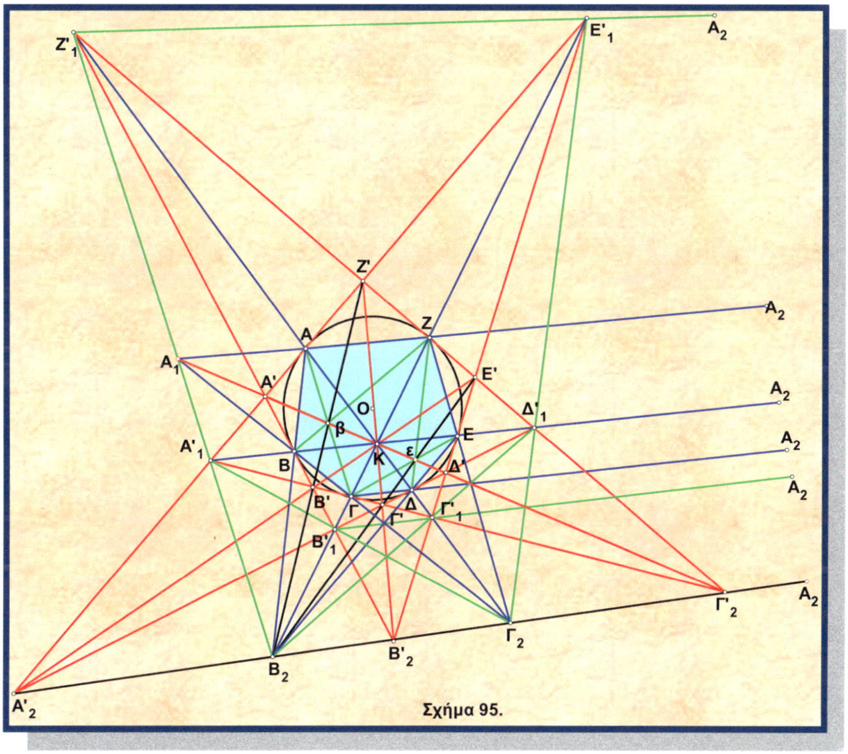

Σχήμα 95.

2nd Solution (Figure 95).

(a). Analysis.

We suppose that we are given an arbitrary triangle $A'_1\Gamma'_1E'_1$ and we are asked to find $B'_1\Delta'_1Z'_1$, such that these triangles would be quadruple homological, hence they have:

$B'_1\Gamma'_1 \cap A'_1\Delta'_1 \cap E'_1Z'_1 \equiv A_2$, $A'_1Z'_1 \cap B'_1E'_1 \cap \Gamma'_1\Delta'_1 \equiv B_2$, $A'_1B'_1 \cap Z'_1\Gamma'_1 \cap E'_1\Delta'_1 \equiv \Gamma_2$, $A'_1\Delta'_1 \cap B'_1E'_1 \cap \Gamma'_1Z'_1 \equiv K$. (1).

If we take into consideration Criteria 140 (§ 19.55), 147 (§ 19.62), 143 (§ 19.58), Propositions 104 (§ 19.11), 146 (§19. 61), Construction 145 (§ 19.60) above and the theory about poles and polars, we are led to the construction below.

(b). Construction.

We draw the inscribed circle (O, ρ) of the given triangle $A'_1\Gamma'_1E'_1$.

If A, Γ, E, are the tacpoints of $A'_1E'_1$, $A'_1\Gamma'_1$, $\Gamma'_1E'_1$, respectively and $B\equiv A'_1E\cap(O,\rho)$, $\Delta\equiv \Gamma'_1A\cap(O,\rho)$, $Z\equiv E'_1\Gamma\cap(O,\rho)$, then from the above way of construction of the hexagon ΑΒΓΔΕΖ, it occurs that it would be harmonic (Construction 145 above § 19.60).

If at B, Δ, we draw tangents that intersect with B'_1, at Δ, Z, that intersect with Δ'_1 and at Z, B, that intersect with Z'_1, then we say that triangle $B'_1\Delta'_1Z'_1$, is the desired one.

(c). Proof.

In fact, according to Criterion 140 above (conclusion), for example $E'_1ZK\Gamma B'_1$, will be a line.

Additionally, according to the above criterion 143, $Z\Gamma B_2$ will be a line ($B_2\equiv AB\cap\Delta E$), as hexagon ΑΒΓΔΕΖ, as we saw above, is harmonic, from its construction.

Moreover, according to Proposition 104, it will also be

$$A\Delta\cap BE\cap\Gamma Z\equiv K.$$

Hence, since lines $E'_1ZK\Gamma B'_1$, $Z\Gamma B_2$, share their points Γ, Z, they will coincide, thus $E'_1ZK\Gamma B'_1B_2$, will also be a line. Similarly we find that the following are lines as well $Z'_1AK\Delta\Gamma'_1\Gamma_2$, $A'_1BKE\Delta'_1A_2$.

The pole of $A'_1Z'_1$ is the intersection β of AΓ, BZ, as AΓ has as pole A'_1 and BZ has as pole Z'_1.

If $E'_1A'_1\cap Z'_1B'_1\equiv A'$, $Z'_1B'_1\cap A'_1\Gamma'_1\equiv B'$, …, $\Delta'_1Z'_1\cap A'_1E'_1\equiv Z'$, $B'\Gamma'\cap Z'E'\equiv\Gamma'_2$, then, according to the known Newton's Theorem, for the circumscribed quadrilateral $Z'A'B'\Gamma'_2$, it will be: $A\Gamma\cap BZ\cap A'\Gamma'_2\cap B'Z'\equiv\beta$. I.e. $A'\Gamma'_2$ runs through pole β of $A'_1Z'_1$. This means that $A'_1Z'_1$ will run through the pole of $A'\Gamma'_2$ too. But, the pole of $A'\Gamma'_2$ is the intersection B_2 of polars AB and ΓZ of points A', Γ'_2, respectively. Thus $A'_1Z'_1$, will run through the intersection B_2 of AB, ZΓ.

Similarly, we also find that $\Gamma'_1\Delta'_1$, runs through B_2, through which $E'_1B'_1$ runs too. I.e., it finally is: $A'_1Z'_1\cap B'_1E'_1\cap\Gamma'_1\Delta'_1\equiv B_2$.

Additionally, if we work similarly to above, we also find that:

$$B'_1\Gamma'_1\cap A'_1\Delta'_1\cap Z'_1E'_1\equiv A_2,\ ,\ A'_1B'_1\cap Z'_1\Gamma'_1\cap\Delta'_1E'_1\equiv\Gamma_2.$$

[If ΓΕ∩ΔΖ≡ε, we find that similarly to above $\varepsilon\Delta'\Gamma'_2$, is a line, whose polar is also B_2, thus $A'\beta\varepsilon\Delta'\Gamma'_2$ is a line. Additionally A'ΚΔ' is a line (Proposition 146 § 19.61), thus also a line is $A'\beta K\varepsilon\Delta'\Gamma'_2$. Similarly we find that the following are lines too $E'KB'A'_2$, $Z'K\Gamma'B'_2$].

Besides, above we have proven that $A\Delta \cap BE \cap \Gamma Z \equiv K \equiv A'_1\Delta'_1 \cap B'_1E'_1 \cap \Gamma'_1Z'_1$. So for the twin triangles $A'_1\Gamma'_1E'_1$, $B'_1\Delta'_1Z'_1$, of hexagon $A'_1B'_1\Gamma'_1\Delta'_1E'_1Z'_1$, (which is not harmonic), relation (1) is true.

This means that, these triangles are actually quadruple homological, with homology centers K, A_2, B_2, Γ_2, hence triangle $B'_1\Delta'_1Z'_1$, is in fact a 2nd solution to the given problem.

(d). Investigation.

Since triangle $B'_1\Delta'_1Z'_1$, is always constructible, here as well we always reach a solution to this Problem.

Conclusion.

So we see that the twin triangles $A'_1\Gamma'_1E'_1$, $B'_1\Delta'_1Z'_1$, of hexagon $A'_1B'_1\Gamma'_1\Delta'_1E'_1Z'_1$, (which is not harmonic), are another solution to the above Problem, as these are quadruple homological.

Remarks.

(a). So we see that there is a solution, to Problem 148 even when hexagon $A'_1B'_1\Gamma'_1\Delta'_1E'_1Z'_1$ is not Harmonic. Though, as seen in figure 95, this is not completely irrelevant to the harmonic hexagon ΑΒΓΔΕΖ. {This in book [1] is called protogenic circumscribic hexagon}.

(b). Easily we find that the twin triangles A'Γ'E', B'Δ'Z', of the respective circumscribed hexagon A'B'Γ'Δ'E'Z', of the harmonic ΑΒΓΔΕΖ hexagon, are also quadruple homological.

In figure 3.2.6.(4).(β).1/, in book [1], we also see other pairs of twin triangles quadruple homological (See also figure in page 714 of the present book).

(c). Points A_2, B_2, Γ_2, A'_2, B'_2, Γ'_2, are collinear, as their polars Γ'Ζ', Α'Δ', Β'Ε', ΑΔ, ΒΕ, ΓΖ, respectively, coincide at K (Proposition 146) (pole of the line of these six points).

(d). Relevant Propositions are 104, 140, 143, 145, 146, 147.

General Conclusion.

If two triangles, are inscribed in the same circle and their apexes are apexes to a Harmonic Hexagon, or if these triangles are circumscribed in the same circle and their tacpoints, are apexes to a Harmonic Hexagon, then in both cases above, the pairs of these triangles are quadruple homological.

19.64. Another Construction of the Simple Harmonic Octagon. (Harmonic Cocyclic Group of Eight Points), Extension to Construction 63, from a line to a circle.

Construction 149 (Figure 96).

On given circle (o), set eight ordered points Α, Β, Γ, Δ, Ε, Ζ, Η, Θ, which are apexes to an arbitrary Simple Harmonic Octagon, or they are a Closed Harmonic Cocyclic Group of Eight Points,

or in a simpler formation:

Construct an arbitrary Simple Harmonic Octagon.

Solution.

(a). Analysis.

If ΑΒΓΔΕΖΗΘ is the desired Harmonic Octagon, then this, according to the relevant alternative definition (Criterion 110 § 19.18), should be convex, inscribed in circle (o), its diagonals (main) are concurrent at point K and for example its following consecutive quadrilaterals are harmonic:

ΑΒΓΖ, ΒΓΔΗ, ΓΔΕΘ, ΔΕΖΑ. (1).

Hence, due to the above, if we also take into consideration Constructions 42 (§ 9.3), or 43 § 9.4 and 63 (§ 15.3), we are led to the construction below.

(b). Construction.

For this construction we work in a way equivalent to the one we followed in way 2 of Construction 63 (§15.3).

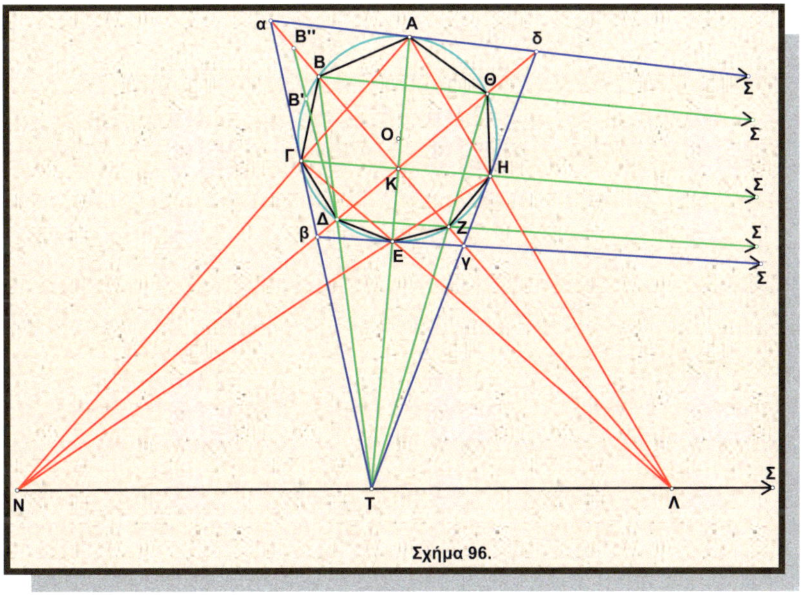

Σχήμα 96.

In the same circle (o) we inscribe, in the known way (Construction 135, § 19.50) an arbitrary Simple Harmonic Quadrilateral ΑΓΕΗ.

Consequently we construct, in the known way (Construction 42, § 9.3, or 43 § 9.4), two pairs of points Δ, Θ and Β, Ζ with which the following two pairs of points are harmonically divided Α, Η - Γ, Ε, and Α, Γ – Ε, Η, respectively. To achieve this we define intersections Κ≡ΑΕ∩ΓΗ, Ν≡ΑΓ∩ΕΗ, Λ≡ΑΗ∩ΓΕ. We draw ΚΝ, ΚΛ, that intersect circle (o) at pairs of points Δ, Θ and Β, Ζ, respectively.

We say that the convex inscribed Octagon ΑΒΓΔΕΖΗΘ is the desired Harmonic one.

(c). Proof.

Octagon ΑΒΓΕΖΗΘ is the desired harmonic one, as from its construction it is convex, inscribed in circle (o), its diagonals are concurrent

K≡AE∩BZ∩ΓH∩ΔΘ, and furthermore, according to the relevant alternative definition (Proposition 110), we are left to prove that its four consecutive quadrilaterals, for example the ones in relation (1), are harmonic.
In fact, from their construction quadrilaterals ABΓZ and ΓΔEΘ are harmonic, as A, Γ, are harmonically divided from B, Z (Construction 42, § 9.3), while similarly Γ, E are also harmonically divided from Δ, Θ.
[Let us note here, that similarly we prove that quadrilaterals EZHB, HΘAΔ are harmonic too. See also below remark (b)].
We are left to prove that quadrilaterals BΓΔH, ΔEZA, are harmonic as well.
To prove that, on quadrilateral AΓEH we circumscribe its polar converse (tangential) quadrilateral αβγδ (where α the intersection of the tangents at A, Γ, β the intersection of the tangents at Γ, E, etc) and we define intersections T≡αβ∩γδ, Σ≡αδ∩βγ.
We prove as above, that quadrilaterals EZHB, HΘAΔ are harmonic as well.
Hence, since quadrilaterals ABΓZ, EZHB, as we saw above, are harmonic, then BZ with the pairs AΓ, EH are conjugate with respect to the circle (o), thus BZ will run through poles α, γ of AΓ, EH, respectively. Similarly we prove, that ΔΘ runs through the poles β, δ of ΓE, AH, respectively.
Thus, ΣγKδαβ is a complete quadrilateral and since AΓEH is harmonic, from its construction, then ΓH is conjugate to AE, with respect to circle (o). This means that ΓH will run through pole Σ of AE. Hence ΓKHΣ is a line and thus a diagonal of the above complete quadrilateral.
Similarly we also find that AKET is a line.
Consequently we will prove that quadrilateral BΓΔH is harmonic. For that to happen we just need to prove that BΔ, ΓH are conjugate, with respect to circle (o), or that BΔ runs through pole T of ΓH. To prove that, we suppose that TΔ does not run through B, but intersects circle (o) at B' and line α-BKZγ at B".
We set the intersection I of lines ΓKΣ and TΔB'B".
Since TΔB' runs through pole T of ΓH, it will be:

$$(TI\Delta B')=1. \qquad (2).$$

After all, from the above complete quadrilateral, it will be:

(ΤΓβα)=1 or Σ(ΤΓβα)=1 or Κ(ΤΓβα)=1 or Κ(ΤΙΔΒ") or (ΤΙΔΒ")=1. (3).

Hence, from relations (2), (3)⇒ Β'≡Β". Hence to ΤΔ belongs a point of αγ, but at the same time of circle (o) as well of to their intersection, which is Β, or at the shared point Β of αγ and circle (o), or that Β'≡Β"≡Β.

So we see that ΒΔΤ is a line too, or that ΒΔ runs through pole Τ of ΓΗ. This means that ΓΗ, ΒΔ are conjugate, with respect to circle (o), or that quadrilateral ΒΓΔΗ is harmonic.

If we work in the exact same way as above, we also prove that quadrilateral ΔΕΖΑ, is harmonic.

Consequently all four quadrilaterals in (1), are in fact harmonic and thus it is proven that octagon ΑΒΓΔΕΖΗΘ is harmonic (alternative definition).

Remarks.

(a). Another way of Construction of Problem 149, is given in Construction 101 (§ 19.7).

(b). In a way, similar to above, we prove that the rest two quadrilaterals ΖΗΘΓ, ΘΑΒΕ are harmonic. The proof of these is carried through with more easily, if we use the auxiliary Proposition 51 in § 9.12.

(c). We easily prove that the «twin» quadrilateral ΒΔΖΘ of the quadrilateral ΑΓΕΗ, is harmonic, based on its way of construction.

(d). It is obvious and easily proven, that the intersections Ν, Λ, Τ, Σ, ΑΖ∩ΒΕ, ΒΗ∩ΓΖ, ΓΘ∩ΔΗ,..., as well as the intersections of the opposite sides of the octagonare collinear points and that this line that these intersections belong to, is the polar of point Κ.

(e). It is obvious that Construction 149 is an extension to Construction 63, from a line to a circle.

(f). As we proved above that ΒΔ runs through Τ, in the exact same way we prove that through Τ, ΖΘ runs as well, while similarly we prove that both ΒΘ, ΔΖ run through Σ.

(g). Many relevant data, concerning Simple Harmonic Octagons, are given in our book «Εγγεγραμμένα - Περιγεγραμμένα Σχήματα» (pages 1769 to 1778), for this reason, we will mention here ONLY what is needed. In the same book the Complexes of Complete Harmonic Octagons are mentioned in detail (Pages 1661 to 1749). In this book, to realize about what amount of work we are talking about, we only need to mention, that to prove the truth of just one of its figures, 175 Geometrical Propositions were formed and proven.

(h). This Construction we believe to be a new one and it firstly appears here.

(i). According to the relevant definitions (§ 16.1c) the Asteroid Octagon ΑΔΗΒΕΘΓΖΑ is also considered Harmonic, since as we saw in previous Propositions, it has many main criteria of harmonic being. Hence, it is harmonic.

(j). Another way of construction is given in Construction 101, while below Construction 154 is mentioned, with the application of the method of Harmonic Transformation.

(k). Relevant Propositions are 42, 43, 63, 101, 110, 135, 154, 205.

19.65. Construction of an Asteroid Harmonic Decagon.

New Easy Way of Construction of an Asteroid Harmonic Decagon.

Extension to Construction 102 to an Asteroid Decagon.

Construction 150 (Figure 78).

Construct an Asteroid Harmonic Decagon ΑΔΗΚΓΖΙΒΕΘΑ, inscribed in given circle (o), based on a Simple Harmonic Decagon ΑΒΓΔΕΖΗΘΙΚ [whose construction is known (Construction 102)], inscribed in the same circle (o).

Solution,

(a). Analysis.

If we take into consideration the relevant definition (§ 16.1e), the way of construction of a convex Harmonic Decagon [Construction 102 (§ 19.8)] and Proposition 108 (§ 19.16) we are led to the construction below.

(b). Construction.

We construct a convex Simple Harmonic Decagon as in Construction 102 and we say that the Asteroid Decagon ΑΔΗΚΓΖΙΒΕΘΑ, that occured, is the desired harmonic one.

(c). Proof.

Since the convex Decagon ΑΒΓΔΕΖΗΘΙΚ, that we constructed is harmonic, according to Proposition 108 then its respective Asteroid Decagon ΑΔΗ-ΚΓΖΙΒΕΘΑ, will be Harmonic too.

Remarks.

(a). The convergence of the diagonals (main), of the Asteroid Decagon, is obvious, as these coincide with the ones of the convex one (See also Proposition 107).

(b). The fact that this Asteroid Decagon is harmonic, is seen in Criterion 116 as well, in which it is proven that this has a Lemoine point, which shows its harmonic being.

(c). Relevant Propositions are 102, 107, 108, 116.

19.66. Harmonic being of Regular Multilaterals.

(Simple but rather useful Proposition).

Property of Regular Multilaterals as Special Harmonic Multilaterals.

Proposition 151 (Figure 61).

Each Regular Multilateral, is Harmonic as well.

Proof.

Let there be a Regular Decagon ΑΒΓΔΕΖΗΘΙΚ, inscribed in circle (o).

We will prove that it is Harmonic as well.

According to the relevant definition (§ 16.1d), for it to be harmonic it should have its following ten quadrilaterals harmonic:

ΑΒΓΗ, ΒΓΔΘ, ΓΔΕΙ, ΔΕΖΚ, ΕΖΗΑ,

ΖΗΘΒ, ΗΘΙΓ, ΘΙΚΔ, ΙΚΑΕ, ΚΑΒΖ. (1).

In fact, for example its quadrilateral ΑΒΓΗ, is harmonic, as its diagonals ΒΗ, ΑΓ are conjugate with respect to circle (o), since due to symmetry ΒΗ runs through the pole of ΑΓ, or since it obviously is ΑΒ.ΓΗ=ΒΓ.ΑΗ.

If we continue, in the exact same way, we prove that the rest nine quadrilaterals in relation (1) are harmonic, hence decagon ΑΒΓΔΕΖΗΘΙΚ is in fact harmonic (Definition § 16.1d).

In the exact same way, easily, we prove that Proposition 151, is true for any other Regular Multilateral too, taking into consideration obviously their definitions (§ 16.1).

Remarks.

(a). Proposition 151, we believe to be new and firstly appear here.

(b). This Proposition obviously is not a Criterion of harmonic being, as it is obvious that its converse is not true.

(c). Regular Multilaterals, as harmonic, have all the properties of the Harmonic Multilaterals. For example, their diagonals coincide at the center of their circumscribed circle, which is also a Lemoine point of the Regular Multilateral, etc.

(d). Relevant Propositions are 79-83.

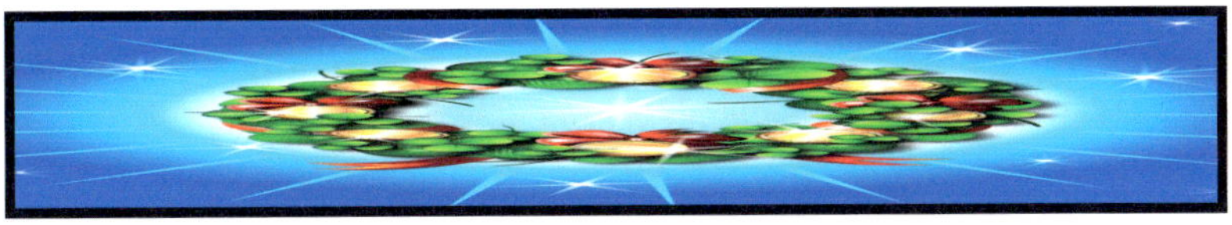

G'. THE PROBLEM OF THE VICENNIUM!!!
AND THE METHOD OF HARMONIC TRANSFORMATION.

20. THE PROBLEM OF THE VICENNIUM

(a). In General.

Section H′, which we enter, is the main part of this book.

In this happens, as we will realize, the SYNTHESIS, of many of those precedented, in a new Method which we call «Harmonic Transformation», and which solved for us the «Problem of the Vicennium», which easily solves for us all relevant problems as well.

We have named Problem of the Vicennium, the Question-Problem below:

«How does a Harmonic Multilateral occur from a Regular Multilateral» (obviously of the same number of sides)?

(b). Historic evolution of the Problem of the Vicennium.

We always believed and it troubled us often, at least for twenty years, that a Harmonic Multilateral, occurs from the projection of a Regular Multilateral, with the same number of sides and we were not proven wrong, as we will see below.

But how? In what way?

The answer to these questions we considered significant and necessary to easily achieve the solution to all relevant problems, even those that their solution did not exist, by then, as for example the Construction of a Harmonic Group of Ten Points, a Harmonic Decagon, etc.

When, after at least twenty years and extensive research-study, we came up with the way of solution of the Problem of the Vicennium, our joy was indescribable and our enthusiasm was huge and that is why we directly expressed it on the website mathematica.gr using the phrase:

AMAZING !!!

The «Problem of the Vicennium» is solved !!!

21. THE AMAZING METHOD OF THE HARMONIC TRANSFORMATION.

(a). Way of thinking.

If a Regular Multilateral is projected (but not in an orthogonal projection), form a point outside of its plane, to another plane (not parallel to the first one), then the Multilateral that will occur, will not be Harmonic, as simply it will not be inscribed in circle, but in a Conic section, as its circle, during the projection, is degenerate and transformed to a conic section.

So this projection should be carried through with in a special way and from a special point, such that this prescribed circle of the Regular Polygon stays unchanged (as a circle, not from a size point of view, since we can choose the size ourselves), which is a necessary prerequisite for a Multilateral to be Harmonic, while only the Regular Multilateral should be transformed to a Harmonic one, of equal number of sides, but not its perscribed circle.

But how and from which point?

This answer – solution came to us, throught the extensive study and SYNTHESIS of a series of other Methods (§ 5, 13, 18), Propositions and Problems, that we have come up with and that have already been mentioned earlier, with their proofs.

Moreover, while achieving this solution, a major role was taken from the «Orthooptical Point» (Relevant terminology is given in paragraph 14), about which we have talked a lot and using which we have solved many relevant Problems.

(b). Interpretation of the Method.

This method in general, is a result of the Problem of the «Vicennium» and it is based on proper Projective Transformations (Projections and Intersections of a line and circles), but (it is based) as well on the so called «Orthooptical Point».

To better interpret and understand in detail this Method, we will solve the Problem below in full detail and of course by applying this method.

The complete General Problem of the Vicennium, is the following:

Based on proper projective transformations and intersections from a line and circles, construct a Harmonic Multilateral based on the Regular Multilateral with the same number of sides, given to us and conversely, similarly and based on a Harmonic Multilateral, given to us construct a Regular Multilateral, with the same number of sides.

The Solution to the Problem.

This solution will be consequently given in detail, for specific multilaterals and of course with the application of the Method of Harmonic Transformation. In this way we will learn this method in depth.

(c). Name of the Method.

Since, from the series of the above transformations, a Collinear Harmonic group of n occurs initially and then a Harmonic Multilateral, this Method, is called «HARMONIC TRANSFORMATION». This method is uniform for every Collinear Harmonic group of n and Multilateral, even those that have an odd number of sides, as we will see below (Triangles, Pentagons , etc), it is a Method, as it has applications to all relevant Problems, many of which we believe cannot be solved in another way. A first taste of this method, is given in paragraph 12.13(d).

(d). Extension and Evolution of the Method.

Through the passing of time and after relevant research, it occurred that with this Method the construction of Harmonic Involutions is possible (collinear and cocyclic), though not based on Regular Multilaterals, which are figures symmetrical with respect to each of their meain diagonals, but based on Inscribed Symmetrical Multilaterals, with respect to one of their diagonals, from which a Simple Harmonic Involution occurs, or with respect to two perpendicular diagonals of theirs, that a Double Harmonic Involution occurs (Corollaries 158, 163-165).

Applications of this Method on Harmonic Involutions, are mentioned in Part C.

Additionally, it occurred that using this method the Construction of Harmonic Lines (collinear and cocyclic) is possible and easy, from Regular Cocyclic Lines. Relevant applications mentioned in section I' below (§ 23).

(e). Value of this Method.

Its value is rather significant, as it is the KEY to the easy proof - solution of almost all the relevant Propositions – Problems, even those that cannot be solved in another way, as the human brain cannot conceive them, due to complexity.

Comment.

The above «Problem of the Vicennium» and the Method of «Harmonic Transformation», we believe to be new and firstly appear here (Relevant terminology is given in paragraphs 11.1, 12.13 και 16.1).

Value of Research.

These amazing things happen through research, this is why we believe in it and we serve it with passion. You are welcome to join us and help us improve using new data this otherwise deserted, mostly from Greeks, mainly Greek science of the Euclidean Geometry.

RESEARCH = Pleasure + Findings + Intense Thrills ⇒ Creation ⇒ Progress ⇒ Culture.

22. APPLICATIONS OF THE METHOD OF HARMONIC TRANSFOMATION.

Solution to the Problem of the «Vicennium», using the Method of the «HARMONIC TRANSFORMATION».

How from a Regular Multilateral a Harmonic one occurs.

To better understand the above Problem of the Vicennium [§ 21(b)] and the new Method of the Harmonic Transformation (§ 21), we will divide this Problem, in smaller ones, for each one of our known Regular Multilaterals (the constructible ones), starting from the simplest one, which is the Regular Quadrialteral (Square), which is elementary and basic.

The complete Problem of the Vicennium is the following:

Based on proper projective transformations and intersections from a line and circles, construct a Harmonic Multilateral based on the Regular Multilateral with the same number of sides, given to us and conversely, similarly and based on a Harmonic Multilateral, given to us construct a Regular Multilateral, with the same number of sides.

(The solution to this Problem, as we will realize later, is carried through with in the same way for all relevant Problems, which is independent to the number of sides of the multilateral, applying of course the Method of Harmonic Transformation).

22.1 . Construction of a Harmonic Quadrilateral from a Square.

(How from a Square a Harmonic Quadrilateral occurs).

Construction 152 (Figure 97).

Based on proper projective transformations and intersections from a line and circles, construct a Harmonic Quadrilateral based on the Square, given to us and conversely, similarly and based on a Harmonic Quadrilateral, given to us construct a Square.

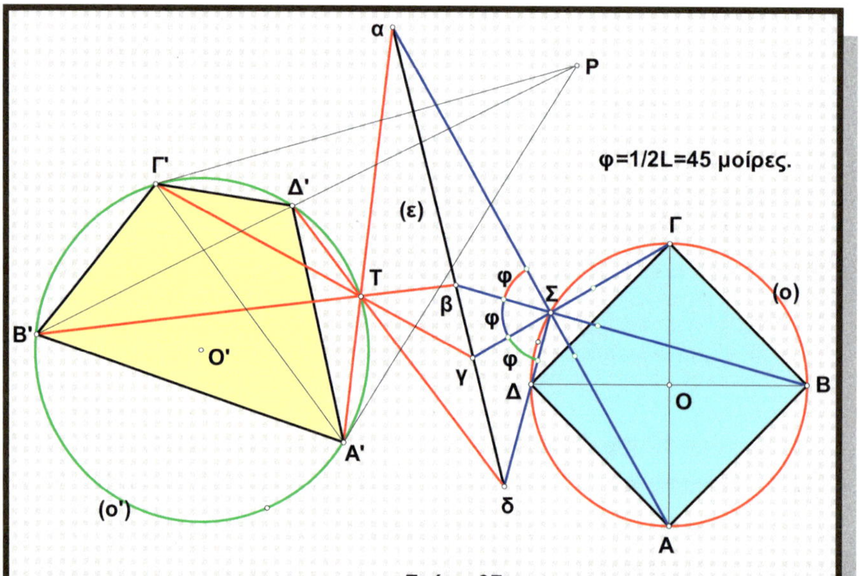

Σχήμα 97.

Solution.

1/. Direct.

(a). Analysis.

Here we are given square ΑΒΓΔ and we are asked, based on that and with proper projective transformations to a get a Harmonic Quadrilateral Α'Β'Γ'Δ'.

If we take into consideration Constructions 79 (§ 15.21) and 99 (§ 19.5) and we compose them properly, we are led to the construction below:

(For convenience and better understanding of the solution we just need to properly compose figures 57 and 68, as for figure 97 to occur).

(b). Construction.

We work similarly to Constructions 79 (converse) and 99 (direct), respectively.

I.e., using an arbitrary point Σ as the center of the circumscribed circle (o) of the square ΑΒΓΔ, we form the central pencil Σ(ΑΒΓΔ), whose beams we suppose that intersect a random line (ε) at points α, β, γ, δ, respectively.

Consequently, if we want to have the desired Harmonic Quadrilateral Α'Β'Γ'Δ', to be inscribed in circle (o'), we work in the following way:

Using an arbitrary point T as the center of circle (o'), we form Pencil T(αβγδ), the beams of which re-intersect circle (o) at points Α', Β', Γ' Δ', respectively.

We say that the desired Harmonic Quadrilateral is Α'Β'Γ'Δ'.

(c). Proof.

If we work similarly to the proof of the converse of Construction 79, we find that the Group of Four Points α, β, γ, δ, is harmonic, as quadrilateral ΑΒΓΔ, is Regular, while if we work afterwards, as in the direct of the proof of Construction 99 and based on this group of six, we find that Quadrialteral Α'Β'Γ'Δ', is harmonic, as it is inscribed in circle (o') and then the group of four points α, β, γ, δ, is harmonic.

Comment.

It is obvious that the method above is another way of construction of Harmonic Group of Four and of a Harmonic Quadrilateral, apart from the ones we ysed in Constructions 61 (§ 15.1) and 135 (§ 19.50).

2/. Converse.

For the converse we follow the opposite way.

(a). Analysis.

I.e. here we are given the Harmonic Quadrilateral Α'Β'Γ'Δ' which is inscribed in circle (o') and we are asked, based on that and with proper projective transformations to construct a square ΑΒΓΔ, inscribed in circle (o).

If we take again into consideration Propositions 99 and 79 and we combine them properly, we are led to the construction below (For our convenience and for a better understanding of the solution, we just need to properly combine figures 68 and 57, in such way that figure 97 occurs):

(b). Construction.

We work similarly to Constructions 99 (converse) and 79 (direct), respectively.

I.e., using an arbitrary point T as center of circle (o'), we form the Central Pencil T(A'B'Γ'Δ') whose beams, we suppose that intersect a random line (ε), at points α, β, γ, δ, respectively.

We find the Orthooptical Point Σ (§ 14), of the group of four α, β, γ, δ (Σ is obviously the intersection of circles with diameters αγ and βδ), from which we suppose that circle (o) runs through and we define on circle (o) the intersections A, B, Γ, Δ, of pencil Σ(αβγδ) [If circle (o), does not run through Σ, then we work as in the direct of Construction 79 (§ 15.21,(a) 2/)].

We say that quadrilateral ΑΒΓΔ is the desired square.

(c). Proof.

If we work similarly to the proof of the converse of Construction 99, we find that the Group of Four Points α, β, γ, δ, is harmonic, as quadrilateral A'B'Γ'Δ', is harmonic, while if we work afterwards, as in the direct of the proof of Construction 79, we find that quadrilateral ΑΒΓΔ, is a Square as it is inscribed in circle (o), the group of four points α, β, γ, δ, is harmonic and Σ, which belongs to circle (o), is the Orthooptical Point of the Harmonic group of Four points α, β, γ, δ (Σ is obviously the intersection of the circles with diameters αγ and βδ).

Remarks.

(a). Problem 152, the way it is given, it is undefined. If we want to define it, then we need to be given sufficient data.

(b). The Problem above and its solution, we believe to be new and firstly appear here.

(c). We notice that with this method, firstly the harmonic collinear group of four occurs and then the desired harmonic quadrilateral.

I.e. using the New Method of the Harmonic Transformation, we simultaneously achieve the Construction of the Harmonic Group of Four as well.

(d). For the case where we are asked to construct the Harmonic Quadrialteral, in such way that it is inscribed in the circumscribed circle of thw square, see figure 48.

(e). Relevant Propositions are 32a, 35, 61, 74, 79, 99, 135.

22.2 . Construction of a Harmonic Hexagon from a Regular Hexagon. (How from a Regular Hexagon a Harmonic Hexagon occurs).

Construction 153 (Figure 98).

Based on proper projective transformations and intersections from a line and circles, construct a Harmonic Hexagon based on the Regular Hexagon, given to us and conversely, similarly and based on a Harmonic Hexagon, given to us construct a Regular Hexagon.

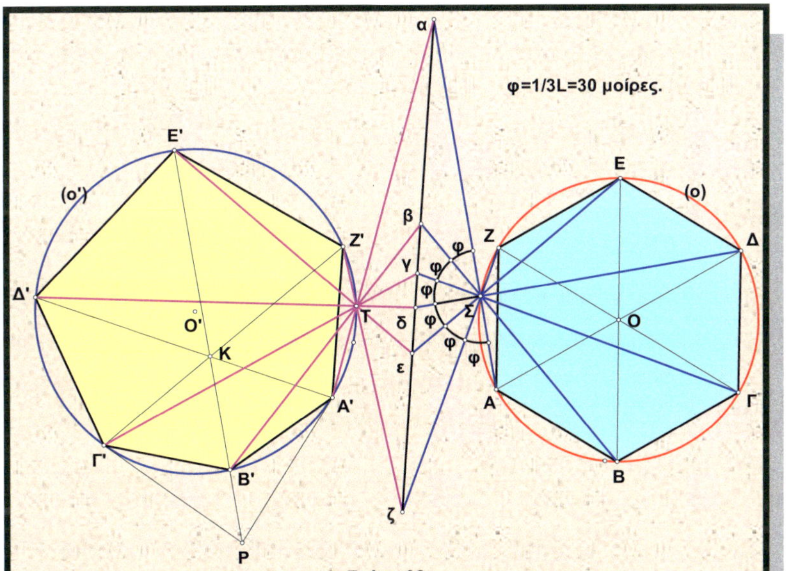

Σχήμα 98.

Solution.

1/. Direct.

(a). Analysis.

Here we are given Regular Hexagon ΑΒΓΔΕΖ and we are asked, based on that and with proper projective transformations to a get a Harmonic Hexagon Α'Β'Γ'Δ'Ε'Ζ'.

If we take into consideration Constructions 80 (§ 15.22) and 100 (§ 19.6) and we compose them properly, we are led to the construction below:

(For convenience and better understanding of the solution we just need to properly compose figures 59 and 69, as for figure 98 to occur).

(b). Construction.

We work similarly to Constructions 80 (converse) and 100 (direct), respectively.

I.e., using an arbitrary point Σ as the center of the circumscribed circle (o) of the Regular Hexagon ΑΒΓΔΕΖ, we form the central pencil Σ(ΑΒΓΔΕΖ), whose beams we suppose that intersect a random line (ε) at points α, β, γ, δ, ε, ζ, respectively.

Consequently, if we want to have the desired Harmonic Hexagon Α'Β'Γ'Δ'Ε'Ζ', to be inscribed in circle (o'), we work in the following way:

Using an arbitrary point T as the center of circle (o'), we form Pencil T(αβγδεζ), the beams of which re-intersect circle (o) at points A', B', Γ' Δ', E', Z', respectively.

We say that the desired Harmonic Hexagon is Α'Β'Γ'Δ'Ε'Ζ'.

(c). Proof.

If we work similarly to the proof of the converse of Construction 80, we find that the Group of Six Points α, β, γ, δ, ε, ζ, is harmonic, as hexagon Α-ΒΓΔΕΖ, is Regular, while if we work afterwards, as in the direct of the proof of Construction 100 and based on this group of six, we find that Hexagon Α'Β'Γ'Δ'Ε'Ζ', is harmonic, as it is inscribed in circle (o') and then the group of four points α, β, γ, δ, ε, ζ, is harmonic.

Comment.

It is obvious that the method above is another way of construction of Harmonic Group of six and of a Harmonic Hexagon, apart from the ones we used in Constructions 62 (§ 15.2) and 145 (§ 19.60).

2/. Converse.

For the converse we follow the opposite way.

(a). Analysis.

I.e. here we are given the Harmonic Hexagon Α'Β'Γ'Δ'Ε'Ζ', which is inscribed in circle (o') and we are asked, based on that and with proper projective transformations to construct a Regular Hexagon ΑΒΓΔΕΖ, inscribed in circle (o).

If we take again into consideration Propositions 100 and 80 and we combine them properly, we are led to the construction below (For our convenience and for a better understanding of the solution, we just need to properly combine figures 69 and 59, in such way that figure 98 occurs):

(b). Construction.

We work similarly to Constructions 100 (converse) and 80 (direct), respectively.

I.e., using an arbitrary point T as center of circle (o'), we form the Central Pencil T(Α'Β'Γ'Δ'Ε'Ζ') whose beams, we suppose that intersect a random line (ε), at points α, β, γ, δ, ε, ζ, respectively.

We find the Orthooptical Point Σ (§ 14), of the group of six α, β, γ, δ, ε, ζ, from which we suppose that circle (o) runs through and we define on circle (o) the intersections Α, Β, Γ, Δ, Ε, Ζ, of pencil Σ(αβγδεζ) [If circle (o), does not run through Σ, then we work as in the direct of Construction 80 (§ 15.22,(a) 2/)].

We say that hexagon ΑΒΓΔΕΖ is the desired Regular one.

(c). Proof.

If we work similarly to the proof of the converse of Construction 100, we find that the Group of Six Points α, β, γ, δ, ε, ζ, η, is harmonic, as hexagon Α'Β'Γ'Δ'Ε'Ζ'Η', is harmonic, while if we work afterwards, as in the direct of the proof of Construction 80, we find that hexagon ΑΒΓΔΕΖ, is a Regular Hexagon as it is inscribed in circle (o), the group of six points α, β, γ, δ, ε, ζ, is harmonic and Σ, which belongs to circle (o), is the Orthooptical Point of the Harmonic group of Six points α, β, γ, δ, ε, ζ.

Remarks.

(a). Problem 153, the way it is given, it is undefined. If we want to define it, then we need to be given sufficient data.

(b). The Problem above and its solution, we believe to be new and firstly appear here.

(c). We notice that with this method, firstly the harmonic collinear group of six occurs and then the desired harmonic hexagon.

I.e. using the New Method of the Harmonic Transformation, we simultaneously achieve the Construction of the Harmonic Group of Six as well.

(d). Relevant Propositions are 35, 62, 76, 80, 96, 100, 145.

22.3 . Construction of a Harmonic Octagon from a Regular Octagon.

(How from a Regular Octagon a Harmonic Octagon occurs).

Construction 154 (Figure 49).

Based on proper projective transformations and intersections from a line and circles, construct a Harmonic Octagon based on the Regular Octagon, given to us and conversely, similarly and based on a Harmonic Octagon, given to us construct a Regular Octagon.

Solution.

1/. Direct.

(a). Analysis.

Here we are given Regular Octagon ΑΒΓΔΕΖΗΘ and we are asked, based on that and with proper projective transformations to a get a Harmonic Octagon Α'Β'Γ'Δ'Ε'Ζ'Η'Θ'.

If we take into consideration Constructions 81 (§ 15.23) and 101 (§ 19.7) and we compose them properly, we are led to the construction below:

(For convenience and better understanding of the solution we just need to properly compose figures 60 and 70, as for figure 49 to occur).

(b). Construction.

We work similarly to Constructions 81 (converse) and 101 (direct), respectively.

I.e., using an arbitrary point Σ as the center of the circumscribed circle (o) of the Regular Octagon ΑΒΓΔΕΖΗΘ, we form the central pencil Σ(ΑΒΓΔΕΖΗΘ), whose beams we suppose that intersect a random line (ε) at points α, β, γ, δ, ε, ζ, η, θ, respectively.

Consequently, if we want to have the desired Harmonic Octagon Α'Β'Γ'Δ'Ε'Ζ'Η'Θ', to be inscribed in circle (o'), we work in the following way:

Using an arbitrary point T as the center of circle (o'), we form Pencil T(αβγδεζηθ), the beams of which re-intersect circle (o) at points A', B', Γ' Δ', E', Z', H', Θ', respectively.

We say that the desired Harmonic Octagon is Α'Β'Γ'Δ'Ε'Ζ'Η'Θ'.

(c). Proof.

If we work similarly to the proof of the converse of Construction 81, we find that the Group of Eight Points α, β, γ, δ, ε, ζ, η, θ, is harmonic, as octagon ΑΒΓΔΕΖΗΘ, is Regular, while if we work afterwards, as in the direct of the proof of Construction 101 and based on this group of eight, we find that octagon Α'Β'Γ'Δ'Ε'Ζ'Η'Θ', is harmonic, as it is inscribed in circle (o') and then the group of eight points α, β, γ, δ, ε, ζ, η, θ, is harmonic.

Comment.

It is obvious that the method above is another way of construction of Harmonic Group of Eight and of a Harmonic Octagon, apart from the ones we used in Constructions 63 (§ 15.3) and 149 (§ 19.64).

2/. Converse.

For the converse we follow the opposite way.

(a). Analysis.

I.e. here we are given the Harmonic Octagon Α'Β'Γ'Δ'Ε'Ζ'Η'Θ', which is inscribed in circle (o') and we are asked, based on that and with proper projective transformations to construct a Regular Octagon ΑΒΓΔΕΖΗΘ, inscribed in circle (o).

If we take again into consideration Propositions 101 and 81 and we combine them properly, we are led to the construction below (For our convenience and for a better understanding of the solution, we just need to properly combine figures 70 and 60, in such way that figure 49 occurs):

(b). Construction.

We work similarly to Constructions 101 (converse) and 81 (direct), respectively.

I.e., using an arbitrary point T as center of circle (o'), we form the Central Pencil T(A'B'Γ'Δ'E'Z'H'Θ') whose beams, we suppose that intersect a random line (ε), at points α, β, γ, δ, ε, ζ, η, θ, respectively.

We find the Orthooptical Point Σ (§ 14), of the group of eight α, β, γ, δ, ε, ζ, η, θ, from which we suppose that circle (o) runs through and we define on circle (o) the intersections A, B, Γ, Δ, E, Z, H, Θ, of pencil Σ(αβγδεζηθ) [If circle (o), does not run through Σ, then we work as in the direct of Construction 81 (§ 15.23,(a) 2/)].

We say that octagon ABΓΔEZHΘ is the desired Regular one.

(c). Proof.

If we work similarly to the proof of the converse of Construction 101, we find that the Group of Eight Points α, β, γ, δ, ε, ζ, η, θ, η, is harmonic, as octagon A'B'Γ'Δ'E'Z'H'Θ' is harmonic, while if we work afterwards, as in the direct of the proof of Construction 81, we find that octagon ABΓΔEZHΘ, is a Regular Octagon as it is inscribed in circle (o), the group of eight points α, β, γ, δ, ε, ζ, η, θ, is harmonic and Σ, which belongs to circle (o), is the Orthooptical Point of the Harmonic group of eight points α, β, γ, δ, ε, ζ, η, θ, from the way it is constructed.

Remarks.

(a). Problem 154, the way it is given, it is undefined. If we want to define it, then we need to be given sufficient data.

(b). The Problem above and its solution, we believe to be new and firstly appear here.

(c). We notice that with this method, firstly the harmonic collinear group of eight occurs and then the desired harmonic octagon.

I.e. using the New Method of the Harmonic Transformation, we simultaneously achieve the Construction of the Harmonic Group of Eight as well.

(d). Relevant Propositions are 35, 63, 77, 81, 97, 101, 149.

22.4 . Construction of a Harmonic Decagon from a Regular Decagon.

(How from a Regular Decagon a Harmonic Decagon occurs).

MAGNIFICENT CONSTRUCTION.

Construction 155 (Figure 99).

Based on proper projective transformations and intersections from a line and circles, construct a Harmonic Decagon based on the Regular Decagon, given to us and conversely, similarly and based on a Harmonic Decagon, given to us construct a Regular Decagon.

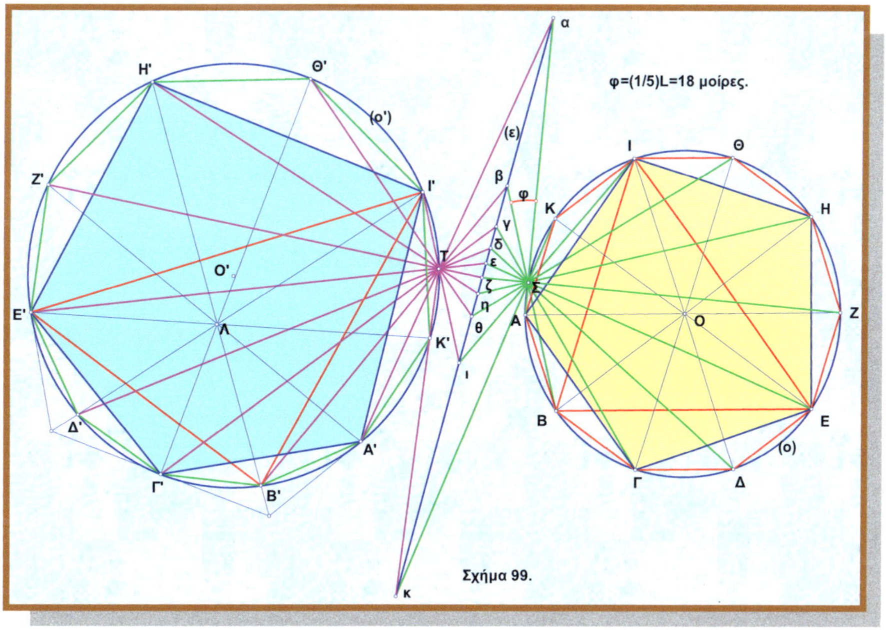

Σχήμα 99.

Solution.

1/. Direct.

(a). Analysis.

Here we are given Regular Decagon ΑΒΓΔΕΖΗΘΙΚ and we are asked, based on that and with proper projective transformations to a get a Harmonic Decagon Α'Β'Γ'Δ'Ε'Ζ'Η'Θ'Ι'Κ'.

If we take into consideration Constructions 82 (§ 15.24) and 102 (§ 19.8) and we compose them properly, we are led to the construction below:

(For convenience and better understanding of the solution we just need to properly compose figures 61 and 71, as for figure 99 to occur).

(b). Construction.

We work similarly to Constructions 82 (converse) and 102 (direct), respectively.

I.e., using an arbitrary point Σ as the center of the circumscribed circle (o) of the Regular Decagon ΑΒΓΔΕΖΗΘΙΚ, we form the central pencil Σ(ΑΒΓΔΕΖΗΘΙΚ), whose beams we suppose that intersect a random line (ε) at points α, β, γ, δ, ε, ζ, η, θ, ι, κ, respectively.

Consequently, if we want to have the desired Harmonic Decagon Α'Β'Γ'Δ'Ε'Ζ'Η'Θ'Ι'Κ', to be inscribed in circle (o'), we work in the following way:

Using an arbitrary point T as the center of circle (o'), we form Pencil T(αβγδεζηθικ), the beams of which re-intersect circle (o) at points Α', Β', Γ' Δ', Ε', Ζ', Η', Θ', Ι', Κ', respectively.

We say that the desired Harmonic Decagon is Α'Β'Γ'Δ'Ε'Ζ'Η'Θ'Ι'Κ'.

(c). Proof.

If we work similarly to the proof of the converse of Construction 82, we find that the Group of Ten Points α, β, γ, δ, ε, ζ, η, θ, ι, κ, is harmonic, as decagon ΑΒΓΔΕΖΗΘΙΚ, is Regular, while if we work afterwards, as in the direct of the proof of Construction 102 and based on this group of ten, we find that decagon Α'Β'Γ'Δ'Ε'Ζ'Η'Θ'Ι'Κ', is harmonic, as it is inscribed in circle (o') and then the group of ten points α, β, γ, δ, ε, ζ, η, θ, ι, κ, is harmonic.

2/. Converse.

For the converse we follow the opposite way.

(a). Analysis.

I.e. here we are given the Harmonic Decagon A'B'Γ'Δ'E'Z'H'Θ'I'K', which is inscribed in circle (o') and we are asked, based on that and with proper projective transformations to construct a Regular Decagon ΑΒΓΔΕΖΗΘΙΚ, inscribed in circle (o).

If we take again into consideration Propositions 102 and 82 and we combine them properly, we are led to the construction below (For our convenience and for a better understanding of the solution, we just need to properly combine figures 71 and 61, in such way that figure 99 occurs):

(b). Construction.

We work similarly to Constructions 102 (converse) and 82 (direct), respectively.

I.e., using an arbitrary point T as center of circle (o'), we form the Central Pencil T(A'B'Γ'Δ'E'Z'H'Θ'I'K') whose beams, we suppose that intersect a random line (ε), at points α, β, γ, δ, ε, ζ, η, θ, ι, κ, respectively.

We find the Orthooptical Point Σ (§ 14), of the group of ten α, β, γ, δ, ε, ζ, η, θ, ι, κ, from which we suppose that circle (o) runs through and we define on circle (o) the intersections A, B, Γ, Δ, E, Z, H, Θ, I, K, of pencil Σ(αβγδεζηθικ) [If circle (o), does not run through Σ, then we work as in the direct of Construction 82 (§ 15.24,(a) 2/)].

We say that decagon ΑΒΓΔΕΖΗΘΙΚ is the desired Regular one.

(c). Proof.

If we work similarly to the proof of the converse of Construction 102, we find that the Group of Ten Points α, β, γ, δ, ε, ζ, η, θ, ι, κ, η, is harmonic, as decagon A'B'Γ'Δ'E'Z'H'Θ'I'K' is harmonic, while if we work afterwards, as in the direct of the proof of Construction 82, we find that decagon ΑΒΓΔΕΖΗ-ΘΙΚ, is a Regular Decagon as it is inscribed in circle (o), the group of ten points α, β, γ, δ, ε, ζ, η, θ, ι, κ, is harmonic and Σ, which belongs to circle (o), is the Orthooptical Point of the Harmonic group of ten points α, β, γ, δ, ε, ζ, η, θ, ι, κ, from the way it is constructed.

Remarks.

(a). Problem 155, the way it is given, it is undefined. If we want to define it, then we need to be given sufficient data.

(b). The Problem above and its solution, we believe to be new and firstly appear here.

(c). We notice that with this method, firstly the harmonic collinear group of ten occurs and then the desired harmonic decagon.

I.e. using the New Method of the Harmonic Transformation, we simultaneously achieve the Construction of the Harmonic Group of Ten as well.

(d). According to what we have written here, another way of construction of a Harmonic Group of Ten and of a Harmonic Decagon, we have not achieved though we tried. So we see how valuable the New Method of Harmonic Transformation is (Constructions 64, 82 and 102, are not independent of the above Construction 155, but they are parts of it and applications of parts of the Harmonic Transformation from the composition of which the above way occurs).

We believe that the Construction above in another way is too difficult, if not impossible. The above way (method) solves for us the problem in an automatic way and easily.

(e). Relevant Propositions are 35, 64, 78, 82, 98, 102.

22.5 . Construction of a Harmonic Pentagon from a Regular Pentagon.
(How from a Regular Pentagon a Harmonic Pentagon occurs).

Construction 156 (Figure 99).

Based on proper projective transformations and intersections from a line and circles, construct a Harmonic Pentagon based on the Regular Pentagon, given to us and conversely, similarly and based on a Harmonic Pentagon, given to us construct a Regular Pentagon.

Solution.

1/. Direct.

(a). Analysis.

Here we are given Regular Pentagon ΑΓΕΗΙ and we are asked, based on that and with proper projective transformations to a get a Harmonic Pentagon Α'Γ'Ε'Η'Ι'.

If we take into consideration Construction 155 (§ 22.4) (direct) and the relevant definition of the Harmonic Pentagon (§ 16.1e), we are led to the construction below:

(b). Construction.

Based on the Regular Decagon ΑΒΓΔΕΖΗΘΙΚ, which is easily constructed as pentagon ΑΓΕΗΙ to be the given regular, inscribed in circle (o) and arbitrary line (ε), we work in the exact same way as in Construction 155 (direct), we firstly construct the Harmonic Group of Ten Points α, β, γ, δ, ε, ζ, η, θ, ι, κ, on line (ε) and consequently on circle (o'), the Harmonic Decagon Α'Β'Γ'Δ'Ε'Ζ'Η'Θ'Ι'Κ'

We say that pentagon Α'Γ'Ε'Η'Ι', is the desired Harmonic one.

(c). Proof.

Since Decagon Α'Β'Γ'Δ'Ε'Ζ'Η'Θ'Ι'Κ' is harmonic, then from its construction pentagon Α'Γ'Ε'Η'Ι' will also be Harmonic, as there are points Β', Δ', Ζ', Θ', Κ' on circle (o'), for whixh Decagon Α'Β'Γ'Δ'Ε'Ζ'Η'Θ'Ι'Κ' is Harmonic [Definition of the Harmonic Pentagon (§ 16.1e)].

2/. Converse.

(a). Analysis.

I.e. here we are given the Harmonic Pentagon Α'Γ'Ε'Η'Ι', which is inscribed in circle (o') and we are asked, based on that and with proper projective transformations to construct a Regular Pentagon ΑΓΕΗΙ, inscribed in circle (o).

If we take again into consideration Propositions 103 and 83 and we combine them properly, we are led to the construction below (For our convenience and for a better understanding of the solution, we just need to properly combine figures 71 and 61, in such way that figure 99 occurs):

(b). Construction.

We work similarly to Constructions 103 (converse) and 83 (direct), respectively.

More specifically, initially we work as in Construction 103 (converse). I.e., using an arbitrary point T as center of circle (o'), we form the Central Pencil T(A'Γ'E'H'I') whose beams, we suppose that intersect a random line (ε), at points α, γ, ε, η, ι, respectively.

Afterwards we work similarly to Construction 83 (direct) and after we construct the Harmonic Group of Ten Points α, β, γ, δ, ε, ζ, η, θ, ι, κ, based on which we construct, on circle (o) decagon ΑΒΓΔΕΖΗΘΙΚ, after we first construct the Orthooptical Point Σ of the Harmonic Group of Ten α, β, γ, δ, ε, ζ, η, θ, ι, κ.

We say that pentagon ΑΓΕΗΙ is the desired Regular one.

[The construction of the above Harmonic Group of Ten Points α, β, γ, δ, ε, ζ, η, θ, ι, κ, can be achieved, if we initially construct the Harmonic Decagon A'B'Γ'Δ'E'Z'H'Θ'I'K', in the following way:

Since we are given that pentagon A'Γ'E'H'I' is harmonic, according to the relevant definition [§ 16,1e(b)], there will be on circle (o′) points B', Δ', Z', Θ', K', such that decagon A'B'Γ'Δ'E'Z'H'Θ'I'K', would be harmonic.

This though, is possible to be constructed.

In fact, since for example then quadrilateral A'B'Γ'H' needs to be harmonic (§ 16.1d) and points A', Γ', H', are known, then B′, as it is known it is easily constructed (Construction 135).

Additionally since quadrilateral Γ'Δ'E'I' should be harmonic and points Γ', E', I' are known, then Δ′ is constructed, as before.

In a similar way, from the Harmonic quadrilateral E'Z'H'A' we define the place of Z′, from the Harmonic quadrilateral H'Θ'I'Γ' we define the place of Θ′ and from the Harmonic quadrilateral I'K'A'E' we define the place of K′.

Hence, after the definition of B', Δ', Z', Θ', K' as well we will have constructed the Harmonic Decagon A'B'Γ'Δ'E'Z'H'Θ'I'K', hence pencil Σ(A'B'Γ'Δ'E'Z'H'Θ'I'K') will be harmonic and thus its intersections α, β, γ, ...κ from line (ε), will constitute the desired Harmonic Group of Ten.

(c). Proof.

If we work similarly to the proof of the converse of Construction 103, we find that the Group of five Points α, γ, ε, η, ι, is harmonic, based on which we constructed the harmonic group of ten points α, β, γ, δ, ε, ζ, η, θ, ι, κ, while if we work afterwards, as in the direct of the proof of Construction 83, we find that decagon ΑΒΓΔΕΖΗΘΙΚ, is a Regular Decagon as it is inscribed in circle (o), the group of ten points α, β, γ, δ, ε, ζ, η, θ, ι, κ, is harmonic and Σ, which belongs to circle (o), is the Orthooptical Point of the Harmonic group of ten points α, β, γ, δ, ε, ζ, η, θ, ι, κ, from the way it is constructed. Therefore, since decagon ΑΒΓΔΕΖΗΘΙΚ is regular, we easily find that pentagon ΑΓΕΗΙ, is regular too.

Remarks.

(a). Problem 156, the way it is given, it is undefined. If we want to define it, then we need to be given sufficient data.

(b). The Problem above and its solution, we believe to be new and firstly appear here.

(c). We notice that with this method, firstly the harmonic collinear group of fivve occurs and then the desired harmonic pentagon.

I.e. using the New Method of the Harmonic Transformation, we simultaneously achieve the Construction of the Harmonic Group of Five as well.

(d). According to what we have written here, another way of construction of a Harmonic Group of Five and of a Harmonic Pentagon, we have not achieved though we tried. So we see how valuable the New Method of Harmonic Transformation is(Constructions 65, 83 and 103, are not independent of the above Construction 155, but they are parts of it and applications of parts of the Harmonic Transformation from the composition of which the above way occurs).

(e). Relevant Propositions are 35, 64, 65, 78, 82, 83, 98, 102, 103, 135, 155.

22.6. Generalization of Constructions 152-156.

If we continue as in applications 152-156, easily and similarly, we can construct any other Regular or Harmonic Multilaterla, solving in this way the

General Problem 157 below, applying the New Method of Harmonic Transformation.

General Construction 157.

Based on proper projective transformations and intersections from line and circles, construct a Harmonic Multilateral based on a Regular Multilateral, given to us, with the same number of sides, and conversely, similarly and based on a Harmonic Multilateral, given to us, construct a Regular Multilateral, with the same number of sides.

(The solution to this Problem, for any other multilateral is carried through with, in the exact same way as in Constructions 152- 156).

Conclusions.

After the above applications 152-156 and the Generalized Problem 157, the following conclusions occur, regarding the Values of the Method of HArmonic Transformation:

(a). During the application of the Harmonic Transformation, at the same time with the Harmonic n-lateral, the Construction of the Harmonic group of n occurs, which is important.

(b). The Construction of a Regular or Harmonic Multilateral (Collinear or Cocyclic), is an easy case using the Method of the Harmonic Transformation (even those whose construction was not known until now), as long as we are given the respective Harmonic or Regular Multilateral, respectively.

Consequently, we are allowed to say:

«Give me a place to stand and I shall move theEarth».

«Give me a Regular n-lateral and I will construct a Harmonic n-lateral (Collinear or Cocyclic), or conversely,

Give me a Harmonic n-lateral (Collinear or Cocyclic), and I will construct a Regular n-lateral».

General Conclusions.

Due to the above the following General Conclusions occur:

(a). Theoretically, it is easy to construct all the Regular Multilaterals (even those whose construction was not known until now), as long as we are given the respective Harmonic Multilateral.

Thereofre:

«If there is another way of constructing any Harmonic n-lateral, then it is obvious that using the Method of Harmonic Transformation, we can construct the respective Regular n-lateral, even that one, that until now we were not able to construct, thus the historical unsolved until now problems will be solved, as the trisection of an angle, etc, this is why we call anyone who knows anything relevant to show it to us».

(b). Using the Method of the Harmonic Transformation, we can say, that in a way, the Euclidean Space is connected to the Harmonic Projective Space and conversely.

General Remarks.

(a). The above General Problem and its solution, we believe to be new and firstly appear here. Since we needed at least twenty years to reach its solution, we called it the "Problem of the Vicennium".

(b). We easily realize that the way we applied above for harmonic quadrilaterals, pentagons, hexagons, octagons, etc, we can also apply it on a triangle and all the multilaterals, as long as their respective regular multilateral is constructible. I.e. for multilaterals, with number of sides 2^2, $2^2.3$, $2^2.5$, $2^v.15$, that as it is known are constructible, then their respective harmonic multilaterals are constructible too.

Hence, the above way, as it is applied on a large number of relevant constructions, it is a method, which we called «Harmonic Transformation».

The detailed interpretation of this method, we tried to provide practically with the above proof of the General Problem 157 (Constructions 152-156).

(c). We notice that using this method, the harmonic collinear group of four,

five, six, eight, ten, twelve, etc points occurs and then the respective harmonic quadrilateral, pentagon, octagon, decagon, dodecagon, etc.

Especially for the triangle we notice the following:

Since each triangle has only some of the properties that harmonic multilaterals have (as for example Lemoine Point), this is why we can call it «pseudoharmonic».

(d). The application of the above Method, can be seen rather graphically and clearly in figures 48, 49, 97-99 above.

(e). Relevant Propositions are 32a, 35, 61-64, 74, 76-82, 97-102, 135, 145, 149, 152-157.

Consequently, we will refer to a series of Corollaries, that stem from Proposition 157 (Rules of the Harmonic Transformation).

These Rules are rather useful, as they help us solve complex problems of Harmonic Geometry.

22.61. Detailed Analysis of the Method of «HARMONIC TRANSFORMATION».

Below we will analyze in detail the Method of Harmonic Transformation, in a series of rules in the form of Corollaries 158 - 168, based on elementary figures, the application of which will help the easier solution of relevant Problems.

Additionally, in Part C we will apply the Method of Harmonic Transformation for the construction of Harmonic Involutions as well (Collinear and Cocyclic), that occur from inscribed in circle convex multilaterals symmetrical with respect to one or two perpendicular diameters of the circle.

22.61 A. Rule 1. Harmonic Transformation of an Inscribed Convex Rhomboid.

1st Corollary of the General Construction 157.

Corollary 158 (Figures 100 and 97) (Rule 1).

Each quadrilateral, inscribed in circle and symmetrical with respect to one of its diagonals (rhomboid), or to both its diagonals (square), is harmonically transformed, to a Harmonic Quadrialteral,

and conversely,

each Harmonic Quadrilateral can be harmonically transformed, to a quadrilateral inscribed in circle and symmetrical with respect to one of its diagonals (rhomboid), or both its diagonals (hence transformed to a square).

Proof.

1st Case (Figure 100).

(a). Direct.

Let there be the inscribed in circle (o) rhomboid quadrilateral ΑΒΓΔ (figure 100), which is symmetrical with respect to its diagonal ΒΔ (terminology § 14). We will prove that this is harmonically transformed to a quadrilateral Α'Β'Γ'Δ', which is harmonic.

We apply, in the known way (Construction 152, direct), the Method of «Harmonic Transformation for a quadrilateral», in the following way:
We set an arbitrary point Σ on circle (o) and construct pencil Σ(ΑΓΒΔ), that intersects an arbitrary line (ε) at points α, β, γ, δ.

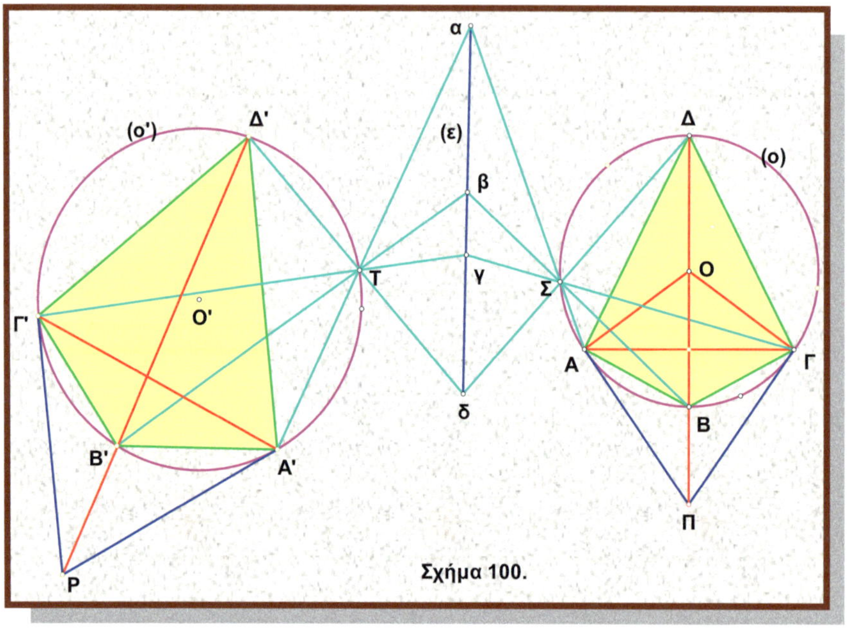

Σχήμα 100.

On random circle (o'), we set an arbitrary point T and form pencil T(αγβδ) which intersects again circle (o') at points A', B', Γ', Δ'.
We say that quadrialteral A'B'Γ'Δ', is Harmonic.
In fact, according to the direct of Criterion 75, pencil Σ(ΑΓΒΔ) or Σ(αγβδ), is harmonic (with angle ΒΣΔ or βΣδ right). Thus, the group of four points α, γ, β, δ, is harmonic, hence pencil T(αγβδ) or T(A'Γ'B'Δ'), is harmonic. Consequently quadrilateral A'B'Γ'Δ' is Harmonic, as it is inscribed in circle (o') (Converse of Criterion 35).

(b). Converse.

Here we are given the Harmonic Quadrilateral A'B'Γ'Δ' and we will prove that it is harmonically transformed to the symmetrical quadrilateral ΑΒΓΔ, with respect to one of its diagonals, inscribed in circle (o) (In the converse of case 2 below, we will see that it is possible for it to be harmonically transformed to a square as well).

To achieve this, we apply here as well, in the known way [as in Construction 152, converse (for quadrialterals)], the Method of «Harmonic Transformation», in the following way:

I.e., we set an arbitrary point T on circle (o') and we construct, pencil T(A'Γ'B'Δ'), that intersects a random line (ε) at points α, β, γ, δ.

Sine we want the quadrilateral ΑΒΓΔ, that will occur, to be symmetrical with respect to for example its diagonal ΒΔ, this is why we draw the circle with diameter βδ, which we suppose intersects a random circle (o) at point Σ and if pencil Σ(αγβδ) intersects circle (o) again at points A, Γ, B, Δ, we say that quadrilateral ΑΒΓΔ is the desired symmetrical one with respect to ΒΔ.

In fact, since quadrialteral A'Γ'B'Δ' is harmonic, according to the direct of Criterion 35, pencil Σ(A'Γ'B'Δ') or Σ(αγβδ), is harmonic, thus the group of four points α, γ, β, δ, will be harmonic, as well as pencil Σ(αγβδ), or Σ(ΑΓΒΔ) will be harmonic, hence quadrilateral ΑΒΓΔ, will also be harmonic (converse of Criterion 35).

Moreover, since the group of four points α, γ, β, δ is harmonic and angle βΣδ is right, in triangle Σαγ, Σβ will be the bisector of angle αΣγ.

Thus ΣΒ is the bisector of angle ΑΣΓ, hence ΑΒ=ΒΓ and since angle ΒΣΔ is right, then ΒΔ is a diameter of circle (o). Therefore we easily prove that quadrilateral ΑΒΓΔ is in fact symmetrical with respect to ΒΔ.

2nd Case (Figure 97).

(a). Direct.

We suppose that the inscribed in circle (o) rhomboid quadrilateral ΑΒΓΔ (figure 97), is symmetrical with respect to its two diagonals (hence it is a square). We will prove that this is harmonically transformed to a quadrilateral A'B'Γ'Δ', which is harmonic.

This case is obviously a special case of the previous one and therefore for the relevant constructions and its proof we work as bove in the direct of case 1 and we prove that this is true (The construction and the proof of this can be based on the proof of the direct of Criterion 152).

(b). Converse.

Here we are given the Harmonic Quadrilateral A'B'Γ'Δ' and we want to prove that this is harmonically transformed to the symmetrical quadrilateral ABΓΔ with respect to both its diagonals, inscribed in circle (o) (square).

We work similarly to the converse of Construction 152 [§ (b)], so we will have figure 97, where based on the Harmonic Quadrilateral A'B'Γ'Δ', the quadrilateral ABΓΔ occurs.

Consequently, to prove that quadrilateral ABΓΔ is a square, we work as in the converse of Construction 152 [§ (c)].

[It is obvious that the Rule above, is also true if the Harmonic Transformation is done on the same circle (See figure 48), using the Orthooptical Point Σ, that occurs as the intersection of circles with diameters αγ and βδ].

Remarks.

(a). The Corollary above and its proof, we believe to be new and firstly appear here.

(b). From figure 100 and paragraphs 3.2 and 8.6, we conclude that, the symmetrical with respect to BΔ, quadrilateral ABΓΔ, is harmonic, or is a Harmonic Involution and that from it, during the Harmonic Transformation, a Collinear Harmonic Groupof Four points occurs or a Collinear Involution of points α, β, γ, δ and a Harmonic Quadrilateral or a Harmonic Cyclic Involution of points A', B', Γ', Δ' (See also § 3,2 and 8.6).

(c). Relevant Propositions are 35, 75, 152-157, 159-168.

22.61. B. Rule 2. Harmonic Transformation of a String Perpendicular to a Diameter of a Circle.

2nd Corollary of the General Construction 157.

Corollary 159 (Figures 100 and 97) (Rule 2).

Each string perpendicular to a diameter of a circle and this diameter, are harmonically transformed, to a pair of conjugate strings of the same circle or another arbitrary circle, and conversely,

each pair of conjugate strings with respect to circle, can be harmonically transformed, to a pair of strings of the same or another circle, perpendicular to each other, from which one or both, are diameters of the second circle.

Proof.

(a). Direct.

I.e. on circle (o) (figure 100), we are given its diameter ΒΔ and random string ΑΓ perpendicular to it (of the same circle).

We will prove that ΑΓ and ΒΔ, are harmonically transformed, to a pair of conjugate strings Α'Γ' and Β'Δ' of the same or another arbitrary circle (o').

In fact, since we easily realize that quadrilateral ΑΒΓΔ is rhomboid (symetrical with respect to ΒΔ), according to the direct of Corollary 158 (Rule 1) (§ 22.61,A), this is Harmonically transformed, to the Harmonic Quadrilateral Α'Β'Γ'Δ', whose diagonals Α'Γ' and Β'Δ', will be as it is known, conjugate with respect to circle (o') (definition §16.1a).

(b). Converse.

Here we are given the conjugate strings Α'Γ' and Β'Δ' of circle (o').

We will prove that these, can be harmonically transformed, in a circle (o), to perpendicular strings ΑΓ, ΒΔ, one of which or both of them are diameters of circle (o) (figures 100 and 97, respectively).

In fact, since quadrilateral Α'Β'Γ'Δ' has conjugate diagonals (definitions, § 16.1a), it will be Harmonic and thus, according to the two converses of Corollary 158 (Rule 1), this is harmonically transformed, to the rhomboid quadrilateral ΑΒΓΔ, inscribed in circle and symmetrical with respect to ΒΔ, or ΑΓ (rhomboid), or to both of them, hence it will be a square.

In both cases ΑΓ, ΒΔ are perpendicular, while one of them or both will be diameters of this circle, respectively.

[It is obvious that the Rule above, is also true if the Harmonic Transformation happens to the same circle (See figure 48)].

Remarks.

(a). The Proposition above and its proof, we believe to be new and firstly appear here.

(b). Relevant Propositions are 152-158, 160, 200.

22. 61. C. Rule 3. Harmonic Transformation of a Pair of Tangents of a Circle Symmetrical with respect to a Diameter of the Same Circle.

3rd Corollary of the General Construction 157.

Corollary 160 (Figures 100 and 97) (Rule 3).

Each pair of symmetrical, with respect to a diameter of a circle, tangents of the same circle, are harmonically transformed, to concurrent string and two tangents respectively of the same or another arbitrary circle,
and conversely,
each string of circle that is concurrent to a pair of tangents of the same circle, can be harmonically transformed, to a diameter and pair of tangents, symmetrical to this diameter, of the same or another circle.

Proof.

(a). Direct.

I.e. on circle (o), we are given its diameter BΔ and the pair of symmetrical, with respect to BΔ, tangents at points A, Γ, of the same circle (It easily occurs that BΔ runs through the intersection Π of the tangents at A and Γ).

We will prove that the diameter BΔ and the pair of these tangents, are harmonically transformed, to concurrent at point P, of a string B'Δ' and pair of two tangents at points A', Γ', of the same or another circle (o').

In fact, since A, Γ will also be symmetrical with respect to BΔ, then BΔ is the perpendicular bisector to AΓ, hence quadrilateral ABΓΔ is rhomboid (symmetrical with respect to BΔ) and thus according to the direct of Corollary 158 (Rule 1) (§22,61,A), this is Harmonically transformed, to the Harmonic Quadrilateral A'B'Γ'Δ'. Thus its diagonals A'Γ' and B'Δ', will be as it is known, conjugate with respect to circle (o'). This means that B'Δ' will also run through the intersection P of the tangents at A', Γ', which is the pole of A'Γ'.

(b). Converse.

Here we are given a string B'Δ' of circle (o') and two tangents at points A',

Γ', that lie on both sides of B'Δ', which with B'Δ' coincide at Point P.

We will prove that B'Δ' and these two tangents, can be harmonically transformed, to a diameter BΔ and two tangents at points A, Γ, respectively of the same of another circle (o), in such way that the above tangents are symmetrical with respect to BΔ (figure 100).

In fact, since diagonal B'Δ' of the quadrilateral A'B'Γ'Δ' runs through pole P of A'Γ', quadrilateral A'B'Γ'Δ' is harmonic and thus according to the converse of Corollary 158 (Rule 1), it is harmonically transformed, to the rhomboid quadrilateral ABΓΔ, inscribed in circle (o), symmetrical with respect to BΔ, or AΓ or with respect to both. In both cases the tangents at A, Γ or B, Δ, are symmetrical with respect to BΔ or AΓ respectively or at the same time with respect to both, thus quadrilateral ABΓΔ is a square (figure 97). This is obvious, since quadrilateral ABΓΔ is symmetrical with respect to BΔ and AΓ.

[It is obvious that the Rule above, is also true when the Harmonic Transformation happens to the same circle (See figure 48)].

Remarks.

(a). It is obvious that the tangents at A, Γ and BΔ, coincide at points Π, or they are parallel.

(b). The Proposition above, and its proof, we believe to be new and firstly appear here.

(c). It is obvious that the Proposition above is also true in the special case where instead of the rhomboid quadrilateral ABΓΔ in figure 100, we have a square [for this case see the second part in Proposition 158 (figure 97)].

(d). In the special case where the tangents at A, Γ are parallel to diameter BΔ, Corollary 160 is true as well.

(e). Relevant propositions are 152, 157, 158, 159, 161, 200.

22.61 D. Rule 4. Harmonic Transformation of a Pair of Symmetrical Strings of Circle with respect to a Diameter of the Same Circle.

4th Corollary of the General Construction 157.

Corollary 161 (Figure 101) (Rule 4).

Each pair of symmetrical, with respect to a diameter of a circle, strings of the same circle (that intersect outside of the circle), the strings and the diameter are harmonically transformed, to three concurrent strings outside of the same circle or another arbitrary circle and the hexagon with apexes the ends of the three strings that occurred, has concurrent diagonals, and conversely,

three concurrent outside of the circle strings, whose ends are apexes to a hexagon, with concurrent diagonals, can be harmonically transformed, to a diameter of another circle, and pair of strings, symmetrical with respect to this diameter.

Proof.

1st Case. The symmetrical strings intersect outside of the circle.

(a). Direct.

I.e. on circle (o), we are given its diameter BE and the pair of its symmetrical, with respect to BE strings AZ and ΓΔ.

We will prove that BE, AZ, ΓΔ, are harmonically transformed, to the concurrent strings B'E', A'Z', Γ'Δ', of the same or another circle (o'), for which it is as well: A'Δ'∩B'E'∩Γ'Z'≡K'. (1).

In fact, since as we easily realize, BE is the perpendicular bisector of both strings AΓ and ΔZ, the two quadrilaterals ABΓE and ΔEZB, are rhomboid (symmetrical with respect to BE). According to Corollary 158 (Rule 1) (§ 22.61,A), these are harmonically transformed in circle (o'), to th Harmonic Quadrilaterals A'B'Γ'E' and Δ'E'Z'B'. Thus, according to the converse of Proposition 51 (§ 9.12), strings B'E', A'Z', Γ'Δ', will be concurrent, at let's say point P', while (1) will also be true.

(b). Converse.

Here we are given the three concurrent strings B'E', A'Z', Γ'Δ', at point P', of circle (o') and for which relation (1) is true.

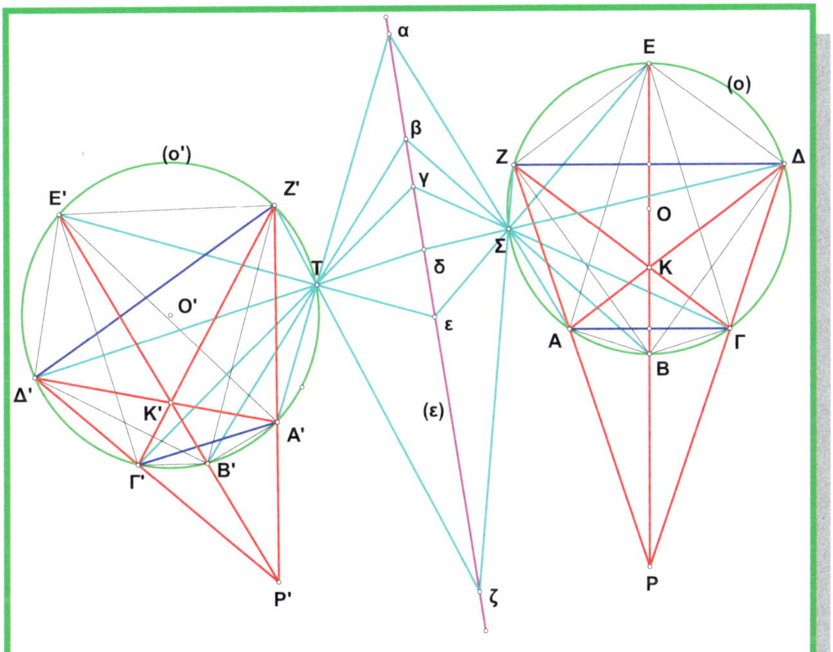

Σχήμα 101.

We will prove that these can be harmonically transformed in circle (o), in such way that B'E', will be harmonically transformed to diameter BE and A'Z', Γ'Δ' to strings AZ, ΓΔ symmetrical with respect to BE.

In fact, since for these strings it is:

$$A'Z' \cap B'E' \cap \Gamma'\Delta' \equiv P'. \qquad (2).$$

and relation (1) is true, according to the direct of Proposition 51, Quadrilaterals A'B'Γ'E' and Δ'E'Z'B' will be harmonic.

This means that these can be harmonically transformed, in circle (o) to the rhomboid quadrialterals ΑΒΓΕ, ΔΕΖΒ, respectively (Corollary 158 Rule 1), whose shared diagonal BE is also a diameter of a circle.

It is obvious that BE will be the perpendicular bisector to ΑΓ and ΔZ, hence strings AZ and ΓΔ will be symmetrical, with respect to BE.

2nd Case. The symmetrical strings intersect inside the circle.

In the exact same way we prove that this Corollary is true when they intersect inside circle (o). For example, instead of AZ and ΓΔ, we have the also symmetrical with respect to BE strings AΔ and ΓZ. In this case, the concave hexagon A'B'Γ'Z'E'Δ', will have its diagonals A'Z', B'E', Γ'Δ', concurrent at point P'.

Remarks.

(a). It is known {Theorem 253 in issue A in Γεωμετρία [28] (page 132, § 253)}, that symmetrical strings AZ, ΓΔ are also concurrent at point P of BE. The same applies to the symmetrical strings AΔ, ΓZ, that are concurrent at point K of BE.

(b). In the special case where strings AZ, ΓΔ, are parallel, to BE, the Corollary above is true. In this special case, we easily prove that strings AΔ and ΓZ as well, are diameters of circle (o), (I.e. they are concurrent at point O of BE, center of this circle), while AZ, BE, ΓΔ, are concurrent at the ideal point.

(c). It is obvious that the tangents at A, Γ and BE, are concurrent at point P″ (Rule 3).

(d). The Corollary above and its proof, we believe to be new and firstly appear here.

(e). Relevant Propositions are 51, 152-160, 162-167, 200.

22.61 E. Rule 5. Harmonic Transformation of Parallel Strings of a Circle.

5th Corollary of the General Construction 157.

Corollary 162 (Figure 102) (Rule 5).

Three or more parallel strings of a circle, are harmonically transformed, to a central pencil of strings, of the same or another arbitrary circle, and conversely,

each central pencil of three or more strings of a circle, can be harmonically transformed, to strings parallel to each other, of the same of another circle.

Proof.

(a). Direct.

Let there be four parallel to each other strings ΑΓ, ΔΚ, ΕΙ, ΖΘ, of circle (o). We suppose that these are harmonically transformed to strings Α'Γ', Δ'Κ', Ε'Ι', Ζ'Θ', of the same or another circle (o'), which also constitute a central pencil, with center P' [outside of circle (o')].

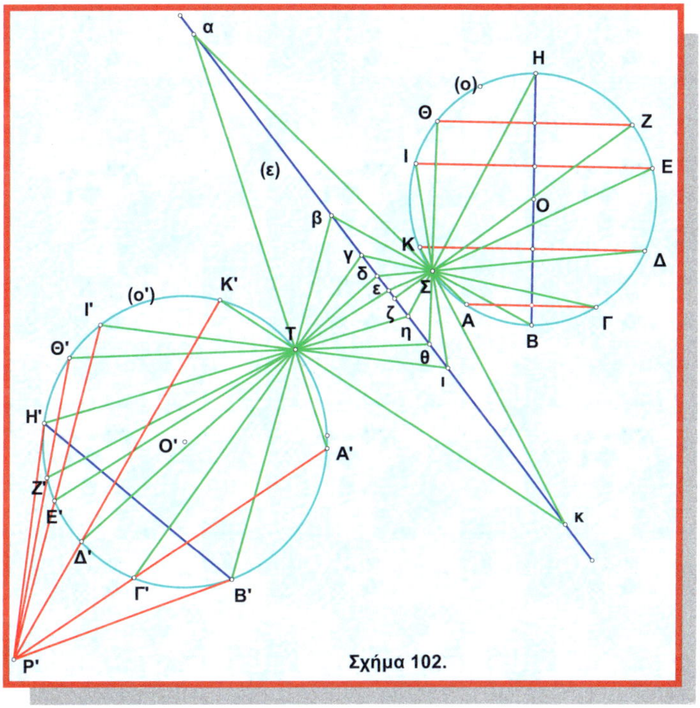

Σχήμα 102.

We easily prove that parallel strings ΑΓ, ΔΚ, ΕΙ, ΖΘ, are all perpendicular to a diameter ΒΗ of circle (o).

Hence, these strings and the diameter are harmonically transformed to strings Α'Γ', Δ'Κ', Ε'Ι', Ζ'Θ' and Β'Η' respectively, of circle (o) or (o'). Thus, according to the direct of Corollary 159 (Rule 2) above, strings Α'Γ', Δ'Κ', Ε'Ι', Ζ'Θ', will be conjugate with respect to string Β'Η'.

This means that Α'Γ', Δ'Κ', Ε'Ι', Ζ'Θ', will be concurrent at pole P' of string Β'Η', which is the intersection of the tangents at Β' and Η'.

(b). Converse.

Let there be four strings A'Γ', Δ'K', E'I', Z'Θ', of circle (o'), that are concurrent at point P' [outside of circle (o')].

We will prove that these, can be harmonically transformed, to strings parallel to each other, of the same of another circle (o).

If from P', we draw the tangents P'B', P'H' at circle (o'), then obviously string B'H' are conjugate to strings A'Γ', Δ'K', E'I', Z'Θ', thus these are harmonically transformed to strings AΓ, ΔK, EI, ZΘ and B'H' to a diameter BH of the same or another circle (o), while these will be perpendicular to diameter BH (Converse of Corollary 159 (Rule 2), hence AΓ, ΔK, EI, ZΘ, will be parallel.

Remarks.

(a). Above we used four parallel strings AΓ, ΔK, EI, ZΘ. We work in the exact same way and prove that Corollary 162 is also true for n parallel strings of circle (o).

(b). It is obvious that AΓ, ΔK, EI, ZΘ are concurrent, bu there at infinity, where also the intersection of tangents at B, H of circle (o) is and which is the pole of BH.

(c). In the general case where instead of the parallel strings we are given parallel lines, then we intersect all the parallel lines with a circle and we use the detached parallel strings, and then we continue as above, in Corollary 162.

(d). The Corollary above and its proof, we believe it is new and firstly appears here.

(e). We easily realize that the three series of points A, B, ...Θ – α, β,...θ -A′, B′.... Θ′, are two Harmonic Involutionary, with Double Points the pairs B, H , β, η and B′, H′, respectively.

(f). Relevant Propositions are 152-161, 163-168, 200.

22.61 F. Rule 6. Properties of the Inscribed in Circle Hexagon, Symmetrical with respect to its Diagonal.

6th Corollary of General Construction 157.

Corollary 163 (Figure 103) (Rule 6).

Each convex inscribed in circle (o), hexagon ΑΒΓΔΕΖ, symmetrical we suppose with respect to its diagonal ΒΕ, is Harmonically transformed, to a convex and inscribed in the same circle (o), or in another circle (o') hexagon Α'Β'Γ'Δ'Ε'Ζ', which has the following properties:

1/. Its diagonals are concurrent.

2/. Similarly for its concave hexagon Α'Β'Γ'Ζ'Ε'Δ', its diagonals are concurrent.

3/. The pairs of its tangents at the pairs of its apexes Α', Γ' and Δ', Ζ', intersect at its diagonals Β'Ε'.

4/. The tangents at points Β', Ε' and Α'Γ' and Δ'Ζ', coincide.

5/. Quadrialterals Α'Β'Γ'Ε', Δ'Ε'Ζ'Β', are Harmonic,

and conversely,

each convex inscribed in circle (o') hexagon Α'Β'Γ'Δ'Ε'Ζ' which itself, as well as the concave hexagon Α'Β'Γ'Ζ'Ε'Δ', have concurrent diagonals, can be Harmonically transformed, to a convex inscribed in circle (o) hexagon ΑΒΓΔΕΖ, that has the following properties:

1/. It should be symmetrical with respect to ΒΕ.

2/. The diagonals of it, as well as of its concave hexagon ΑΒΓΖΕΔ, should be concurrent, in each hexagon.

3/. The pairs of tangents at pairs of points Α, Γ and Δ, Ζ, should intersect at diameter ΒΕ.

4/. The tangents at Β, Ε and ΑΓ, ΔΖ, should coincide at infinity (they should be parallel).

5/. Quadrilaterals ΑΒΓΕ, ΔΕΖΒ, should be symmetrical with respect to ΒΕ.

Proof.

1st Way.

(a). Direct.

In fact:

§1/ and 2/. Since strings ΑΖ, ΓΔ are symmetrical with respect to diameter ΒΕ, according to Corollary 161 (direct) (§ 22.61, D), during the Harmonic

Transformation, it would be: A'Δ'∩B'E'∩Γ'Z'≡K' και A'Z'∩B'E'∩Γ'Δ'≡P'. (1)

3/. Since the tangents at A and Γ are symmetrical with respect to BE, according to Corollary 160 (§ 22.61,C), during the Harmonic Transformation, tangents at A', Γ' and B'E' will be concurrent at point α'.

In a similar way we prove that tangents at Δ', Z' and B'E' coincide at point β'.

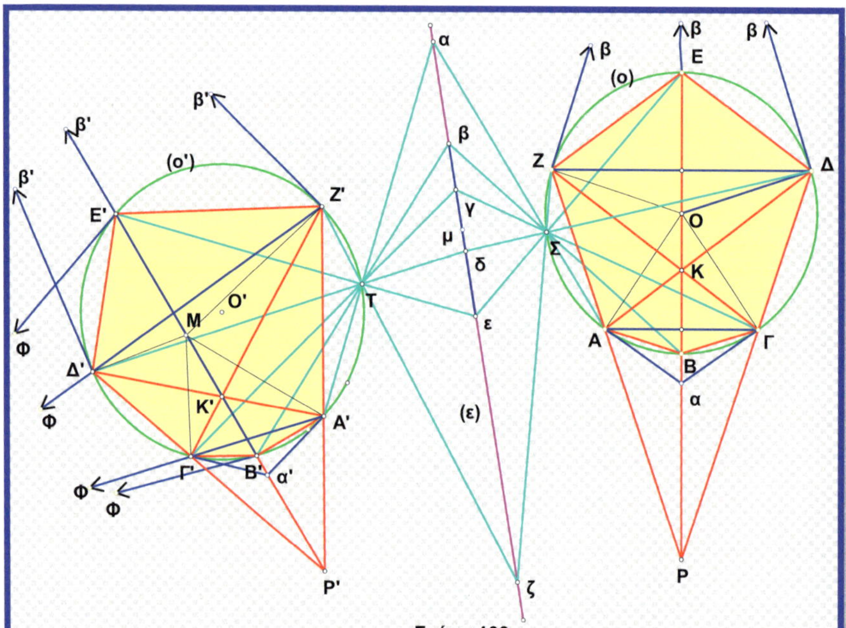

Σχήμα 103.

4/. Since AΓ and ΔZ are perpencidular to diameter BE, according to Corollary 159 (direct) (§ 22.61, B), during the Harmonic Transformation, A'Γ' and Δ'Z', will be conjugate to B'E'.

This means that these will run through pole Φ of B'E', which is the intersection of the tangents at points B' and E'.

5/. Since quadrilaterals ABΓE and ΔEZB are rhomboid, during the Harmonic Transformation, quadrilaterals A'B'Γ'E' and Δ'E'Z'B' will be harmonic [Corollary 158 (direct) (§ 22.61,A)].

(b). Converse.

1/. According to the converse of Corollary 161, during the Harmonic Transformation, hexagon ΑΒΓΔΕΖ, will be symmetrical with respect to ΒΕ.

2/. Hence, since the hexagon is symmetrical with respect to ΒΕ, ΑΖ and ΓΔ will be symmetrical with respect to ΒΕ. Thus, ΑΖ, ΓΔ, ΒΕ will coincide at point P {Theorem 253 in Big Geometry [28] (issue A, page 132)}.

Similarly we prove that the symmetrical with respect to ΒΕ, strings ΑΔ, ΓΖ and ΒΕ, coincide at point K.

3/. Additionally, since the pairs of points Α, Γ and Δ, Ζ are symmetrical with respect to ΒΕ and the pairs of tangents respectively at the pairs of these points, will be symmetrical lines and thus according to Theorem 253 above these tangents, in pairs and ΒΕ, will coincide at two points α and β respectively.

(This can also be proven as follows:

Since for example triangle αΑΓ is isosceles and ΒΕ a perpendicular bisector to ΑΓ, then ΒΕ will run through. We work similarly to prove that ΒΕ runs through the intersection β).

4/. Due to symmetry, ΑΓ, ΔΖ are perpendicular to the diameter ΒΕ. Thus, ΑΓ, ΔΖ and the tangents at Β, Ε, will be parallel, hence all these are concurrent at the ideal point.

5/. ΑΒΓΕ and ΔΕΖΒ are rhomboids, form their construction, as hexagon ΑΒΓΔΕΖ is symmetrical with respect to ΒΕ as well.

(c). Investigation.

If we also want string ΔΖ or ΑΓ to be the diameter of circle (o), then we should define point Σ as the intersection of circles with diameters βε and δζ or αγ, respectively.

2nd Way.

(a). Direct.

We prove, as in the first way of proof above, that during the Harmonic Transformation, the tangents at Β', Ε' and Α'Γ', Δ'Ζ' coincide at Φ and that Β'Ε' is the polar of Φ.

Hence, since α' and β' are the polars of A'Γ' and Δ'Z' respectively, the intersection of A'Γ' and Δ'Z', that is Φ, has as polar α'β', then since B'E' is the polar of Φ, it will be α'β'≡B'E', or that α', B', E' β', are collinear.
Moreover, since K'P' is the polar of Φ≡ A'Γ'∩Δ'Z' and K'P'≡B'E', or points P', α′, B', K', E', β', are collinear, then P'α'K'E'β' is a line.
From the above it easily occurs that quadrilaterals A'B'Γ'E' and Δ'E'Z'B' are harmonic.

(b). Converse.

The converse is easily proven, if we take into consideration the above mentioned.

Remarks.

(a). The Corollary above and its proof, we believe to be new and firstly appear here.

(b). The above 5th desired feature is the direct, can also be proven in the following way:
It is obvious that, since B'E' is conjugate to A'Γ' and Δ'Z', as we saw above, quadrilaterals A'B'Γ'E' and Δ'E'Z'B' are harmonic.

(c). It is obvious that AΓ, ΔZ coincide, but here at infinity, where the intersection of tangents at B, E of circle (o) is as well and where the pole of BE is.

(d). From figure 103 and Corollary 163, we conclude that, the symmetrical with respect to BE, hexagon ABΓΔEZ is a Harmonic Involution, and that this, during the Harmonic Transformation, is initially transformed to a Harmonic Collinear Involution, of two Pairs of Conjugate Points α, γ and δ, ζ and Double Points β, ε and finally to a Harmonic Cyclic Involution of two Pairs of Conjugate Points A′, Γ′ and Δ′, Z′ and Double Points B′, E′.

(e). Relevant Propositions are 157-162, 164-168.

22.61 G. Rule 7. Properties of an Inscribed in Circle Octagon, Symmetrical with respect to its Diagonal.

7th Corollary of General Construction 157.

Corollary 164 (Figure 104) (Rule 7).

Every convex inscribed in circle (o) octagon ΑΒΓΔΕΖΗΘ, symmetrical for example with respect to its diagonal ΒΖ, is Harmonically transformed to a convex inscribed in the same circle (o) or another circle (o') octagon Α'Β'Γ'Δ'Ε'Ζ'Η'Θ', with the following properties:

1/. Α'Γ', Δ'Θ', Ε'Η', coincide at the intersection Φ of the tangents at points Β', Ζ'.

2/. Intersections Α'Ε'∩Γ'Η'≡Κ', Α'Δ'∩Γ'Θ'≡α, Α'Θ'∩Γ'Δ'≡γ, Α'Η'∩Γ'Ε'≡δ, Δ'Η'∩Ε'Θ'≡α', Δ'Ε'∩Η'Θ'≡γ' and intersections β, β', ε, of the tangents at the pairs of points Α'-Γ', Ε'-Η', Δ'-Θ' respectively, belong to Β'Ζ'.

3/. Quadrilaterals Α'Β'Γ'Ζ', Ε'Ζ'Η'Β', Β'Δ'Ζ'Θ' are Harmonic,

and conversely,

each convex inscribed in circle (o') octagon Α'Β'Γ'Δ'Ε'Ζ'Η'Θ', whose three quadrialterals Α'Β'Γ'Ζ', Ε'Ζ'Η'Β', Β'Δ'Ζ'Θ', are Harmonic, can be Harmonically transformed, to a convex inscribed in circle (o) octagon ΑΒΓΔΕΖΗΘ, with the following properties:

1/. It should be symmetrical with respect to ΒΖ.

2/. ΑΓ, ΔΘ, ΕΗ, should coincide at infinity (they are parallel).

3/. Intersections of the pairs of strings ΑΕ-ΓΗ, ΑΔ-ΓΘ, ΑΘ-ΓΔ, ΑΗ-ΓΕ, ΔΗ-ΕΘ, ΔΕ-ΗΘ as well as the intersections of the tangents at tha pairs of symmetrical points Α-Γ, Ε-Η, Δ-Θ, should belong to ΒΖ.

4/. Quadrilaterals ΑΒΓΖ, ΕΖΗΒ, ΒΔΖΘ, should be symmetrical, with respect to ΒΖ.

Proof .

(a). Direct.

1/. Since octagon ΑΒΓΔΕΖΗΘ is symmetrical with respect to ΒΖ, ΒΖ is the perpendicular bisector to strings ΑΓ, ΔΘ, ΕΗ, therefore parallel to each other and ΒΖ is a diameter of circle (o). Thus, according to the direct of Corollary 162 (§ 22.61,E), these are Harmonically transformed, to three strings Α'Γ', Δ'Θ', Ε'Η' of circle (o'), that are conjugate to strings Β'Ζ', of this circle. This means that Α'Γ', Δ'Θ', Ε'Η' coincide at the pole of Β'Ζ', which is the intersection Φ of the tangents at points Β', Ζ'.

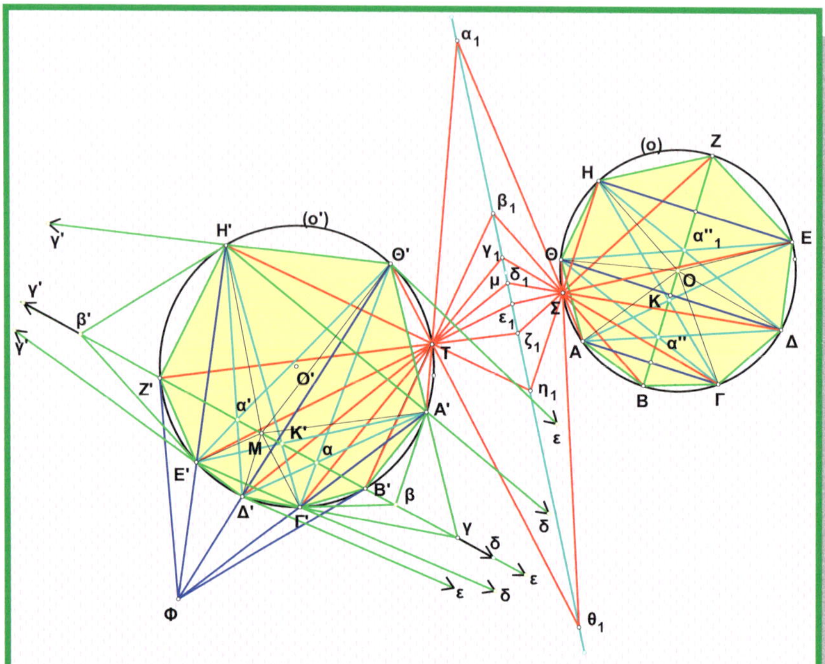

Σχήμα 104.

This can also be proven in the following way:

Since the convex hexagon ΑΒΓΕΖΗ is also symmetrical with respect to ΒΖ, according to Corollary 163 (§ 22.61, F) (direct), this is Harmonically transformed to the convex hexagon A'B'Γ'E'Z'H', in which A'Γ', E'H' and the tangents at points B', Z' coincide at point Φ.

Moreover, since quadrilateral ΒΔΖΘ is symmetrical with respect to ΒΖ (rhomboid), this is Harmonically transformed to B'Δ'Z'Θ' which will be harmonic [Corollary 158 (direct) (§ 22.61,A)]. Hence Δ'Θ', will be conjugate to B'Z', thus Δ'Θ', will also coincide at Φ (pole of B'Z').

2/. Since, as we saw above hexagon A'B'Γ'E'Z'H', is Harmonic, according to Corollary 163 (direct), intersections K', β, β', δ, belong to B'Z'.

Moreover, since hexagon ΑΒΓΔΖΘ is symmetrical with respect to ΒΖ,

according to Corollary 163 (direct), in hexagon A'B'Γ'Δ'Z'Θ', intersections α, γ, ε, β, will belong to B'Z'.
Additionally, similarly we find that for hexagon E'Z'H'Θ'B'Δ' too, intersections α', γ', ε, β', will belong to B'Z'.
Hence, due to the above, it occurs that all intersections K', α, β, γ, δ, ε, α', β', γ', actually belong to B'Z'.
3/. Since quadrilaterals ABΓZ, EZHB, BΔZΘ, are obviously rhomboid, they are Harmonically transformed, to the Harmonic quadrilaterals A'B'Γ'Z', E'Z'H'B', B'Δ'Z'Θ' (Corollary 158 (direct) (§ 22.61,A).

(b). Converse.

1/. Since quadrilaterals A'B'Γ'Z', E'Z'H'B', B'Δ'Z'Θ', are given to be Harmonic A'Γ', E'H', Δ'Θ', they will be conjugate to B'Z', thus these will constitute a pencil and thus according to the converse of Corollary 162 and according to the Method of the Harmonic Transformation and if Σ is received on the circle with diameter $\beta_1\zeta_1$ of (ε), strings AΓ, EH, ΔΘ, will be parallel to each other, while, according to Corollary 159, these will be perpendicular to BZ, which will be a diameter of circle (o). Thus, it easily occurs that Octagon ABΓΔEZHΘ, is in fact symmetrical with respect to BZ.
This can also be proven in the following way:
Hence, since quadrilaterals A'B'Γ'Z', E'Z'H'B' are harmonic, it will be A'E'∩Γ'H' B'Z'≡K' and A'H'∩B'Z'∩Γ'E'≡δ (Criterion 51, converse).
Since hexagons A'B'Γ'E'Z'H', A'B'Γ'H'Z'E', have concurrent diagonals K', δ, (for each hexagon) respectively, the first one of them can be Harmonically transformed to hexagon ABΓEZH, symmetrical with respect to BZ (Corollary 163, converse).
Moreover, since quadrilateral B'Δ'Z'Θ' is Harmonic too, then this can be transformed to quadrilateral BΔZΘ, symmetrical with respect to BZ (rhomboid) [Corollary 158 (converse)].
Therefore, since hexagon ABΓEZH , as well as quadrilateral BΔZB are symmetrical, with respect to BZ, we easily find that symmetrical with respect to BZ, will also be the following Octagon ABΓΔEZHΘ.
2/. Due to the above, since Octagon ABΓΔEZHΘ is symmetrical with respect

to BZ then AΓ, ΔΘ, EH will be perpendicular to BZ and thus parallel to each other, hence they actually coincide at infinity, where the intersection of the tangents at points B and Z, which is the pole of BZ.

3/. Since Octagon ΑΒΓΔΕΖΗΘ is symmetrical with respect to BZ then pairs of strings AE-ΓH, AΔ-ΓΘ, AΘ-ΓΔ, AH-ΓE, ΔH-EΘ, ΔE-HΘ are symmetrical with respect to BZ. Thus their intersections in pairs as well as the intersections of the tangents at the following pairs of points, A-Γ, E-H, Δ-Θ, will belong to BZ (Theorem 253 in volume A of Big Geometry in [28]).

4/. Since, as we saw above, octagon ΑΒΓΔΕΖΗΘ is symmetrical with respect to BZ, then obviously quadrilaterals ΑΒΓΖ, EZHB, ΒΔΖΘ, will also be symmetrical with respect to BZ.

(c). Investigation.

If we want string for example ΔΘ to be a diameter of a circle (o) as well, then we should set point Σ as the intersection of circles with diameters $\beta_1\zeta_1$ and $\delta_1\theta_1$, of (ε).

Conclusion.

We notice that in Harmonic Transformations, the transformed multilaterals (Regular, collinear and cocyclic) have equal number of sides, diagonals and angles, while to each point of one multilateral one point of the other matches (apexes, intersections, etc).

The same applies to line segments and and by extension to lines as well.

Remarks.

(a). The Corollary above and its proof, we believe to be new and firstly appear here.

(b). The 3rd desired fact in the direct above, can also be proven in the following way:

It is obvious that, since B'Z' is conjugate to A'Γ' Δ'Θ', Z'H', as we saw above, quadrilaterals A'B'Γ'Z', E'Z'H'B', B'Δ'Z'Θ' are harmonic.

(c). From figure 104 and Corollary 164, we conclude that, the symmetrical with respect to BZ, octagon ΑΒΓΔΕΖΗΘ is a Harmonic Involution and that this, during the Harmonic Transformation, is transformed to a Harmonic Collinear Involutory series of points α_1, β_1, γ_1, δ_1, ε_1, ζ_1, η_1, θ_1 and a Harmonic Cyclic Involutory series of points A', B', Γ', Δ', E', Z', H', Θ'.

(d). Relevant Propositions are 157-163, 165, 166-168.

22.61 H. Rule 8. Properties of an Inscribed in Circle Decagon, Symmetrical with respect to one of its Diagonals.

8th Corollary of General Construction 157.

Corollary 165 (Figure 105) (Rule 8).

Every convex inscribed in circle (o) decagon ΑΒΓΔΕΖΗΘΙΚ, symmetrical for example with respect to its diagonal ΒΗ, is Harmonically transformed to a convex inscribed in the same circle (o) or another circle (o') decagon Α'Β'Γ'Δ'Ε'Ζ'Η'Θ'Ι'Κ', with the following properties:

1/. Α'Γ', Κ'Δ', Ι'Ε', Θ'Ζ', coincide at the intersection Φ of the tangents at points Β', Η'.

2/. Intersections Α'Ζ'∩Γ'Θ'≡Λ, Δ'Ι'∩Ε'Κ'≡Λ', Δ'Θ'∩Κ'Ζ'≡α, Ε'Θ'∩Ι'Ζ'≡β, Ε'Ζ'∩Ι'Θ'≡δ, Δ'Ζ'∩Κ'Θ'≡ε, Κ'Ι'∩Δ'Ε'≡θ, Α'Ε'∩Γ'Ι'≡α', Α'Δ'∩Γ'Κ'≡β', Α'Κ'∩Γ'Δ'≡δ', Α'Ι'∩Γ'Ε'≡ε', Α'Θ'∩Γ'Ζ'≡η' and intersections γ, ζ, γ', ζ', of the tangents at the pairs of points Θ'- Ζ', Ι'- Ε', Α'- Γ', Κ'- Δ', respectively, belong to Β'Η'.

3/. Quadrilaterals Α'Β'Γ'Η', Ζ'Η'Θ'Β', Β'Δ'Η'Κ', Β'Ε'Η'Ι' are Harmonic,

and conversely,

each convex inscribed in circle (o') decagon Α'Β'Γ'Δ'Ε'Ζ'Η'Θ'Ι'Κ', whose four quadrilaterals Α'Β'Γ'Η', Ζ'Η'Θ'Β', Β'Δ'Η'Κ', Β'Ε'Η'Ι', are Harmonic, can be Harmonically transformed, to a convex inscribed in circle (o') or (o) decagon ΑΒΓΔΕΖΗΘΙΚ, with the following properties:

1/. ΑΓ, ΚΔ, ΙΕ, ΘΖ, should coincide at infinity (they are parallel).

2/. The decagon should be symmetrical with respect to ΒΗ.

3/. Intersections of the pairs of strings ΑΖ-ΓΘ, ΔΙ-ΕΚ, ΔΘ-ΚΖ, ΕΘ-ΙΖ, ΕΖ-ΙΘ, ΔΖ-ΚΘ, ΚΙ-ΔΕ, ΑΕ-ΓΙ, ΑΔ-ΓΚ, ΑΚ-ΓΔ, ΑΙ-ΓΕ, ΑΘ-ΓΖ, as well as the intersections of the tangents at the pairs of symmetrical points Θ- Ζ, Ι- Ε, Α- Γ, Κ- Δ, should belong to ΒΗ.

4/. Quadrilaterals ΑΒΓΗ, ΖΗΘΒ, ΒΔΗΚ, ΒΕΗΙ, should be symmetrical, with respect to ΒΗ.

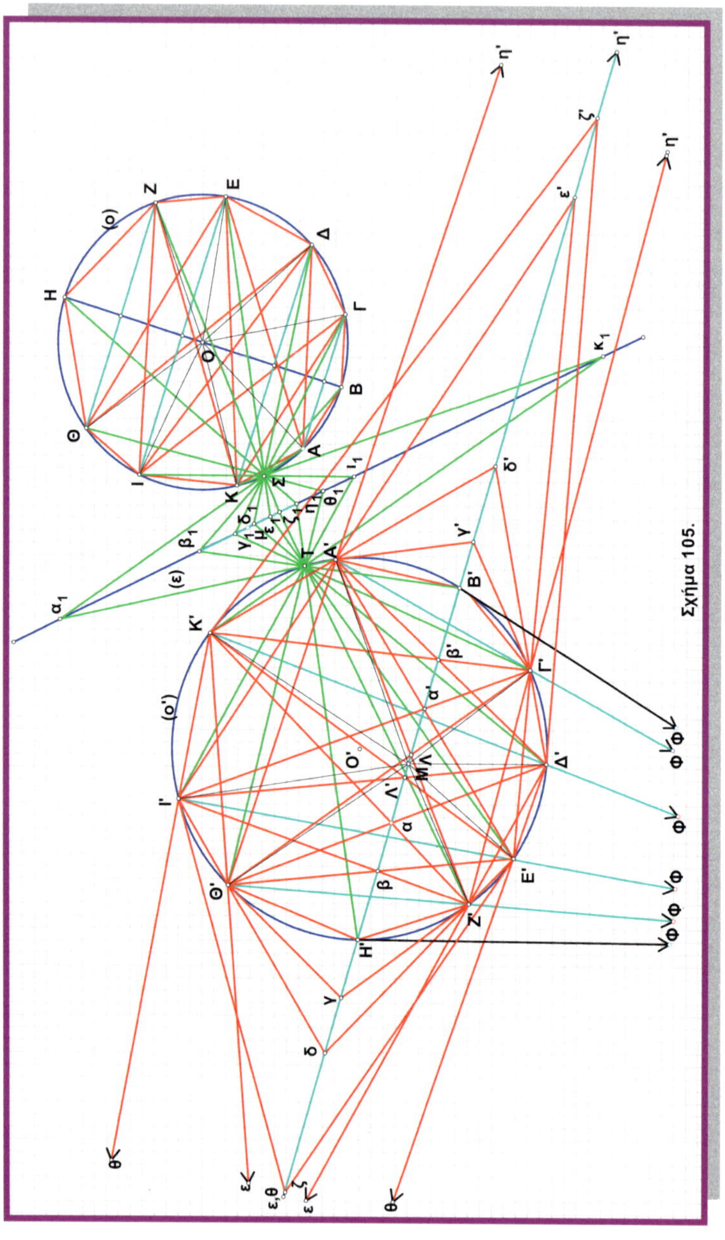

Σχήμα 105.

Proof.

(a). Direct.

1/. Since the four strings ΑΓ, ΚΔ, ΙΕ, ΘΖ, are obviously perpendicular to diameter ΒΗ of circle (o), them and ΒΗ, are harmonically transformed to strings Α'Γ', Κ'Δ', Ι'Ε', Θ'Ζ', conjugate to their shared string Β'Η', of the same or another circle (o') [Corollary 159 (direct), (§ 22.61,B)].

Besides since ΑΓ', Κ'Δ', Ι'Ε', Θ'Ζ', as we saw above, are conjugate to Β'Η', they will coincide at the pole of Β'Η', which is the intersection Φ of the tangents at Β', Η'.

2/. We consider its octagon ΑΒΓΔΕΗΙΚ, which is symmetrical with respect to its diameter ΒΗ. This, according to Corollary 164 (direct) (§ 22.61, G), is Harmonically transformed to octagon Α'Β'Γ'Δ'Ε'Η'Ι'Κ', for which intersections: Λ', α', β', γ' δ', ε', ζ', θ, ζ, (1).

will belong to Β'Η'.

Similarly, for the same reason for octagon Β'Δ'Ε'Ζ'Η'Θ'Ι'Κ', intersections:

Λ', α, β, γ, δ, ε, ζ, θ, ζ', (2).

belong to Β'Η'.

After all, we notice that hexagon ΑΒΓΖΗΘ is also symmetrical, with respect to ΒΗ, hence this is as well Harmonically transformed to hexagon Α'Β'Γ'Ζ'Η'Θ', for which, according to Corollary 163 (direct) (§ 22.61, F) intersections, Λ, γ, γ', η', (3).

will belong to Β'Η', as well.

Hence, from (1), (2), (3), it occurs that all 16 intersections

Λ, Λ', α, β, γ, δ, ε, ζ, α', β', γ' δ', ε', ζ', η', θ, (4).

belong to Β'Η'.

3/. Since, as we saw above Α'Γ', Κ'Δ', Ι'Ε', Θ'Ζ', they are all conjugate to Β'Η', quadrilaterals Α'Β'Γ'Η', Ζ'Η'Θ'Β', Β'Δ'Η'Κ', Β'Ε'Η'Ι', will be harmonic.

(b). Converse.

1/. Since from the implication quadrilaterals Α'Β'Γ'Η', Ζ'Η'Θ'Β', Β'Δ'Η'Κ', Β'Ε'Η'Ι', are harmonic, strings Α'Γ', Κ'Δ', Ι'Ε', Θ'Ζ', will be conjugate to Β'Η', with respect to circle (o'), hence these coincide at point Φ of Β'Η'.

Thus, according to the converse of Corollary 162 (§ 22.61, E), these strings, can be transformed to strings ΑΓ, ΚΔ, ΙΕ, ΘΖ, of the same or another circle (o), parallel to each other, as long as point Σ, lies on the circle with diameter $\beta_1\eta_1$.

2/. Due to the above paragraph 1/, decagon Α'Β'Γ'Δ'Ε'Ζ'Η'Θ'Ι'Κ', is harmonically transformed to decagon ΑΒΓΔΕΖΗΘΙΚ, inscribed in circle (o), whose diameter is ΒΗ and strings ΑΓ, ΚΔ, ΙΕ, ΘΖ, that are parallel to each other and perpendicular to ΒΗ (converse of Corollary 159), hence it easily occurs that decagon ΑΒΓΔΕΖΗΘΙΚ, is symmetrical with respect to ΒΗ.

3/. Thus, since decagon ΑΒΓΔΕΖΗΘΙΚ, is symmetrical with respect to ΒΗ, the following pairs of strings ΑΖ-ΓΘ, ΔΙ-ΕΚ, ΔΘ-ΚΖ, ΕΘ-ΙΖ, ΕΖ-ΙΘ, ΔΖ-ΚΘ, ΚΙ-ΔΕ, ΑΕ-ΓΙ, ΑΔ-ΓΚ, ΑΚ-ΓΔ, ΑΙ-ΓΕ, ΑΘ-ΓΖ, as well as the tangents at pairs of symmetrical points Θ- Ζ, Ι- Ε, Α- Γ, Κ- Δ, will be concurrent at ΒΗ (axis of symmetry) (Theorem 253 in volume A of Big Geometry in [28]).

4/. Since, decagon ΑΒΓΔΕΖΗΘΙΚ is symmetrical with respect to ΒΗ, it easily occurs that quadrilaterals ΑΒΓΗ, ΖΗΘΒ, ΒΔΗΚ, ΒΕΗΙ, will be symmetrical, with respect to ΒΗ.

(c). Investigation.

If we want for example string ΕΙ to be a diameter of circle (o) as well, then we should set a point Σ as the intersection with diameters $\beta_1\eta_1$ and $\iota_1\varepsilon_1$, of (ε).

Conclusions.

1/. We notice that in Harmonic Transformations, the transformed multilaterals (Regular, collinear and cocyclic) have equal number of sides, diagonals and angles, while to each point of one multilateral one point of the other matches (apexes, intersections, etc).

2/. Due to the above Corollary 165 and with a simple application of it, we can prove rather easily Proposition 201 (§29.2) as well (we will mention the proof later).

The same can also happen for hexagons, octagons, etc, with a simple application of Corollaries 163 (§ 22.61, F), 164 (§ 22.61, G), respectively, etc.

3/. Due to Corollaries 158, 163-165, above, the Generalized Corollary 166 (Rule 9) below occurs easily.

Remarks.

(a). The Corollary above and its proof, we believe to be new and firstly appear here.

(b). From figure 105 and paragraphs 2 and 8 in the present book, we conclude that, the symmetrical with respect to ΒΗ, octagon ΑΒΓΔΕΖΗΘΙΚ is a Harmonic Involution and that this, during the Harmonic Transformation, is transformed to a Harmonic Collinear Involutory Series of Points α_1, β_1, γ_1, δ_1, ε_1, ζ_1, η, θ_1, ι_1, κ_1 and a Harmonic Cyclic Involutory Series of Points Α', Β', Γ', Δ', Ε', Ζ', Η', Θ', Ι', Κ'.

(c). Relevant Propositions are 157-164, 166. 167, 168.

22.61 J. General Rule 9. Properties of the Inscribed in Circle 2n-lateral, Symmetrical with respect to one of its Diagonals.

9th General Corollary of General Construction 157.

(It is the Generalization of Corollaries 158, 163-165).

Corollary 166 (Rule 9).

Each convex inscribed in circle (o) 2n-lateral, symmetrical with respect to one of its diagonals, is Harmonically transformed to a convex inscribed in circle (o) or (o') 2n-lateral, which has analogous properties to the ones of quadrilaterals, hexagons, octagons, decagons, (Corollaries 158, 163-165), and conversely.

Proof.

The proof of this can happen in analogous way, to the one we used for Corollaries 158, 163-165.

22.61 I. Rule 10. Properties of Rectangles Inscribed in Circle and with Common Axes of Symmetry.

10th Corollary of Construction 157.

Corollary 167 (Figure 106) (Rule 10).

Two rectangles ΑΓΕΗ and αγεη or even more (rectangles), inscribed in the same circle (o) and which have shared axes of symmetry ΒΖ, ΔΘ, are Harmonically Transformed, to two convex quadrilaterals Α'Γ'Ε'Η' and α'γ'ε'η' respectively, inscribed in the same circle (o) or another circle (o'), that have the following properties:

1/, The two groups of four of their opposite sides Α'Γ', Ε'Η', α'γ', ε'η' and Α'Η', Γ'Ε', α'η', γ'ε' coincide, the first (group of four) at point Φ and the second at point Ρ. Similarly strings Δ'Θ', Β'Ζ' coincide at Φ, Ρ, respectively that are the shared axes of symmetry of the rectangles ΔΘ, ΒΖ, Harmonically transformed.

2/. Their diagonals Α'Ε', Γ'Η', α'ε', γ'η', coincide at the point of intersection Κ of the two strings Β'Ζ', Δ'Θ',

and conversely,

two quadrilaterals Α'Γ'Ε'Η' and α'γ'ε'η' or more (quadrilaterals), inscribed in the same circle (o'), whose diagonals coincide at point Κ and thw two groups of four of their opposite sides Α'Γ', Ε'Η', α'γ', ε'η' and Α'Η', Γ'Ε', α'η', γ'ε' are concurrent, the first (group of four) at point Φ and the second at point Ρ, can be Harmonically Transformed, to two rectangles ΑΓΕΗ and αγεη, inscribed in the same circle (o), that share their two axes of symmetry ΒΖ, ΔΘ, wich run through the center Ο of the circle (o) and which have occrued through the Harmonic Transformation, respectively (It is easy to realize that the Harmonic Transformation, can be carried through with on the same circle).

Proof.

(a). Direct.

1/. We apply, in the known way [Proposition 157 (§ 22.6)] the Method of Harmonic Transformation.

It is easily occurs that the shared axes ΒΖ, ΔΘ of these rectangles, that are

also diameters of circle (o), are perpendicular to each other. Hence, they will be Harmonically transformed to axes B'Z', Δ'Θ', which will be conjugate

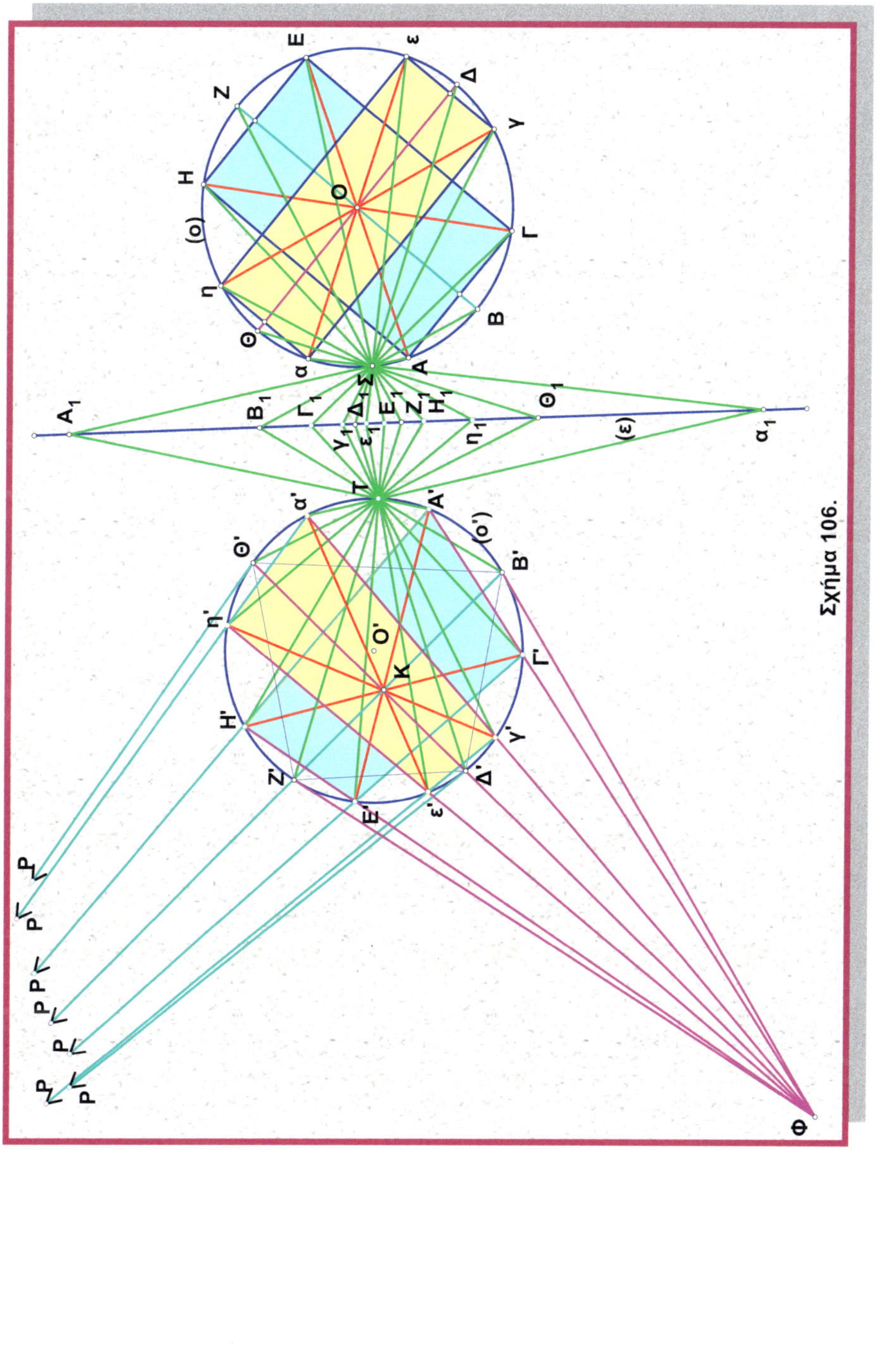

Σχήμα 106.

(Corollary 159 (direct) (§ 22.61, B)] and therefore they will coincide at poles P, Φ respectively, of Δ'Θ', B'Z', (P, Φ are intersections of the tangents at Δ', Θ' andı B', Z', respectively).

The exact same thing happens with the two groups of four of their parallel sides AH, ΓE, αη, γε and AΓ, EH, αγ, εη, which are, each group, perpendicular to diameters ΔΘ, BZ respectively, and which are Harmonically transformed to the two groups of four strings A'H', Γ'E', α'η', γ'ε' and A'Γ', E'H', α'γ', ε'η', respectively from which the first group of four coincides at P and the second at Φ [Corollary 162, direct (§ 22.61, E)].

2/. It easily occurs that diagonals AE, ΓH as well, of the rectangle AΓEH, are symmetrical with respect to the two axes ΔΘ and BZ, thus they will be Harmonically Transformed to A'E', Γ'H' of quadrilateral A'Γ'E'H', whose intersection will belong to both strings Δ'Θ' and B'Z' [Corollary 161 (§ 22.61,D)). This means that the intersection of A'E', Γ'H' coincides with the intersection K of B'Z', Δ'Θ' or that it is: $A'E' \cap \Gamma'H' \cap B'Z' \cap \Delta'\Theta' \equiv K$. (1).

In the exact same way we find that it is as well:

$$\alpha'\varepsilon' \cap \gamma'\eta' \cap B'Z' \cap \Delta'\Theta' \equiv K. \quad (2).$$

Consequently from (1), (2), it occurs that:

$$A'E' \cap \Gamma'H' \cap \alpha'\varepsilon' \cap \gamma'\eta' \cap B'Z' \cap \Delta'\Theta' \equiv K. \quad (3).$$

(b). Converse.

We prolong KΦ and KP which intersect circle (o') at Δ', Θ' and B', Z', respectively (B', Δ' between pairs A', Γ' and γ', ε', respectively).

We notice that what we are asked for can be true, if during the Harmonic Transformation, we draw a circle (o) that runs through point Σ, which has occurred as the intersection of circles with diameters B_1Z_1 and $\Delta_1\Theta_1$ (B_1Z_1 and $\Delta_1\Theta_1$ will have occurred obviously from B'Z', Δ'Θ').

Hence then, BZ, ΔΘ that will occur, will be diameters of circle (o) and perpendicular to each other, as B'Z', Δ'Θ' are conjugate with respect to circle (o') [Corollary 159 (converse)].

Besides, since A'Γ', Δ'Θ', E'H', α'γ', ε'η', are conjugate to B'Z', then AΓ, ΔΘ, EH, αγ, εη will be perpendicular to BZ, thus parallel to each other [Corollary 162 (converse)].

In the exact same way it also occurs that AH, BZ, ΓE, αη, γε will be parallel to each other.

So, we see that quadrilaterals ΑΓΕΗ, αγεη, have their opposite sides parallel. This means that these are parallelograms and they are also inscribed in circle (o), they will be rectangle, hence their diagonals AE, ΓΗ, αε, γη, will run through O. But, since as we saw above BZ, ΔΘ will run through O, it easily occurs that the pairs of sides AH, ΓE and αη, γε are symmetrical with respect to BZ, while pairs ΑΓ, EH and αγ, εη are symmetrical with respect to ΔΘ.

Remarks.

(a). This Corollary is significant, as based on it an analytical proof of Proposition 106 (§ 19.13) occurs, about the concurrent diagonals of the Harmonic Decagon, which we will give later (Proposition 200).

(b). The Corollary above and its proof, we believe to be new and firstly appear here.

(c). From figure 106 and paragraphs 2.1 , 6.5, 6.24, 8.1 and 8.65, we conclude that, the symmetrical with respect to BZ and ΔΘ, rectangles ΑΓΕΗ and αγεη constitute a Double Harmonic Involutionand that these, during the Harmonic Transformation, are transformed to a Double Harmonic Collinear Involutory Series of Points A_1, Γ_1, E_1, H_1, α_1, γ_1, ε_1, η_1, with double points, the following two pairs of points B_1, Z_1- Δ_1, Θ_1, and a Double Harmonic Cyclic Involutory Series of Points A', Γ', E', H', α', γ', ε', η', with double points the following two pairs of points B', Z'- Δ', Θ'.

(d). It is obvious that decagon ΑΒΓγΔεΕΖΗηΘα is symmetrical with respect to its two perpendicular diagonals BZ, ΔΘ (main).

(e). Relevant Propositions are 157-166.

22.61 K. Rule 11. Invariant the Relations (Equalities), during the Harmonic Transformation.

11th Corollary of General Construction 157.

Corollary 168 (Rule 11).

During the Harmonic Transformation, all relations (equalities, etc), that are true for Harmonic Involutions (collinear and cocyclic), for Harmonic Multilaterals (collinear and cocyclic), as well as for Harmonic Lines (collinear and cocyclic), are transformed Invariant.

Proof.

1. For Harmonic Involutions (Collinear and cocyclic).

(a). Harmonic Involutions of n Pairs of Conjugate Points

1/. According to relevant definitions (§ 2.1 and 8.1).

We will only refer to Harmonic Involutions of one and four Pairs of Conjugate Points, as in an equivalent way it is proved that the same applies to n Pairs of conjugate points.

a/. Of one pair (Figure 100).

It is (ΑΓΒΔ)=(αγβδ)=(Α′Γ′Β′Δ′)=1. (1).

From relation (1), comes the invariant of this case.

The fact that (ΑΓΒΔ)=1, is obvious as ΑΒ=ΒΓ and ΑΔ=ΔΓ. The remaining two relations are true due to the relevant definitions § 2.11 and 8.11, respectively.

β/. Of four pairs (Figure 102).

It is: (ΒΗΑΓ)= (ΒΗΔΚ)=(ΒΗΕΙ)=(ΒΗΖΘ)=1. (2).

(βηαγ)= (βηδκ)=(βηει)=(βηζθ)=1. (3).

(Β′Η′Α′Γ′)= (Β′Η′Δ′Κ′)=(Β′Η′Ε′Ι′)=(Β′Η′Ζ′Θ′)=1. (4).

From relations (2)-(4), comes the invariant of this case.

Relations (2) are obvious, as ΒΑ=ΒΓ and ΗΑ=ΗΓ, etc, relations (3) and (4) are true due to the relevant definitions § 2.11 and 8.11, respectively.

[It will obviously be: Σ(βηαγ)= Σ(βηδκ)=Σ(βηει)=Σ(βηζθ)=1. (3′)].

2/. According to Newton's Theorem Criterion 33 and § 8.61 - 8.65, as well as the relevant definitions (§ 2.1 and 8.1).,

We will only refer to Harmonic Involutions of one and four Pairs of Conjugate Points, as in an equivalent way it is proved that the same applies to n Pairs of conjugate points.

a/. Of one pair (Figure 100).

It is: $OB^2=O\Delta^2=OA.O\Gamma$, $\mu\beta^2=\mu\delta^2=\mu\alpha.\mu\gamma$, $MB'^2=M\Delta'^2=MA'.M\Gamma'$. (5).

(Where μ the mid point of βδ and M the midpoint of B′Δ′).

From the three equalities in (5), comes the invariant of this case (Hence we will have here the invariant of Newton's Theorem).

The first equality in (5) is obvious, as OA=OB=OΓ=OΔ=R, the second is true due to the known Newton's Theorem and the third due to the extension of it (Criterion 33).

b/. Of four pairs (Figure 102).

It is: $OB^2=OH^2=OA.O\Gamma=O\Delta.OK=OE.OI=OZ.O\Theta$. (6).

$\mu\beta^2=\mu\eta^2=\mu\alpha.\mu\gamma=\mu\delta.\mu\kappa=\mu\varepsilon.\mu\iota=\mu\zeta.\mu\theta$. (7).

$MB'^2=MH'^2=MA'.M\Gamma'=M\Delta'.MK'=ME'.MI'=MZ'.M\Theta'$ (8).

(Where μ the midpoint of βη and M the midpoint of B′H′).

From relations (6)-(8), comes the invariant of this case.

Relations (6) are obvious, as OA=OB=OΓ=...,=OI=OK=R, etc, relation (7) is proven if we apply Newton's Theorem (§ 6.5 remarks) and (8) if we work as in § 8.64 [relation (3)].

3/. According to Criteria 4 and 37 (Figure 107).

We will only refer to Harmonic Involutions of two Pairs of Conjugate Points, as only then Corollary 168 is true.

It is: $$\frac{AB}{B\Gamma}\cdot\frac{\Gamma\Delta}{\Delta E}\cdot\frac{E\Gamma}{\Gamma A}=1. \quad (9).$$

$$\frac{\alpha\beta}{\beta\gamma}\cdot\frac{\gamma\delta}{\delta\varepsilon}\cdot\frac{\varepsilon\gamma}{\gamma\alpha}=1. \quad (10).$$

$$\frac{A'B'}{B'\Gamma'}\cdot\frac{\Gamma'\Delta'}{\Delta'E'}\cdot\frac{E'\Gamma'}{\Gamma'A'}=1. \quad (11).$$

From relations (9)-(11), comes the invariant of this case.

Relation (9) is obvious, as it is AB=ΔE, BΓ=ΓΔ, AΓ=ΓE, relation (10) is true due to Criterion 4 [relation (1)] and (11) if we take into consideration Criterion 37 [relation (2′)].

[Since angle γΣζ is right and in triangle Σαε, Σβ and Σδ are isogonal, as it is known, it will also be: $\frac{\alpha\beta}{\beta\gamma}\cdot\frac{\gamma\delta}{\delta\varepsilon}\cdot\frac{\varepsilon\Sigma}{\Sigma\alpha}=1$ (10′)].

4/. According to Criteria 5 and 38 (Figure 107).

We will only refer to Harmonic Involutions of two Pairs of Conjugate Points, as only then Corollary 168 is true.

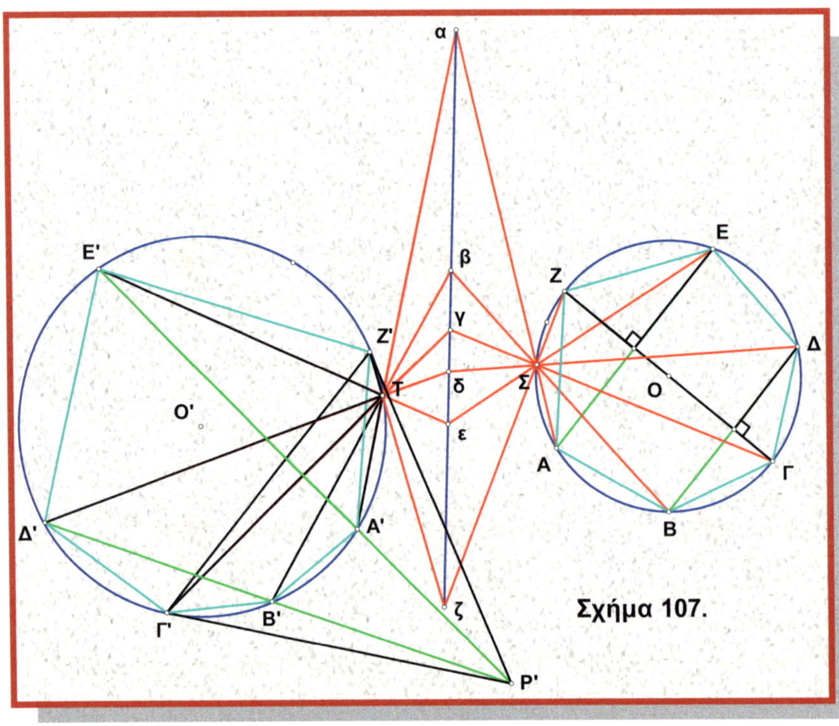

Σχήμα 107.

It is: $$\frac{A\Gamma^2}{\Gamma E^2}=\frac{AZ^2}{ZE^2}=\frac{AB}{BE}\cdot\frac{A\Delta}{\Delta E}. \qquad (12).$$

$$\frac{\alpha\gamma^2}{\gamma\varepsilon^2}=\frac{\alpha\zeta^2}{\zeta\varepsilon^2}=\frac{\alpha\beta}{\beta\varepsilon}\cdot\frac{\alpha\delta}{\delta\varepsilon}. \qquad (13).$$

$$\frac{A'\Gamma'^2}{\Gamma'E'^2}=\frac{A'Z'^2}{Z'E'^2}=\frac{A'B'}{B'E'}\cdot\frac{A'\Delta'}{\Delta'E'}. \qquad (14).$$

From relations (12)-(14), comes the invariant of this case.

Relation (12) is obvious, as ΑΒ=ΔΕ, ΑΔ=ΒΕ, ΑΓ=ΓΕ, ΑΖ=ΖΕ, relation (13) is true according to Criterion 5 [relation (1)] and (14) if we take into consideration Criterion 38 [relations (1) and (1′).

[Since angle γΣζ is right and in triangle Σαε, Σβ and Σδ are isogonal, as it is known (Theorem of isogonals),it will also be:

$$\frac{\Sigma\alpha^2}{\Sigma\varepsilon^2}=\frac{\alpha\gamma^2}{\gamma\varepsilon^2}=\frac{\alpha\zeta^2}{\zeta\varepsilon^2}=\frac{\alpha\beta}{\beta\varepsilon}\cdot\frac{\alpha\delta}{\delta\varepsilon}. \qquad (13')].$$

5/. According to Propositions 8 and 41 (Figures 100 and 102).

We will only refer to Harmonic Involutions of one and four Pairs of Conjugate Points, as in an equivalent way it is proved that the same applies to n Pairs of conjugate points.

a/. Of one pair (Figure 100).

The above relation (1), can also be written in the following way:

$$\frac{AB}{B\Gamma}\cdot\frac{\Gamma\Delta}{\Delta A}=1,\ \frac{\alpha\beta}{\beta\gamma}\cdot\frac{\gamma\delta}{\delta\alpha}=1,\ \frac{A'B'}{B'\Gamma'}\cdot\frac{\Gamma'\Delta'}{\Delta'A'}=1, \qquad (15).$$

From relation (15), comes the invariant of this case.

The fact that $\frac{AB}{B\Gamma}\cdot\frac{\Gamma\Delta}{\Delta A}=1$, is obvious as ΑΒ=ΒΓ and ΑΔ=ΔΓ (The remaining two equalities, are known to us).

b/. Of four pairs (Figure 102).

It is:

$$\frac{AB}{B\Gamma}\cdot\frac{\Gamma\Delta}{\Delta E}\cdot\frac{EZ}{ZH}\cdot\frac{H\Theta}{\Theta I}\cdot\frac{IK}{KA}=1. \qquad (16).$$

$$\frac{\alpha\beta}{\beta\gamma}\cdot\frac{\gamma\delta}{\delta\varepsilon}\cdot\frac{\varepsilon\zeta}{\zeta\eta}\cdot\frac{\eta\theta}{\theta\iota}\cdot\frac{\iota\kappa}{\kappa\alpha}=1. \qquad (17).$$

$$\frac{A'B'}{B'\Gamma'}\cdot\frac{\Gamma'\Delta'}{\Delta'E'}\cdot\frac{E'Z'}{Z'H'}\cdot\frac{H'\Theta'}{\Theta'I'}\cdot\frac{I'K'}{K'A'}=1. \qquad (18).$$

From relations (16)-(18), comes the invariant of this case.

Relation (16) is obvious, as ΒΑ=ΒΓ, ΓΔ=ΚΑ, etc, relation (17) and (18) are true due to Propositions 8 and 41, respectively.

2. For Harmonic Quadrilaterals (Collinear and cocyclic),.

(a). Harmonic 2n-laterals

1/. According to the relevant definitions (§ 11.1a, 16.1a and § 11.1d, 16.1d).

We will only refer to Harmonic Quadrilaterals and Decagon, as in an equivalent way it is proven that the same applies to 2n-laterals.

a/. Quadrilateral (Figure 97).

It is (ΑΓΒΔ)=(αγβδ)=(Α′Γ′Β′Δ′)=1. (1).

From relation (1), comes the invariant of this case.

The fact that (ΑΓΒΔ)=1, is obvious as ΑΒ=ΒΓ=ΓΔ=ΔΑ (The rest second and third, are known to us, § 11.1a and 16.1a, respectively).

b/. Decagon (Figure 99).

It is: (ΑΓΒΗ)= (ΒΔΓΘ)=(ΓΕΔΙ)=...=(ΚΒΑΖ)=1. (2).

(αγβη)= (βδγθ)=(γεδι)=...=(κβαζ)=1. (3).

(Α′Γ′Β′Η′)= (Β′Δ′Γ′Θ′)=(Γ′Ε′Δ′Ι′)=...=(Κ′Β′Α′Ζ′)=1. (4).

From relations (1)-(3), comes the invariant of this case.

Relation (2) is obvious as ΑΒ=ΒΓ=ΓΔ=...και ΑΗ=ΒΘ=ΓΙ=... , relations (3) and (4) are true due to the relevant definitions § 11.1d and 16.1d, respectively.

[It will obviously be: Σ(αγβη)= Σ(βδγθ)=Σ(γεδι)=...=Σ(κβαζ)=1. (3′)].

2/. According to Criteria 86 and 121 (Figures 97 and 99).

We will only refer to Harmonic Quadrilaterals and Decagon, as in an equivalent way it is proven that the same applies to 2n-laterals.

a/. Quadrilateral (Figure 97).

The above relation (1), is also written in the following way:

$$\frac{ΑΒ}{ΒΓ}\cdot\frac{ΓΔ}{ΔΑ}=1,\ \frac{αβ}{βγ}\cdot\frac{γδ}{δα}=1,\ \frac{Α'Β'}{Β'Γ'}\cdot\frac{Γ'Δ'}{Δ'Α'}=1, \qquad (4').$$

From relation (4′), comes the invariant of this case.

The fact that $\frac{ΑΒ}{ΒΓ}\cdot\frac{ΓΔ}{ΔΑ}=1$, is obvious as ΑΒ=ΒΓ=ΓΔ=ΔΑ, the second and third parts, are known to us.

b/. Decagon (Figure 99).

It is:

$$\frac{ΑΒ}{ΒΓ}\cdot\frac{ΓΔ}{ΔΕ}\cdot\frac{ΕΖ}{ΖΗ}\cdot\frac{ΗΘ}{ΘΙ}\cdot\frac{ΙΚ}{ΚΑ}=1, \qquad (5).$$

$$\frac{αβ}{βγ}\cdot\frac{γδ}{δε}\cdot\frac{εζ}{ζη}\cdot\frac{ηθ}{θι}\cdot\frac{ικ}{κα}=1, \qquad (6).$$

$$\frac{Α'Β'}{Β'Γ'}\cdot\frac{Γ'Δ'}{Δ'Ε'}\cdot\frac{Ε'Ζ'}{Ζ'Η'}\cdot\frac{Η'Θ'}{Θ'Ι'}\cdot\frac{Ι'Κ'}{Κ'Α'}=1. \qquad (7).$$

From relations (5)-(7), comes the invariant of this case. Relation (5) is obvious, as ΑΒ=ΒΓ=ΓΔ=..., relation (6) and (7) are true due to Propositions 86 and 121, respectively.

3/. According to Propositions 87, 88 and 144 (Figures 97 and 98).

We will only refer to Harmonic Quadrilaterals and Decagon, as in an equivalent way it is proven that the same applies to 2n-laterals.

a/. Quadrilateral (Figure 97).

It is: $ΑΒ.ΓΔ=ΒΓ.ΔΑ=\frac{ΑΓ.ΒΔ}{2}$,

$$αβ.γδ=βγ.δα=\frac{αγ.βδ}{2},\ \ Α'Β'.Γ'Δ'=Β'Γ'.Δ'Α'=\frac{Α'Γ'.Β'Δ'}{2}. \quad (8).$$

From relation (8), comes the invariant of this case.

The second relation in (8) is true due to Proposition 87, while the first and third one are true due to the known Proposition in [32] page 53 IV, as quadrilaterals ΑΒΓΔ and Α′Β′Γ′Δ′, are harmonic (Proposition 151).

b/. Hexagon (Figure 98).

It is:
$$ΑΒ.ΓΔ.ΕΖ=ΒΓ.ΔΕ.ΖΑ=\frac{ΑΔ.ΒΕ.ΓΖ}{8}, \quad (9).$$

$$αβ.γδ.εζ=βγ.δε.ζα=\frac{αδ.βε.γζ}{8}, \quad (10).$$

$$Α'Β'.Γ'Δ'.Ε'Ζ'=Β'Γ'.Δ'Ε'.Ζ'Α'=\frac{Α'Δ'.Β'Ε'.Γ'Ζ'}{8}. \quad (11).$$

From relations (9)-(11), comes the invariant of this case.

Relation (10) is true due to Proposition 88, while (9) and (10) are true due to Proposition 144, as hexagons ΑΒΓΔΕΖ and Α′Β′Γ′Δ′Ε′Ζ′ are harmonic (Proposition 151).

4/. According to Criterion 114.

We will only refer to Harmonic Quadrilaterals and Decagon, as in an equivalent way it is proven that the same applies to 2n-laterals.

a/. Quadrilateral (Figure 97). It is: $\frac{α'}{α}=\frac{β'}{β}=\frac{γ'}{γ}=\frac{δ'}{δ}$, (12).

$$\frac{\alpha''}{\alpha_1}=\frac{\beta''}{\beta_1}=\frac{\gamma''}{\gamma_1}=\frac{\delta''}{\delta_1}, \qquad (13).$$

(Where α′, β′, γ′, δ′, is the distance of O, from each side AB=α, BΓ=β, ΓΔ=γ, ΔA=δ, respectively and α″, β″, γ″, δ″, the distance of intersection K of A′Γ′ and B′Δ′, from each side $A'B'=\alpha_1$, $B'\Gamma'=\beta_1$, $\Gamma'\Delta'=\gamma_1$, $\Delta'A'=\delta_1$, respectively).

From relations (12) and (13), comes the invariant of this case.

It is obvious that relation (12), is true as it is α′=β′=γ′=δ′ and α=β=γ=δ, while relation (13) is true due the Proposition in page 54 IV in [32].

b/. Decagon (Figure 99).

It is:

$$\frac{\alpha'}{\alpha}=\frac{\beta'}{\beta}=\frac{\gamma'}{\gamma}=\ldots=\frac{\kappa'}{\kappa}, \qquad (14).$$

$$\frac{\alpha''}{\alpha_1}=\frac{\beta''}{\beta_1}=\frac{\gamma''}{\gamma_1}=\ldots=\frac{\kappa''}{\kappa_1}. \qquad (15).$$

(Where α′, β′, γ′, ...,κ′, is the distance of O, from each side AB=α, BΓ=β, ΓΔ=γ, ..., KA=κ, respectively and α″, β″, γ″, ..., κ″, the distance of intersection Λ of A′Z′, B′H′ , ... from each side $A'B'=\alpha_1$, $B'\Gamma'=\beta_1$, $\Gamma'\Delta'=\gamma_1$, ..., $K'A'=\kappa_1$, respectively).

From relations (14) and (15), comes the invariant of this case.

It is obvious that relation (14) is true, as it is α′=β′=γ′=...=κ′ and α=β=γ=...=κ, while relation (15) is true due to Criterion 114.

5/. According to Criteria 122 and 125.

We will only refer to Harmonic Quadrilaterals and Decagon, as in an equivalent way it is proven that the same applies to 2n-laterals.

a/. Quadrilateral (Figure 97).

It is:

$$\frac{\alpha'}{\beta'}\cdot\frac{\gamma'}{\delta'}=1, \qquad (16).$$

$$\frac{\alpha''}{\beta''}\cdot\frac{\gamma''}{\delta''}=1. \qquad (17).$$

(Where α′, β′, γ′, δ′, is the distance of O, from each side AB=α, BΓ=β, ΓΔ=γ, ΔA=δ, respectively and α″, β″, γ″, δ″, the distance of intersection K of A′Γ′

and $B'\Delta'$, from each side $A'B'=\alpha_1$, $B'\Gamma'=\beta_1$, $\Gamma'\Delta'=\gamma_1$, $\Delta'A'=\delta_1$, respectively).
From relation (16) and (17), comes the invariant of this case.
It is obvious that relation (16) is true, as it is $\alpha'=\beta'=\gamma'=\delta'$, while relation (17) is true due to Criterion 122.

b/. Decagon (Figure 99).

It is: $$\frac{\alpha'}{\beta'}\cdot\frac{\gamma'}{\delta'}\cdot\frac{\varepsilon'}{\zeta'}\cdot\frac{\eta'}{\theta'}\cdot\frac{\iota'}{\kappa'}=1, \quad (18).$$

$$\frac{\alpha''}{\beta''}\cdot\frac{\gamma''}{\delta''}\cdot\frac{\varepsilon''}{\zeta''}\cdot\frac{\eta''}{\theta''}\cdot\frac{\iota''}{\kappa''}=1. \quad (19).$$

(Where α', β', γ', ...,κ', is the distance of O, from each side $AB=\alpha$, $B\Gamma=\beta$, $\Gamma\Delta=\gamma$, ..., $KA=\kappa$, respectively and α'', β'', γ'', ..., κ'', the distance of intersection Λ of $A'Z'$, $B'H'$, ... from each side $A'B'=\alpha_1$, $B'\Gamma'=\beta_1$, $\Gamma'\Delta'=\gamma_1$, ..., $K'A'=\kappa_1$, respectively).
From relation (18) and (19), comes the invariant of this case.
It is obvious that relation (18) is true, as it is $\alpha'=\beta'=\gamma'=...=\kappa'$, while relation (19) is true due to Criterion 125.

3. For Harmonic Lines (Collinear and Cocyclic).

(a). According to Relevant Definitions (§ 23.1, 23.2 and 23.3) (Figure 111).

It is:

$(A\Gamma BB') =(B\Delta\Gamma\Gamma')=(\Gamma E\Delta\Delta')=...= (A'\Gamma'B'B) =(B'\Delta'\Gamma'\Gamma)=(\Gamma'E'\Delta'\Delta)=...=1.$ (1).

$(\alpha\gamma\beta\beta') =(\beta\delta\gamma\gamma')=(\gamma\varepsilon\delta\delta')=...=(\alpha'\gamma'\beta'\beta) =(\beta'\delta'\gamma'\gamma)=(\gamma'\varepsilon'\delta'\delta)=...=1.$ (2).

$(A_1\Gamma_1B_1B'_1)=(B_1\Delta_1\Gamma_1\Gamma'_1)=(\Gamma_1E_1\Delta_1\Delta'_1)=...$
$=(A'_1\Gamma'_1B'_1B_1)=(B'_1\Delta'_1\Gamma'_1\Gamma_1)=(\Gamma'_1E'_1\Delta'_1\Delta_1)....=1.$ (3).

From relations (1) - (3), comes the invariant of this case.
Relation (1) above it obvious, as it is for example $AB=B\Gamma$ and $AB'=B'\Gamma$, etc (definition § 23.3).
Relations (2) and (3) are true due to the relevant definitions § 23.1 and § 23.2, respectively.

(b). According to Criterion 178 (Figure 111).

If π, ρ, σ, τ, ..., is the distance of O from each one of the following $AB=\alpha$,

$B\Gamma=\beta$, $\Gamma\Delta=\gamma$, $\Delta E=\delta$, ..., if π', ρ', σ', τ', ... is the distance of O from each one of the following $A'B'=\alpha'$, $B'\Gamma'=\beta'$, $\Gamma'\Delta'=\gamma'$, $\Delta'E'=\delta'$,... and if π_1, ρ_1, σ_1, τ_1,..., is the distance of Λ from each one of the following $A_1B_1=\alpha_1$, $B_1\Gamma_1=\beta_1$, $\Gamma_1\Delta_1=\gamma_1$, $\Delta_1E_1=\delta_1$..., αν π'_1, ρ'_1, σ'_1, τ'_1,..., is the distance of Λ from each one of the following $A'_1B'_1=\alpha'_1$, $B'_1\Gamma'_1=\beta'_1$, $\Gamma'_1\Delta'_1=\gamma'_1$, $\Delta'_1E'_1=\delta'_1$..., respectively, then it will be:

$$\frac{\pi}{\alpha}=\frac{\rho}{\beta}=\frac{\sigma}{\gamma}=\frac{\tau}{\delta}=...=\frac{\pi'}{\alpha'}=\frac{\rho'}{\beta'}=\frac{\sigma'}{\gamma'}=\frac{\tau'}{\delta'}=.... \quad (4).$$

$$\frac{\pi_1}{\alpha_1}=\frac{\rho_1}{\beta_1}=\frac{\sigma_1}{\gamma_1}=\frac{\tau_1}{\delta_1}=...=\frac{\pi'_1}{\alpha'_1}=\frac{\rho'_1}{\beta'_1}=\frac{\sigma'_1}{\gamma'_1}=\frac{\tau'_1}{\delta'_1}=.... \quad (5).$$

From relations (4) and (5), comes the invariant of this case.
It is obvious that relation (4) is true, as it is $\pi=\rho=\sigma=\tau=$... $\pi'=\rho'=\sigma'=\tau'=$ ... and $\alpha=\beta=\gamma=\delta=$... $\alpha'=\beta'=\gamma'=\delta'=$ ..., while (5) is true due to Criterion 178.

Remarks.

(a). It is obvious that during the Harmonic Transformation, the circumscribed circle of the Regular or Harmonic Multilateral stays invariant as well (Except of course of the size of the circle, which we choose at will).

(b). Relevant Propositions are 4, 5, 8, 33, 37, 38, 41, 86-89, 114, 121, 122, 125, 144, 151, 178.

H' PAIR OF HARMONIC COLLINEAR AND COCYCLIC LINES.

23. In General.

It has to do with Pairs of Open Harmonic Collinear and Cocyclic Series of Points, n points of each series of points, or Pairs of Polygonal Harmonic Lines, which have some harmonic properties, to which we will refer below.

We could not find features, regarding the Harmonic Lines, in bibliography, even though we asked for relevant information at the mathematica.gr website, no such information has been found until today (except for Harmonic Groups of Four Points and Harmonic Quadrilaterals, that we have forementioned as known to us).

So, at first we approached the matter according to our own view and intuition, while later on we developed and used relevant methods

Through the passing of time and after research and study, we realized that the Pairs of Open Collinear Groups of n, is an extreme case of Pairs of Cocyclic Harmonic Groups of n, during which their circumscribed circle is degenerate to a line. Hence the circle to which they belong has a radius of infinite length (With zero curvature obviously).

23.1. Pairs of Open Harmonic Collinear Series of Points of 2n Points.

(Harmonic Polygonal Line degenerate to a line).

Definitions.

We consider, on line (ε), the group of n ordered points A, B, Γ, Δ, E,...,N (Figure 108) and consequently, the group of n ordered points A', B', Γ', Δ', E',...,N'.

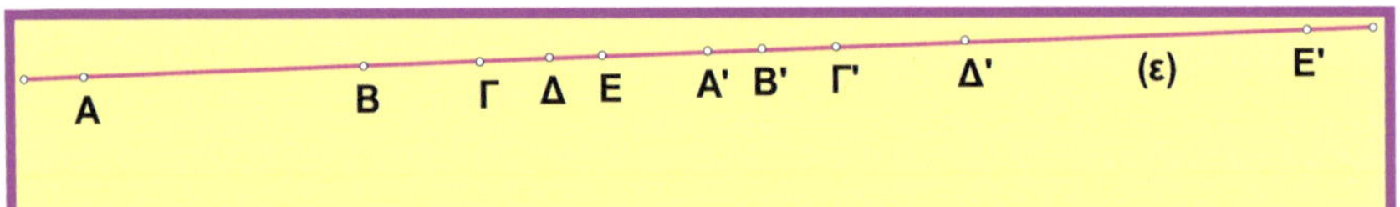

Σχήμα 108.

The set of the above 2n points will be called «Harmonic Open Collinear Series of Points of 2n Points or more simply Harmonic Polygonal Collinear Line, or Harmonic Collinear Line», if all the groups of four points below, are harmonic:

A, B, Γ, B', - B, Γ, Δ, Γ', - Γ, Δ, E, Δ', ,n-2 groups
A', B', Γ', B - B', Γ', Δ', Γ, - Γ', Δ', E', Δ,.., n-2 -//- } or 2(n-2) groups, (1).

the above 2(v-2) groups of four points in relation (1), we will call them « Group of 2(n-2) Groups of Four Points, of the Harmonic Open Collinear Series of Points of 2n Points». The following line segments AB, ΒΓ, ΓΔ, ΔE,...and A'B', B'Γ', Γ' Δ', Δ' E',...we will call them sides and points A, B, Γ, Δ, E,...N,... A', B', Γ', Δ', E',... N',... we will call them apexes of this Harmonic Open Series of Points of 2n Points, while the n segments, AA', BB',ΓΓ', ΔΔ', EE',....,NN', we will call them «Main Diagonals» of it.

Remarks.

(a). We named the above group of 2n points «Collinear», since, as we will see below, there is a matching Harmonic Open «Cocyclic» Series of Points of 2n Points.

(b). Each such series of points of 2n points, consists of two series of points of n points each.

(c). The above collinear group of 2n points, can be considered as degenerate «Cocyclic» Series of Points of 2n Points, whose the circumscribed circle is degenerate to a line (ε), as it has a radius of infinite length (zero curvature).

(d). Below we will also mention various properties that each Harmonic Open Collinear Series of Points of 2n Points has, as well as some Criteria of its Harmonic being.

Moreover we will refer to the ways of construction of an arbitrary Harmonic Open Collinear Series of Points of 2n Points (Construction 171, etc), among those also using the Method of Harmonic Transformation (Construction 179).

23.2. Pair of a Harmonic Open Convex Cocyclic Series of Points of 2n Points (Apexes). (Harmonic Open Polygonal Line).

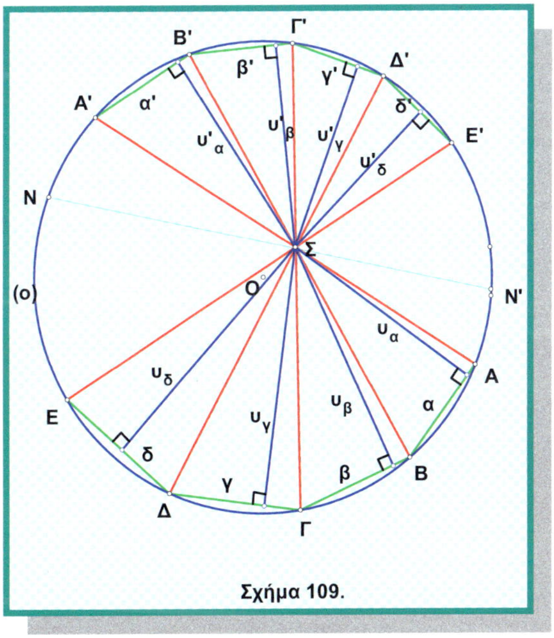

Σχήμα 109.

Definitions.

We consider on circle (o), the group of n ordered points A, B, Γ, Δ, E,...,N (Figure 109) and consequently the group of n ordered points A', B', Γ', Δ', E',...,N'.

The set of the above 2n points, we will call «Pair of Harmonic Open Convex Cocyclic Series of Points of 2n Points, or more simply Harmonic Polygonal Cocyclic Line, or Harmonic Cocyclic Line», if all its 2(n-2) quadrilaterals below are harmonic:

ΑΒΓΒ', ΒΓΔΓ', ΓΔΕΔ', ...,n-2 quadrilaterals
Α'Β'Γ'Β, Β'Γ'Δ'Γ, Γ'Δ'Ε'Δ,...,n-2 quadrilaterals } 2(n-2) quadrilaterals, (2).

The above quadrilaterals in relation (2), we will call them «2(n-2) Quadrilaterals of the Harmonic Open Cocyclic Series of Points of 2n Points».

The line segments AB, ΒΓ, ΓΔ, ΔΕ,...and A'B', Β'Γ', Γ'Δ', Δ'Ε',...we will call them sides, the points A, B, Γ, Δ, Ε,...,Ν,..., - Α', Β', Γ', Δ', Ε',..., Ν',... we will call them apexes of the Harmonic Open Cocyclic Series of Points of 2n Points, while the n segments, ΑΑ', ΒΒ', ΓΓ', ΔΔ', ΕΕ',....,ΝΝ', we will name them its «Main Diagonals».

Remarks.

(a). We called the above group of 2n points «Cocyclic», since as we saw above, there is a matching Open Harmonic «Collinear» Group of 2n Points.

(b). Each such series of points of 2n points (apexes), consists of two series of points of n points each.

(c). It is obvious that if the radius of the circumscribed circle of this Open Harmonic group of 2n, is of infinite length or if the circle has zero curvature, then this group of 2n is degenerate to an Open Harmonic «Collinear» group of 2n Points.

(d). Below we will also mention various properties that each Open Harmonic Cocyclic Series of Points of 2n Points has, in the form of Propositions and Criteria of its Harmonic being. For example the convergence of its diagonals (Proposition 177), etc.

Moreover we will refer to the ways of construction of an arbitrary such group of 2n points, among which also using the Method of Harmonic Transformation.

23.3. Pair of a Regular Open Cocyclic Series of Points of 2n Points.(Regular Open Polygonal Cocyclic Line).

Definitions.

We consider on circle (o), the group of n ordered points A, B, Γ, Δ, Ε,...,Ν (Figure 110) for which it is: ΑΒ=ΒΓ=ΓΔ=ΔΕ=...=τ, (3).

and the group of the n points A', B', Γ', Δ', Ε',...,Ν', symmetrical to the above first group of n with respect to a center, the center O of circle (o) respectively.

Here as well, it will obviously be: A'B'= B'Γ'= Γ'Δ'= Δ'E'=...= τ, (4).
and A', B', Γ', Δ', E',...,N', will belong to circle (o).
The set of the above 2n-points we would call «Pair of a Regular Open Cocyclic Series of Points of 2n Points or more simply Regular Polygonal Cocyclic Line, or Regular Cocyclic Line».
We easily realize that all the 2(n-2) quadrilaterals in (5) below, that belong to the group of 2n above, are rhomboid and therefore they are harmonic, as for example the rhomboid quadrilateral ΑΒΓΒ' has ΑΒ.ΓΒ'=ΒΓ.Β'Α, thus it will be Harmonic.
In the same way we also find that all the rhomboid quadrilaterals in (5) below, are Harmonic.

ΑΒΓΒ', ΒΓΔΓ', ΓΔΕΔ', ……, n-2 quadrilaterals
A'B'Γ'B, B'Γ'Δ'Γ, Γ'Δ'E'Δ,...,n-2 quadrilaterals } 2(n-2) quadrilaterals, (5).

The line segments AB, BΓ, ΓΔ, ΔΕ,...and A'B', B'Γ', Γ'Δ', Δ'E',...we will call them sides, points A, B, Γ, Δ, E,...N,... - A', B', Γ', Δ', E',... N',... we will call them apexes of the Harmonic Open Cocyclic Series of Points of 2n Points, while the n segents, AA', BB', ΓΓ', ΔΔ', EE' ,...., NN', we will call them its «Main Diagonals» [here they are also diameters of circle (o)].

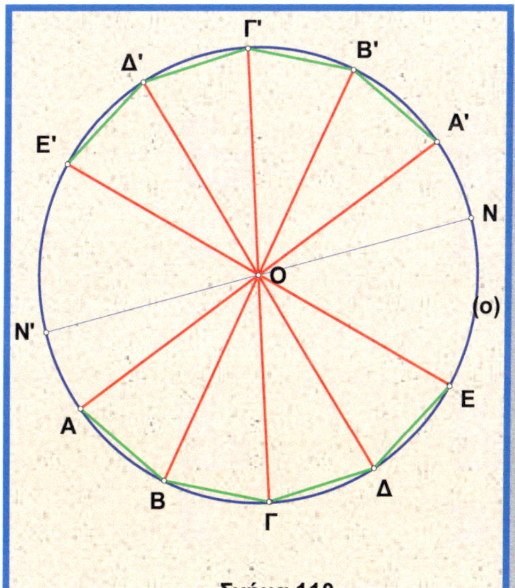

Σχήμα 110.

Comments.

(a). In all three cases of lines above, the pairs of points B-B', Γ-Γ', Δ-Δ', ...harmonically divide the pairs of segments ΑΓ-Α'Γ', ΒΔ-Β'Δ', ΓΕ-Γ'Ε', ..., respectively (In the Third case they are obviously perpendicular).

(b). It is obvious that in all three cases of series of points above, some apexes of each series of points can be located among other apexes of the same series of points, or all of them together to not cover the whole circle (Hence the term «open»).

(c). The construction of the above series of points is easy, in comparison to the construction of the Harmonic Multilaterals, as the above series of points are open and thus the difficulty in closing them does not exist (Construction 172, etc).

24. Pairs of Open Central Flat Harmonic and Regular Pencils of 2n Beams and their Orthooptical Points.

24.1. In General.

Below we will refer to the Harmonic and Regular Pairs of Open Flat Central Pencils of 2n Beams, as well as to their Orthooptical Points.

24.2. Harmonic Pairs of an Open Flat Central Pencil of 2n Beams (Figure 111).

a. **Definition**.

Based on what we know so far and from all we have mentioned here (for example paragraphs 12.13, 23.1, etc), we can give the definition of these pencils below:

We call Pair of a Harmonic Open Central Pencil of 2n Beams, that pencil whose intersections to an arbitrary line, consist a Pair of Harmonic Open Collinear Series of Points of 2n Points» (Relevant Terminology in paragraph 23.1).

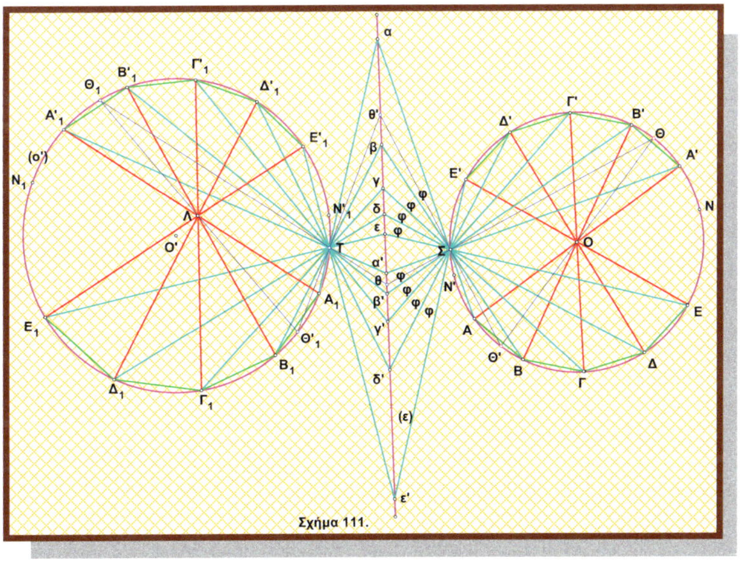

Σχήμα 111.

b. **Properties.**

These Open Pencils have many and significant properties, some of which we will mention below:

1/. The Harmonic Open Central Flat Pencil of 2n Beams T(αβγδε...ν, α'β'γ'δ'ε'...ν') in figure 111], if we take into consideration paragraph 23.1, we easily find that it includes 2(n-2) Harmonic Central Flat Pencils of four beams.

These are: T(αγββ'), T(βδγγ'), ..., T(α'γ'β'β), T(β'δ'γ'γ), ...

2/. If we intersect this pencil with a line it gives us 2n points, that constitute an Open Harmonic Collinear Series of Points of 2n Points, or a Collinear Harmonic Line (Relevant terminology in paragraph 23.1).

3/. A Harmonic Open Central Pencil of 2n Beams, if we inscribe them in circle (o'), then tis beams re-intersect the circle at points that obviously are the points of a Harmonic Open Cocyclic Series of Points of 2n Points. Thus from the above Harmonic Pencil T(αβγδε...ν, α'β'γ'δ'ε'...ν') in figure 111, occurred the Harmonic Open Cocyclic Series of Points of 2n Points A_1, B_1, Γ_1, Δ_1, E_1,...,N_1, A'_1, B'_1, Γ'_1, Δ'_1, E'_1,...,N'_1, which for the random point T of circle (o') has 2(n-2) Harmonic Pencils of four of its beams:

$$T(A_1\Gamma_1B_1B_1'),\ T(B_1\Delta_1\Gamma_1\Gamma'_1),\ \ldots,\ T(A'_1\Gamma'_1B'_1B_1),\ T(B'_1\Delta'_1\Gamma'_1\Gamma_1),\ \ldots.$$

24.3. Pair of a Regular Open Central Flat Pencil of 2n Beams (Figure 111).

a. **Definition**.

«We call Regular, a Pair of Open Central Flat Pencil of 2n Beams Σ(αβγδε...ν, α'β'γ'δ'ε'...ν') , if the angles of this pencil, that are formed between the consecutive beams, of each pencil of its above n beams, are equal to each other, with random measure φ and f for these pencils, the pairs of their homologous beams, are perpendicularly intersected.

I.e., for the Pair of this Regular Pencil above: Σ(αβγδε...ν, - α'β'γ'δ'ε'...ν') it is: $\angle\alpha\Sigma\beta=\angle\beta\Sigma\gamma=\angle\gamma\Sigma\delta=\angle\delta\Sigma\varepsilon=\angle\varepsilon\Sigma\zeta=\ldots$

$$=\angle\alpha'\Sigma\beta'=\angle\beta'\Sigma\gamma'=\angle\gamma'\Sigma\delta'=\angle\delta'\Sigma\varepsilon'=\angle\varepsilon'\Sigma\zeta'=\ldots=\varphi, \quad (1).$$

and ∠αΣα'=∠βΣβ'=∠γΣγ'=∠δΣδ'=∠εΣε' =......=∠νΣν'=1L. (2).
(The question is raised. Does such a pencil exist? The answer is positive, as it is constructed easily, if we depend in relations (1) and (2). This after all can also be seen in figure 111. Proof in Construction 179).
Due to relation (2), these pencils can also be called Orthooptical.

Conclusion.

Each such Regular Pencil, is also Orthooptical, as relation (2) is true.

b. **Properties.**

These Open Pencils have many and significant properties, some of which we will mention below:

1/. The Regular Open Central Flat Pencil of 2n Beams Σ(αβγδε...ν, - α'β'γ'δ'ε'...ν') in figure 111, has 2(n-2) Regular Pencils of four beams. These are:

Σ(αβγβ'), Σ(βγδγ'), ... Σ(α'β'γ'β), Σ(β'γ'δ'γ), ...

2/. It is easily proven (Proposition 169) that, if we intersect this pencil with a line, it gives us 2n points, that are a Pair of a Harmonic Collinear Series of Points of 2n Points (Relevant terminology in paragraph 23.1). This is the reason why we call this pencil Harmonic. Hence, if we also take into consideration the definition of Closed Harmonic Pencils in paragraph 12.13, we reach the conclusion that: «All these Regular Pencils, are also Harmonic». Relevant constructions are 61-64, 79, 97 and 171, that will follow.

3/. A Regular Open Pencil of 2n beams, if we inscribe them in circle (o), then its beams re-intersect the circle at points which are the points of a Regular Open Cocyclic Series of Points of 2n Points (Criterion 173). Hence from the above Regular Pencil Σ(αβγδε...ν, α'β'γ'δ'ε'...ν') in figure 111, occured the Regular Open Cocyclic Series of Points of 2n Points Α, Β, Γ, Δ, Ε,...,Ν, - Α' Β' Γ' Δ' Ε' ...,Ν', which for point Σ of circle (o), has 2(n-2) Regular Pencils of four beams: Σ(ΑΒΓΒ'), Σ(ΒΓΔΓ'), ... Σ(Α'Β'Γ'Β), Σ(Β'Γ'Δ'Γ), ...

4/. Additionally, when we intersect this pencil with a line, then its center Σ is an «Orthooptical Point» for some segments (diagonals) at which the line is

intersected, while if we intersect it with a circle that runs through its center Σ, then its center is an «Orthooptical Point» for some diameters of the circle in which it is inscribed.
Thus, in figure 111 Σ, is an «Orthooptical Point», for segments (diagonals) αα', ββ', γγ', δδ'.... , as well as for diameters AA', BB', ΓΓ', ΔΔ',...

24.4. Pair of Orthooptical Points, of a Pair of Open Collinear Harmonic Lines (Figure 111).

As we will later prove, each Collinear Harmonic Line, has two symmetrical with respect to this line Orthooptical Points, from which its diagonals, are seen in right angles. Such points are these we mentioned above (§ b 4/).

24.5. Description-Analysis of Figure 111.

Based on the above, we can easily describe figure 111.
In this figure we can see the Open Regular Flat Central Pencil of 2n Beams Σ(αβγδε...ν, - α'β'γ'δ'ε'...ν') with equal angles of random measure φ, which intersects line (ε) at the Open Harmonic Collinear Series of Points of 2n Points α, β, γ, δ, ε,...ν, - α', β', γ', δ', ε', ...,ν', but which also intersects circle (o) at points A, B, Γ, Δ, E,...N, - A', B', Γ', Δ', E' ,...,N', which are obviously apexes of an Open Regular Cocyclic Series of Points of 2n Points.
Here quadrilaterals ABΓB', BΓΔΓ', ..., A'B'Γ'B, B'Γ'Δ'Γ ,.., are rhomboids with axes of symmetry the diameters BB', ΓΓ'...., that are obviously harmonic.
Additionally, in this figure, we can see the Open Central Harmonic Pencil of 2n Beams T(αβγδε...ν, - α'β'γ'δ'ε'...ν'), which intersects line (ε) at th Open Harmonic Series of Points of 2n Points α, β, γ, δ, ε,...,ν, - α', β', γ', δ', ε', ...,ν' (known from the previous).
Moreover, if we set an arbitrary point T on a random circle (o'), then pencil Σ(αβγδε...ν, - α'β'γ'δ'ε'...ν') re-intersects circle (o') at the Open Harmonic

Cocyclic Series of Points of 2n Points A_1, B_1, Γ_1, Δ_1, E_1,...,N_1, - A'_1, B'_1, Γ'_1, Δ'_1, E'_1,...,N'_1, which are apexes to the Open Harmonic Cocyclic Series of Points of 2n Points.

It is obvious that, if the sides of the Open Regular Cyclic Series of Points of 2n Points, are many and in comparison to the circle big, then some apexes can be located among other of its apexes. Hence, for example in figure 111, apex Θ lies between A' and B', while apex Θ' lies between A and B.

Remarks.

(a). We will prove the above in Propositions and Problems that will follow or that have already been mentioned.

(b). Below based on figure 111, as we mentioned above, we will give the way and the interpretation of the way that an Open Collinear and a Cocyclic Harmonic Series of Points of 2n Points occurs, from an Open Cocyclic Regular Series of Points of 2n Points, with an equal number of sides, using the new Method of the «Harmonic Transfomation» (Construction 179), that is the solution to a Problem that troubled us often, for at least twenty years. (A first taste of this method, was given above in paragraph 24.5).

25. APPLICATIONS RELEVANT TO THE PAIRS OF COLLINEAR AND CO-CYCLIC HARMONIC LINES AND PENCILS.

In General.

Consequently we will proceed to applications of the method of Harmonic Pencils, that is mentioned in paragraphs 5 and 14, adjusted though here to the pencils of paragraph 24, in the form of Geometrical Propositions and Constructions, with their proofs, as we will need them to base on them the proofs of other Propositions, etc.

Moreover on these and those in paragraphs 13, 17 and 18, we will also depend the New Method of the Harmonic Transformation (§ 21), to which we will refer later, adjusted though here to the pencils of paragraph 24.

25.1. Intersection of a Pair of a Regular Open Flat Pencil of 2n Beams, from a Line.

Proposition 169 (Figure 112).

Each line that intersects a Pair of an Open Regular Flat Pencil of 2n Beams, is intersected at 2n points, which constitute a Pair of a Harmonic Open Collinear Series of Points of 2n Points.

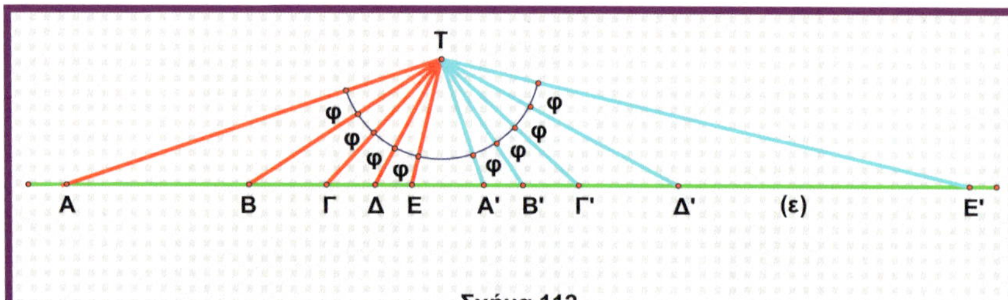

Σχήμα 112.

Proof.

Let there be the Pair of the Open Regular Flat Pencil of 2n Beams

$$T(AB\Gamma\Delta E...N - A'B'\Gamma'\Delta'E'...N'), \qquad (1).$$

which is intersected from line (ε) at points:

$$A, B, \Gamma, \Delta, E,...,N - A', B', \Gamma', \Delta', E',...,N', \qquad (2).$$

respectively. We will prove that the pair of the series of points (2), is Harmonic (§ 23.1).

Since the given pencil (1) is Regular, according to the relevant definition (§ 24.3a), it will have: ∠ATB= ∠BTΓ= ∠ΓTΔ= ∠ΔTE= ∠ETZ= ...,=∠A'TB'= ∠B'TΓ'= ∠Γ'TΔ'= ∠Δ'TE'= ∠E'TZ'= ... =φ, (3)

and ∠ATA'=∠BTB'=∠ΓTΓ'=∠ΔTΔ'=∠ETE' =...=∠NTN' =1L. (4).

Hence, since TB is the bisector of ∠ATΓ and TB' is perpendicular to TB, then B, B' will harmonically divide A and Γ or it will be (AΓBB')=1.

Additionally, since TΓ is the bisector of angle BTΔ and TΓ' is perpendicular to TΓ, then Γ, Γ' will harmonically divide B, Δ, or it will be (BΔΓΓ')=1.

We continue in a similar way and we find that: (ΓEΔΔ')=1, (ΔZEE')=1, and so on and so forth.

Similarly, since TB' is the bisector of angle A'TΓ' and TB is perpendicular to TB', then B', B will harmonically divide A', Γ', or it will be: (A'Γ'B'B)=1.

In a similar way we continue and find that it is: (B'Δ'Γ'Γ)=1, (Γ'E'Δ'Δ)=1, (Δ'Z'E'E)=1, and so on and so forth.

Due to the above and according to the relevant definitions (§ 23.1), the pair of the series of points A, B, Γ, Δ, E,...,N - A', B', Γ', Δ', E',...,N', is actually harmonic.

Conclusion.

From the above Proposition and the relevant definition of the Pair of the Open Regular Flat Pencil of 2n Beams (§ 24), we reach the following conclusion:

Each Pair of an Open Regular Flat Pencil of 2n Beams, is Harmonic as well.

Thus, to easily deifne, on a line, an arbitrary Pair of an Open Regular Flat Series of Points of 2n Points, we just need to intersect this line with a Pair of an Open Regular Flat Pencil of 2n Beams.

Remarks.

(a). Proposition 169 we believe to be new and firstly appear here.

(b). The above Central Pencil in figure 112, is symmetrical too, with respect to the bisector of angle ATN', or A'TN, and so on and so forth.

(c). Relevant Constructions are 11, 32a, 61-65.

25.2. Orthooptical Points of a Pair of a Harmonic Open Series of Points of 2n Points.

(Significant Property. Extenstion to Corollary 1 of Proposition 72).

Proposition 170 (Figure 113).

We are given the Pair of the Harmonic Open Collinear Series of Points of 2n Points A, B, Γ, Δ, E,...,N, - A', B', Γ', Δ', E',...,N' and we are asked to prove that the circles with diameters AA', BB', ΓΓ', ΔΔ', EE', ..., NN', are concurrent, hence they are a pencil of circles of the first kind.

(a). Proof.

Let (ε) be the line of the above points A, B, Γ, Δ, E,...,N, - A', B', Γ', Δ', E',...,N'.

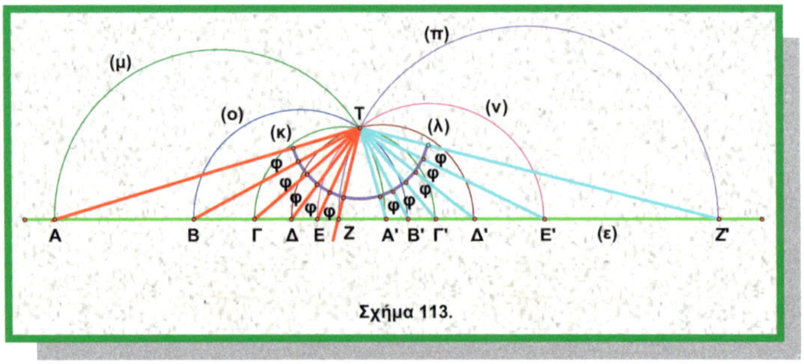

Σχήμα 113.

According to the relevant definition of the Harmonic Collinear Series of Points of 2n Points (§ 23.1), this should have harmonic the following groups of four points:

A, B, Γ, B'- B, Γ, Δ, Γ'- Γ, Δ, E, Δ'-,........,n-2 groups
A', B', Γ', B- B', Γ', Δ', Γ- Γ', Δ', E', Δ-,...,n-2 groups } 2(n-2) groups. (1).

To prove Proposition 170, we use the new Method of «Symmetrical Central

Pencils» (§ 5), adjusted to the pencils in paragraph 24, and taking into consideration the definition of the Pairs of Regular Open Central Pencils of 2n Beams (§ 24.3), that are harmonic too.

To apply this method, we draw circles (ο), (κ) with diameters BB' and ΓΓ' respectively, that intersect at points T, T', symmetrical with respect to (ε).

We draw TA, TB, TΓ, TΔ, TE, TZ,......, TA', TB', TΓ', TΔ', TE', TZ',......

If we take into consideration relation (1) above and we suppose that for triangle ATΓ since (AΓBB')=1 and $\angle BTB'=90^0$, circle (o) is Apollonian, then

$\angle ATB=\angle BT\Gamma$. (2).

Moreover, in triangle BTΔ, since it is (BΔΓΓ')=1 and $\angle \Gamma T\Gamma'=90^0$, circle (κ) is Apollonian, hence $\angle BT\Gamma=\angle \Gamma T\Delta$. (3).

Thus, from the relations in (2), (3), it occurs that:.

$\angle ATB=\angle BT\Gamma= \angle \Gamma T\Delta=\varphi$. (4).

We continue as above and if we consider triangle A'TΓ' for which since (Γ'A'B'B)=1 and $\angle BTB'=90^0$, circle (o) is Apollonian, then

$\angle A'TB'=\angle B'T\Gamma'$. (5).

Moreover, for triangle B'TΔ', since it is (Δ'B'Γ'Γ)=1 and $\angle \Gamma T\Gamma'=90^0$, circle (κ) is Apollonian, then $\angle B'T\Gamma'=\angle \Gamma'T\Delta'$. (6).

Hence, from the relations in (5), (6), it occurs that:

$\angle A'TB'=\angle B'T\Gamma'= \angle \Gamma'T\Delta'=\omega$. (7).

After all it is $\angle BT\Gamma=\varphi=\angle B'T\Gamma'=\omega$, or $\omega=\varphi$, (8).

ως συμπληρώματα της γωνίας ΓTB'.

Από τις σχέσεις (4), (7), (8) προκύπτει ότι:.

$\angle ATB=\angle BT\Gamma=\angle \Gamma T\Delta=\angle A'TB'=\angle B'T\Gamma'=\angle \Gamma'T\Delta'=\varphi=\omega$. (9).

Since it is $\angle \Gamma T\Gamma'=90^0$ and since, due to (9), it also is $\angle \Gamma T\Delta=\angle \Gamma'T\Delta'$, it will be $\angle \Gamma T\Gamma'-\angle \Gamma T\Delta+\angle \Gamma'T\Delta'=90^0$ or $\angle \Delta T\Delta'=90^0$, thus then circle (λ), with diameter ΔΔ', will run through T, as well as through T', its symmetrical with respect to (ε).

Additionally, for triangle ΓTE, since it is (ΓEΔΔ')=1 and since $\angle \Delta T\Delta'=90^0$, circle (λ) is Apollonian, thus due to (9), it will be:

$\angle \Gamma T\Delta=\angle \Delta TE=\varphi$. (10).

Similarly, for triangle Γ'TE', since it is (Γ'E'Δ'Δ)=1 and since $\angle \Delta'T\Delta=90^0$, circle (λ) is Apollonian, thus due to (9), it will be:

$$\angle\Gamma'T\Delta'=\angle\Delta'TE'=\varphi. \quad (11).$$

Since it is $\angle\Delta T\Delta'=90^0$ and due to (10), (11) it is $\angle\Delta TE=\angle\Delta'TE'$, and it will also be $\angle\Delta T\Delta'-\angle\Delta TE+\angle\Delta'TE'=90^0$ or $\angle ETE'=90^0$, thus circle (v), with diameter EE', will run through T, as well as through T', its symmetrical with respect to (ε).

In the exact same way we continue and we find that circle (π) with diameter ZZ', runs through T, T', as well, and so on and so forth, until circle with diameter MM'. But when we try to prove that the circle with diameter NN' runs through T, T' we encounter difficulty. To achieve this we add in the end of each one of the two given series of points one more point Ξ, Ξ' respectively, such that it would be (MΞNN')=1, (M'Ξ'N'N)=1 and we work as above. This can easily happen as points M, N, N', exist, so we define in the known way point Ξ, while since points M', N', N, exist, we define in the known way Ξ'.

Finally since we have not proven that circle (μ) with diameter AA', runs through T, T', we will prove that too. To prove that we work in the following way:

Since it is $\angle BTB'=90^0$ and since, due to (9), it is $\angle ATB=\angle A'TB'$, it will be $\angle BTB'-\angle A'TB'+\angle ATB=90^0=\angle ATA'$ or $\angle ATA'=90^0$, hence circle (μ), with diameter AA', will also run through T, as well as through T', its symmetrical with respect to (ε).

Conclusions.

(a). From the above it finally occurs that:

$$\angle ATA'=\angle BTB'=\angle\Gamma T\Gamma'=\angle\Delta T\Delta'=\angle ETE'=.....=\angle NTN'=90^0. \quad (12).$$

This means that the circles with diameters AA'=BB'=ΓΓ'=ΔΔ'=EE'=.....=NN', do in fact run through the pair of points T, T'. This means that they constitute a pencil of circles of the first kind and since these segments, are seen in right angles from T, T', they are «Orthooptical Points» of the given harmonic series of points.

(b). Additionally from (9), (10), (11) it also occurs that:

$$\angle ATB=\angle BT\Gamma=\angle\Gamma T\Delta=\angle\Delta TE=....$$
$$=\angle A'TB'=\angle B'T\Gamma'=\angle\Gamma'T\Delta'=\angle\Delta'TE'=....=\varphi. \quad (13).$$

(c). Hence, since (12), (13), are true, pencil T(ΑΒΓΔΕ...Ν - Α'Β'Γ'Δ'Ε'...Ν'), is regular (definition § 24.3).

Remarks.

(a). Proposition 170, is obviously an extension to Corollary 1 of Proposition 72, which refers to a Harmonic Group of Ten Collinear Points.

(b). The Proposition above we believe to be new and firstly appear here.

(c). We have to do with a significant Proposition, as it will be useful in many cases below.

(d.) Relevant Constructions are 32a, 63. 68, 70-72, 169, 171.

25.3. Easy Construction of a Pair of a Harmonic Open Collinear Series of Points of 2n Points.

(Harmonic Open Polygonal Collinear Line).

Construction 171 (Figure 113).

Construct an arbitrary Pair of a Harmonic Open Collinear Series of Points of 2n Points, on a given line (ε).

Solution.

1st Way (Analytical).

(a). Analysis.

We suppose that the desired series of points is constructed on line (ε) and it is this one Α, Β, Γ, Δ, Ε,...,Ν, - Α', Β', Γ', Δ', Ε',...,Ν'.

If we take into consideration the definitions of Regular Central Pencils (§ 12.13 and 24, especially their paragraphs a1/ and 24.3a/, respectively) and the definition of the Pair of a Harmonic Open Collinear Series of Points of 2n Points (§ 23.1), we are easily led to the construction below.

(b). Construction.

Outside of line (ε), we set a point Τ, from which we draw a line that intersects (ε), at let's say point Α.

Consequently from Τ we draw a second line, that intersects (ε) at let's say Β and forms with ΤΑ an acute angle φ (φ is preferably of a small measure).

We repeat that with lines TΓ, TΔ, TE,..,TN [points Γ, Δ, E, ...,N belong to (ε)].
To TA, TB, TΓ, TE,...,TN at point T we draw perpendiculars that intersect (ε), at points A', B', Γ', Δ', ...,N', respectively.
We say that the desired series of points is A, B, Γ, Δ, E,...,N - A', B', Γ', Δ', E',...,N'.

(c). Proof.

We easily find that it is as well:

$$\angle A'TB' = \angle B'T\Gamma' = \angle \Gamma'T\Delta' = \angle \Delta'TE' = \angle E'TZ' = \ldots = \varphi. \quad (2).$$

Hence, since TB is the bisector of ∠ATΓ and TB' is perpendicular to TB, then B, B' will harmonically divide A and Γ or it will be (AΓBB')=1.
Moreover, since TΓ is the bisector of angle BTΔ and TΓ' is perpendicular to TΓ, then Γ, Γ' will harmonically divide B, Δ, or it will be (BΔΓΓ')=1.
We continue, and in a similar way we also find that: (ΓEΔΔ')=1, (ΔZEE')=1, and so on and so forth.
Similarly, since TB' is the bisector of angle A'TΓ' and TB is perpendicular to TB', then B', B will harmonically divide A', Γ', or it will be: (A'Γ'B'B)=1.
We continue in a similar way and we find that it is: (B'Δ'Γ'Γ)=1, (Γ'E'Δ'Δ)=1, (Δ'Z'E'E)=1, and so on and so forth.
Due to the above and according to the relevant definition (§ 23.1), the following series of points: A, B, Γ, Δ, E,...,N - A', B', Γ', Δ', E',...,N'. (3).
is in fact the desired harmonic one.

2nd Way (Based on Proposition 169).

(a). Analysis.

If we take into consideration Proposition 169, we are easily led to the construction below.

(b). Construction.

We construct an arbitrary Pair of a Regular Open Harmonic Pencil of 2n Points, in the following way:
Using an arbitrary point T, outside of line (ε), as the center, we construct the central pencil of n beams T(ABΓΔE,...,N), with equal angles, hence:

$$\angle ATB = \angle BT\Gamma = \angle \Gamma T\Delta = \angle \Delta TE = \angle ETZ = \ldots = \varphi. \quad (4).$$

Consequently, we draw perpendiculars TA′, TB′, TΓ′, TΔ′, TE′,....,TN′, to TA,

TB, TΓ, TΔ, TE,....,TN, respectively, thus the Open Regular Pencil of 2n Beams occurs: T(ABΓΔE,...,N - A'B'Γ'Δ'E',....,N',), (5).
which intersects (ε), at let's say points A, B, Γ, Δ, E,...,N - A', B', Γ', Δ', E',...,N'. which we say constitue the desired Pair of a Harmonic Open Collinear Series of Points of 2n Points.

(c). Proof.

We easily find that the angles of the second part of pencil (5) are equal to φ [Relation (2)].
Hence, from its construction, pencil (5) we see that it is Open Regular, thus according to Proposition 169, the intersection of pencil (5) from (ε), will in fact constitute a Pair of a Harmonic Open Collinear Series of Points of 2n Points.

Investigation.

This construction can always be carried through with.
It is obvious that this construction is undefined, it can become defined though, if we are given sufficient data about it.

Remarks.

(a). The above Problem and its solution, we believe to be new and firstly appear here.
(b). Below we will also mention some properties that each Pair of an Open Harmonic Collinear Series of Points of 2n Points, in the form of Criteria of its harmonic being.
(d). The construction of the above series of points is rather easy, in comparison to the construction of the Harmonic Multilaterals, as the above series of points is open and thus the difficulty of closing it does not exist.
(e). Extenstion to Construction 171 to a circle, is exactly below, in paragraph 25.4 (Construction 172).
(f). Relevant Propositions are 64, 100 - 102, 169, 172.

25.4. Easy Construction of a Pair of a Convex Harmonic Open Cocyclic Series of Points of 2n Points.

(Harmonic Open Polygonal Line. Extension to Construction 171 to a circle as well).

Construction 172 (Figure 114).

Construct an arbitrary Pair of a Convex Harmonic Open Cocyclic Series of Points of 2n Points, on a given circle (o).

Solution.

(a). Analysis.

We suppose that the desired series of points is constructed and it is the following one A, B, Γ, Δ, E,...,N, - A', B', Γ', Δ', E',...,N'.

According to the relevant definition (§ 23.2), its 2n quadrilaterals below, should be Harmonic:

ABΓB', BΓΔΓ', ΓΔEΔ', ..., n-2 quadrilaterals
A'B'Γ'B, B'Γ'Δ'Γ, Γ'Δ'E'Δ,....,n-2 quadrilaterals } 2(n-2) quadrilaterals. (1).

This can be easily achieved, if we take into consideration Proposition 34 (§ 8.71) or 207 (§ 29.8).

(b). Construction.

Inside circle (o), we set point Σ, at which we want the diagonals (main), of the series of points, to coincide [Remark (d) in § 23.2].

From Σ we draw an arbitrary line, which intersects the circle at points A, A'.

From an arbitrary point α on line AA', that is located outside of circle (o), we draw a tangent αB.

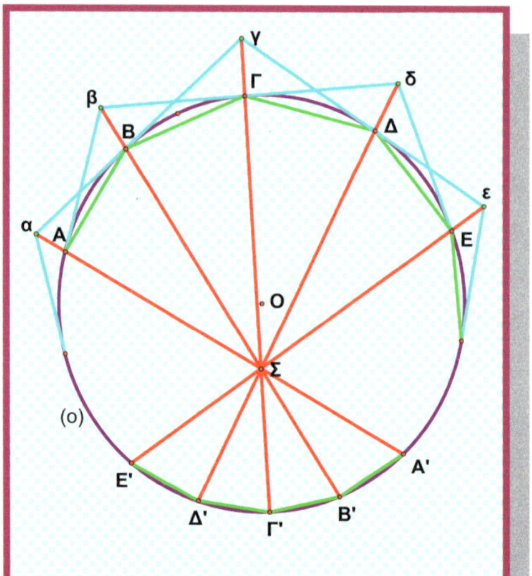

Σχήμα 114.

Consequently we draw ΣB, which intersects at point β, the tangent at A and re-intersects the circle at B'.

From β we draw the 2nd tangent βΓ to circle (o) (the 1st one is βA).

Consequently we draw ΣΓ, that intersects αB at γ and re-intersects the circle at Γ'.

From γ we draw the 3rd tangent γΔ to circle (o).

Afterwards we draw ΣΔ, which intersects βΓ at δ ad re-intersects the circle at Δ'.

From δ we draw the 4th tangent δE to the circle (o).

Afterwards we draw ΣE, that intersects γΔ at ε and re-intersects the circle at E'.

We continue as above and we define points Z, Z', ζ- H, H', η-...- N, N', ν.

We say that the two series of the 2n points in total below, is the desired series of points: A, B, Γ, Δ, E,...,N and A', B', Γ', Δ', E',...,N'.

(c). Proof.

It is obvious that quadrilateral ABΓB', is Harmonic, as its diagonal BB', runs through pole β of its other diagonal AΓ, from its construction.

Similarly for quadrilateral BΓΔΓ', it is Harmonic, since its diagonal ΓΓ', runs through pole γ of its other diagonal BΔ, from its construction.

In the exact same way we find that quadrilateral ΓΔEΔ', is Harmonic.

We continue, as above until we prove that quadrilateral ΛMNM', is Harmonic.

Moreover, since for example hexagon ABΓA'B'Γ' has concurrent diagonals at Σ and its quadrilateral ABΓB' is Harmonic, according to the above Proposition 34 or 307 (direct) its quadrilateral A'B'Γ'B, will also be Harmonic.

We continue in the exact same way as above and find that quadrilaterals B'Γ'Δ'Γ, Γ'Δ'E'Δ,..., Λ'M'N'M are Harmonic too, thus the above series of points is in fact the desired one (definition § 23.2).

(d). Investigation.

This construction can always be carried through with.

It is obvious that this construction is undefined, it can become defined though, if we are given sufficient data about it.

Remarks.

(a). The above Problem and its solution, we believe to be new and firstly appear here.

(b). Below we will also mention some properties that each Pair of an Open Harmonic Cocyclic Series of Points of 2n Points, in the form of Criteria of its harmonic being [for example convergence of its diagonals (Proposition 177), etc].

(c). The construction of the above series of points is rather easy, in comparison to the construction of the Harmonic Multilaterals, as the above series of points is open and thus the difficulty of closing it does not exist.

(d). Construction 172 is obviously an extenstion to Construction 171, from a line to circle.

(e). Relevant Propositions are 34, 64, 80-82, 100 -102, 171, 177, 207.

25.5. Criterion of Regularity of a Pair of an Open Cocyclic Series of Points of 2n Points.

(Extension to Criterion 78 to a Pair of a Regular Open Cocyclic Series of Points of 2n Points).

Criterion 173 (Figure 115).

Each Pair of an Open Cocyclic on circle (o) Series of Points of 2n Points Α, Β, Γ, Δ, Ε,...,Ν - Α', Β', Γ', Δ', Ε',...,Ν', is Regular, if and only if, pencil Σ(ΑΒΓΔΕ...Ν - Α'Β'Γ'Δ'Ε' ...Ν'), with its center an arbitrary point Σ of circle (o), is Regular.

Proof.

a). Direct.

I.e., if the Pair of the Open Cocyclic in circle (o) Series of Points of 2n Points: Α, Β, Γ, Δ, Ε,...,Ν - Α', Β', Γ', Δ', Ε',...,Ν', (1).

Is Regular (definition § 23.3), then pencil:

Σ(ΑΒΓΔΕ...Ν - Α'Β'Γ'Δ'Ε'...Ν'), (2).

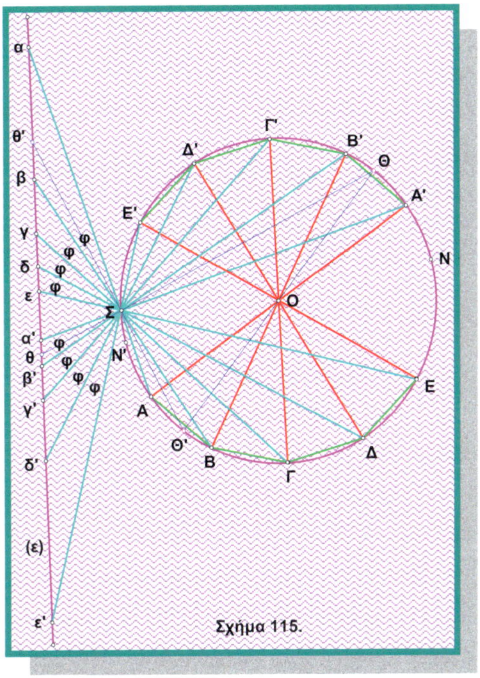

Σχήμα 115.

is Regular as well (Terminology § 24.3). According to the relevant definition of the above Regular Series of Points (1), AA', BB', ΓΓ', ΔΔ', EE', ..., NN', are diameters of circle (o), while it is as well:

AB=BΓ=ΓΔ=ΔE=...=A'B'=B'Γ' = Γ'Δ'=Δ'E'=...=τ, thus we easily find that: ∠AΣB = ∠BΣΓ = ∠ΓΣΔ = ∠ΔΣE = ∠EΣZ = ...= ∠A'ΣB' = ∠B'ΣΓ' = ∠Γ'ΣΔ' = ∠Δ'ΣE' = ∠E'ΣZ'= ... = φ. (3).

and ∠AΣA'= ∠BΣB'= ∠ΓΣΓ'= ∠ΔΣΔ'= ∠EΣE' =...=1L (4).

Hence, since relations (3) and (4) are true, according to the relevant definition of the above Regular Pencil (§ 24.3), it occurs that pencil (2), is in fact Regular.

(b). Converse.

I.e., if pencil (2) is Regular, then the Cocyclic Series of Points (1), is Regular too.

According to the relevant definition of the above Regular Pencil (2), (3) and (4) will be true.

Thus, easily from (3), it occurs that AB= BΓ= ΓΔ= ΔE= ...= A'B'= B'Γ'= Γ'Δ'= Δ'E' =... =τ, while from (4), we easily find that as well AA', BB', ΓΓ', ΔΔ', EE', ..., NN', are diameters of circle (o), hence according to the relevant definition of the Cocyclic Regular Series of Points, the above Series of Points (1), is in fact Regular.

Conclusion.

The Criterion above is rather significant, as it has many applications, which we will find later on and because it allows us to give the alternative definition of the Pair of a Regular Open Central Flat Pencil of 2n Beams below, apart from the, until now known to us, definition (§ 24.3).

«A Pair of a Central Open Pencil of 2n Beams, is called Regular, if we inscribe this pencil in a circle and its intersections to this circle are apexes to a Pair of an Open Regular Cocyclic Series of Points of 2n Points.

Remarks

(a). As we have already seen, earlier in this paper, equivalent Criteria, are true for Regular Polygons, as the Regular Hexagon, Octagon, Decagon, etc (Criteria 76, 77, 78, etc, respectively).

(b). As we saw in the conclusion above, with this Criterion the definition of the Pair of Open Regular Pencils is extended from a line to a circle.

(c). This Criterion we believe to be new, and firstly appear here.

(d). We notice that if an arbitrary line (ε) intersects the Regular Pencil (2) above (figure 115), then the pair of the Harmonic Series of Points α, β, γ, δ, α,...,ν - α', β', γ', δ', ε',...,ν' occurs [Proposition 169 (§ 25.1].

(e). Relevant Propositions are 35, 76, 77, 78, 96, 97, 98, 169, 174.

25.6. Criterion of Harmonic being of a Pair of a Convex Open Cocyclic Series of Points of 2n Points.

(Extension to Criterion 98 to a Pair of a Harmonic Open Cocyclic Series of Points of 2n Points)

Criterion 174 (Figure 116).

Each Pair of a Convex Open Cocyclic in circle (o) Series of Points of 2n Points A, B, Γ, Δ, E,...,N, - A', B', Γ', Δ', E',...,N' is Harmonic, if and only if, pencil T(ABΓΔE...N - A'B'Γ'Δ'E' ...N'), whose center is an arbitrary point T of circle (o), is Harmonic.

Proof.

(a). Direct.

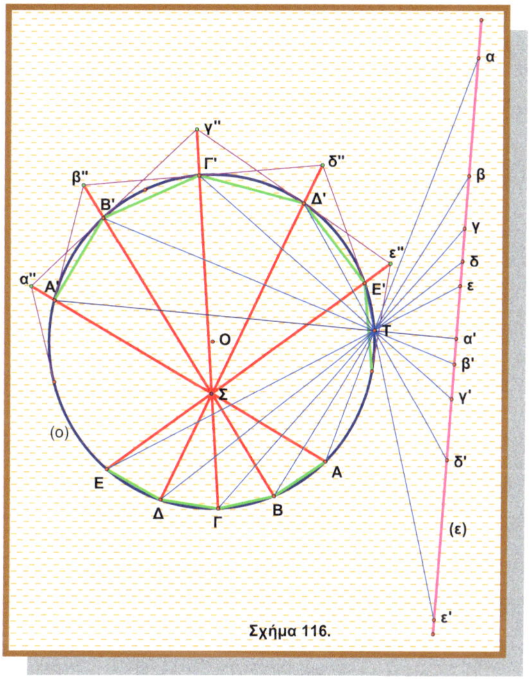

Σχήμα 116.

I.e., if the convex Open Cocyclic in circle (o) Series of Points of 2n Points: A, B, Γ, Δ, E,...,N, - A', B', Γ', Δ', E', ...,N', is Harmonic (§ 23.2), then will also prove that pencil: T(ΑΒΓΔΕ...Ν, - Α'Β'Γ'Δ'Ε'...Ν'), is harmonic (Terminology § 24.2).

In fact, since the following Series of Points: A, B, Γ, Δ, E,...,N, - A', B', Γ', Δ', E', ...,N', is harmonic, according to the relevant definition of harmonic being (§ 23.2), the above cocyclic series of points, has all its 2n quadrilaterals below harmonic:

ΑΒΓΒ', ΒΓΔΓ', ΓΔΕΔ', ...,n-2 quadrilaterals
Α'Β'Γ'Β, Β'Γ'Δ'Γ, Γ'Δ'Ε'Δ,...,n-2 quadrilaterals } 2(n-2) quadrilaterals. (1).

Hence, according to the direct of criterion 35 (§ 8.72), the matching 2(n-2) pencils of four beams below, will also be harmonic, as the 2(n-2) quadrilaterals below are inscribed too:

Τ(ΑΒΓΒ'), Τ(ΒΓΔΓ'), Τ(ΓΔΕΔ'), ..., n-2 pencils
Τ(Α'Β'Γ'Β), Τ(Β'Γ'Δ'Γ), Τ(Γ'Δ'Ε'Δ),......,n-2 pencils } 2(n-2) pencils. (2).

Thus, if a line (ε) intersects pencil Τ(ΑΒΓΔΕ...Ν, - Α'Β'Γ'Δ'Ε'...Ν') of 2n beams, then it will intersect (ε) at points α, β, γ, δ, ε,...ν, - α', β', γ', δ', ε',...,ν', respectively and pencil Τ(αβγδε...ν, - α'β'γ'δ'ε'...ν') will occur, for which according to what we know, the 2(n-2) groups of four points below will be harmonic:

α, β, γ, β' – β, γ, δ, γ' – γ, δ, ε, δ'...... n-2 groups
α', β', γ', β – β', γ', δ', γ – γ', δ', ε', δ..... n-2 groups } 2(n-2) groups. (3).

Therefore, according to the relevant definition of harmonic being (§ 23.1), the group of 2n points α, β, γ, δ, ε,...ν, - α', β', γ', δ', ε',...,ν', is harmonic. This, according to the relevant definition of the Harmonic Central Pencil of 2n Beams (§ 24.2), means that pencil T(αβγδε...ν, α'β'γ'δ'ε'...ν'), or pencil T(ΑΒΓΔ,Ε...Ν, Α'Β'Γ'Δ'Ε'...Ν'), will in fact be harmonic.

(b). Converse.

I.e., if pencil T(ΑΒΓΔΕ...,Ν - Α'Β'Γ'Δ'Ε'...Ν'), is harmonic, then we will prove that the Series of Points of 2n Points A, B, Γ, Δ, E,...,N, - A', B', Γ', Δ', E', ...,N', is harmonic.

To prove that, we will follow the exactly reverse route.

Hence:

Since pencil T(ΑΒΓΔΕ..., - Α'Β'Γ'Δ'Ε'...Ν'), or pencil T(αβγδε...ν, - α'β'γ'δ'ε'...ν'), is harmonic, if a line (ε) intersects it, according to the relevant definition (§ 24.2), the collinear Series of Points of 2n points α, β, γ, δ, ε,...ν, - α', β', γ', δ', ε',...,ν', will also be harmonic, thus according to the relevant definition (§ 23.1) all 2(n-2) groups of four collinear points below, will be harmonic:

α, β, γ, β' – β, γ, δ, γ' – γ, δ, ε, δ'...... n-2 groups
α', β', γ', β – β', γ', δ', γ – γ', δ', ε', δ..... n-2 groups } 2(n-2) groups. (3).

Thus, due to what we know all 2(n-2) pencils of four beams below, will be harmonic:

T(ΑΒΓΒ'), T(ΒΓΔΓ'), T(ΓΔΕΔ'), ...,n-2 pencils
T(Α'Β'Γ'Β), T(Β'Γ'Δ'Γ), T(Γ'Δ'Ε'Δ),...,n-2 pencils } 2(n-2) pencils (2).

Thus, since these 2(n-2) pencils of four beams are harmonic, according to criterion 35 (§ 8.72), they will intersect circle (o), at 2n points A, B, Γ, Δ, E,...,N, A', B', Γ', Δ', E', ...,N', for which, the 2(n-2) groups of four of their cocyclic points below, will be harmonic (or they will be apexes to harmonic quadrilaterals), as these are also inscribed in circle (o):

A, B, Γ, B' - B, Γ, Δ, Γ' - Γ, Δ, E, Δ'- ......n-2 groups
A', B', Γ', B - B', Γ', Δ', Γ - Γ', Δ', E', Δ- n-2 groups } 2(n-2) groups. (1).

This, according to the relevant definition of harmonic being (§ 23.2), means that the convex Open Cocyclic in circle (o) Series of Points of 2n Points A, B, Γ, Δ, E, - ,N, A', B', Γ', Δ', E', ...,N', is in fact Harmonic.

(c). Investigation.

This criterion is also true in the special case where T coincides with an apex, of the above series of points.

In fact, if T coincides for example with Δ', then beam TΔ' of pencil T(ABΓΔE...,N - A'B'Γ'Δ'E'...N') will obviously coincide with the tangent γ"Δ'ε" at point Δ', hence: ∠ATB = ∠AΔ'B, ∠BTΓ = ∠BΔ'Γ, ∠ΓTΔ = ∠ΓΔ'Δ, ∠ΔTE = ∠ΔΔ'E, ,...∠A'TB' = ∠A'Δ'B', ∠B'TΓ' = ∠B'Δ'Γ', ∠Γ'TΔ' = ∠Γ'Δ'γ'', ∠Δ'TE' = ∠ε''Δ'E',.... (4).

This means that, all the angles formed among the consecutive beams of two pencils T(ABΓΔE...,N, - A'B'Γ'Δ'E'...N') and Δ'(ABΓΔE...,N, - A'B'Γ'Δ'E'...N') are equal, one by one. Thus, since the first one of the pencils is harmonic, then the second will be harmonic too.

Conclusion.

The Criterion above is rather significant, as it has many applications, which we will find later on and because it allows us to give the alternative definition of the Pair of a Harmonic Open Central Flat Pencil of 2n Beams below, apart from the, until now known to us, definition (§ 24.2).

«A Pair of an Open Harmonic Central Flat Pencil of 2n Beams, is called Harmonic, if we inscribe this pencil in a circle and its intersections with this circle are apexes to a Pair of an Open Harmonic Cocyclic Series of Points of 2n Points.

Remarks

(a). Equivalent Criteria, are true for Harmonic Polygons, as the Harmonic Hexagon, Octagon, Decagon, etc (Criteria 96, 97, 98, etc, respectively).

(b). As we saw in the conclusion above, with this Criterion the definition of the Pair of Open Harmonic Pencils is extended from a line to a circle.

(c). This Criterion we believe to be new, and firstly appear here.

(d). Relevant Propositions are 35, 76, 77, 78, 96, 97, 98, 173.

25.7. Construction of a Pair of an Open Cocyclic Series of Points of 2n Points from a Pair of a Harmonic Open Collinear Series of Points of 2n Points and Conversely.

(Pair of a Regular Cocyclic Polygonal Line).

Construction 175 (Figure 115).

Construct a Pair of a Regular Open Cocyclic Series of Points of 2n Points, based on a Pair of a Harmonic Open Collinear Series of Points of 2n Points, given to us, and conversely, construct a Pair of a Harmonic Open Collinear Series of Points of 2n Points, based on a Pair of a Regular Open Cocyclic Series of Points of 2n Points, given to us.

Solution.

(a). Direct.

1/. Analysis.

I.e. we are given on a line (ε), the Pair of a Harmonic Open Collinear Series of Points of 2n Points: α, β, γ, δ, ε,...,ν - α', β', γ', δ', ε',...,ν' (1).

and we are asked to construct the Pair of a Regular Open Cocyclic Series of Points of 2n Points: A, B, Γ, Δ, E,...,N - A', B', Γ', Δ', E',...,N', (2).

Inscribed in given circle (o).

If we take into consideration Criterion 170 (§ 25.2) and Proposition 173 (§ 25.5), we are led to the construction below.

2/. Construction.

We draw circles using segments ββ' and γγ' as diameters, that generally intersect at point Σ «Orthooptical Point», of the given series of points (Proposition 170).

We suppose that Σ by chance belongs to the given circle (o). We form pencil Σ(αβγδε...ν – α'β'γ'δ'ε'...ν'), the beams of which, if they re-intersect circle (o) at points A, Γ, B, Δ, E, ...N - A', B', Γ', Δ', E',...,N', respectively, then we say that the series of points Α, Γ, Β, Δ, Ε, ...Ν - Α', Β', Γ', Δ', Ε',...,Ν', is the desired Pair of a Regular Open Cocyclic Series of Points of 2n Points.

In the case where the given circle (o) does not run through Σ, then we draw, in the known way, a circle equal to the given (o), that does run through Σ, continue, as above and after we construct the Pair of the Regular Open Cocyclic Series of Points of 2n Points Α, Γ, Β, Δ, Ε, ...Ν - Α', Β', Γ', Δ', Ε',...,Ν', as above, we move it, in the known way, on the given circle (o).

3/. Proof.

Since the series of points α, β, γ, δ, ε,..., ν – α', β', γ', δ', ε', ν', is Harmonic, through Σ will also run the circles with diameters αα', δδ', εε',...,νν', «Orthooptical Point» and the Central Pencil Σ(αβγδε...ν – α'β'γ'δ'ε'...ν) or Σ(ΑΓΒΔΕ ...Ν - Α'Β'Γ'Δ'Ε'...Ν') is Regular [Conclusion (c) in Proposition 170].

Thus, according to Criterion 173 (converse) intersections Α, Γ, Β, Δ, Ε, ...Ν - Α', Β', Γ', Δ', Ε',...,Ν' of the Regular Pencil, with the given circle (o), will be apexes to the Pair of the Regular Open Cocyclic Series of Points of 2n Points Α, Γ, Β, Δ, Ε, ...Ν - Α', Β', Γ', Δ', Ε',...,Ν', thus this is the desired Regular pair of series of points, as it is also inscribed in circle (o).

(b). Converse.

1/. Analysis.

I.e. we are given the inscribed in circle (o), Pair of a Regular Open Cocyclic Series of Points of 2n Points:

Α, Β, Γ, Δ, Ε,...,Ν - Α', Β', Γ', Δ', Ε',...,Ν' (2).

and we are asked to construct on a given line (ε) the Pair of a Harmonic Open Collinear Series of Points of 2n Points:

α, β, γ, δ, ε,...,ν - α', β', γ', δ', ε',...,ν'. (1).

If we take into consideration Criterion 173 and Proposition 169 (§ 25.1), we are led to the construction below:

2/. Construction.

On circle (o), we easily construct the Pair of the Regular Open Cocyclic Series of Points of 2n Points A, B, Γ, Δ, E,...,N - A', B', Γ', Δ', E',...,N' (§ 23.3) (if it is not given) and after we set an arbitrary point Σ, we form the regular pencil Σ(ΑΓΒΔΕ ...N - Α'Β'Γ'Δ'Ε'...N'), the beams of which, we suppose that intersect line (ε) at points α, β, γ, δ, ε,...,ν - α', β', γ', δ', ε',...,ν'. respectively.

We say that the desired Pair of a Harmonic Open Collinear Series of Points of 2n Points, is the α, β, γ, δ, ε,...,ν - α', β', γ', δ', ε',....,ν'.

3/. Proof.

Since the Pair of the Open Cocyclic Series of Points of 2n Points A, B, Γ, Δ, E,...,N - A', B', Γ', Δ', E',...,N', is Regular and the Central Pencil Σ(ΑΓΒΔΕ ...N - Α'Β'Γ'Δ'Ε'...N'), or Σ(αβγδε...ν – α'β'γ'δ'ε'...ν'), is Regular (Criterion 173) and the Pair of an Open Collinear Series of Points of 2n Points: α, β, γ, δ, ε,...,ν - α', β', γ', δ', ε',...ν', is Harmonic (Proposition 169).

Thus, the Pair of an Open Collinear Series of Points of 2n Points: α, β, γ, δ, ε,...,ν - α', β', γ', δ', ε',...,ν' is the desired one, as it is also located on line (ε).

Remarks.

(a). The Problem above and its solution, we believe to be new and firstly appear here.

(b). Problem 175, the way it is given, it is undefined. If we want to define it we need to be given sufficient data.

(c). In an equivalent way we have constructed on a line closed harmonic groups of four, eight, ten, etc, or on a circle Squares, Regular hexagons, octagons, etc, Constructions 79, 80, 81, etc, respectively.

(d). Relevant Propositions are 35, 61, 62, 72, 74, 77, 79, 80, 81, 82, 96 - 103, 169, 170, 173, 176, 179.

25.8. Construction of a Pair of a Convex Harmonic Open Cocyclic Series of Points of 2n Points, from a Pair of a Harmonic Open Collinear Series of Points of 2n Points and Conversely. (Extension to Construction 175).

(Pair of a Convex Harmonic Cocyclic Polygonal Line).

Construction 176 (Figure 116).

Construct a Pair of a Convex Harmonic Open Cocyclic Series of Points of 2n Points, based on a Pair of a Harmonic Open Collinear Series of Points of 2n Points, given to us, and conversely, construct a Pair of a Harmonic Open Collinear Series of Points of 2n Points, based on a Pair of a Convex Harmonic Open Cocyclic Series of Points of 2n Points, given to us.

Solution.

(a). Direct.

1/. Analysis.

I.e., we are given on a line (ε), the Pair of the Harmonic Open Collinear Series of Points of 2n Points:

α, β, γ, δ, ε,...ν, - α', β', γ', δ', ε',...,ν', (1)

and we are asked to construct on the given circle (o), the Pair of the Harmonic Open Cocyclic Series of Points of 2n Points:

A, B, Γ,Δ, E, ,N, - A', B', Γ', Δ', E', ... ,N'. (2).

If we take into consideration Criterion 174 (§ 25.6) and the relevant definitions (paragraphs 23.1, 23.2 and 24.2), we are led to the construction below.

2/. Construction.

On circle (o) we set an arbitrary point T and form pencil T(αβγδε,...ν, - α'β'γ'δ'ε',...,ν'), the beams of which, we suppose re-intersect circle (o) at points A, B, Γ, Δ, E,...,N, - A', B', Γ', Δ', E',...,N' , respectively [If relation (1) is not given to us, we construct it ourselves (Construction 177)].

We say that the series of points A, B, Γ, Δ, E,...,N, - A', B', Γ', Δ', E',...,N', is the desired Open Harmonic one.

3/. Proof.

Since the Open Collinear Series of Points of 2n Points, α, β, γ, δ, ε,...ν, - α', β', γ', δ', ε',...,ν', is Harmonic, according to the relevant definition (§ 24.2), Central Pencil T(αβγδε...ν- α'β'γ'δ'ε'...ν'), or T(ΑΒΓΔΕ...Ν - Α'Β'Γ'Δ'Ε' ...Ν') will also be Harmonic.

Thus, according to Criterion 174 (converse) the cocyclic series of points A, B, Γ, Δ, E,...,N,   A', B', Γ',Δ', E',...,N' , will also be Harmonic, thus it is the desired one, as it belongs to circle (o).

(b). Converse.

1/. Analysis.

I.e. we are given on circle (o), the Pair of the Harmonic Open Cocyclic Series of Points of 2n Points: A, B, Γ, Δ, E,...,N, - A', B', Γ',Δ', E', ... ,N' and we are asked to construct on a given line (ε) the Pair of a Harmonic Open Collinear Series of Points of 2n Points:

α, β, γ, δ, ε,...ν, - α', β', γ', δ', ε',...,ν',   (1).

If we take into consideration Criterion 174 and the relevant definitions (§ 23.1, 23.2 and 24.2) , we are led to the construction below.

2/. Construction.

On circle (o), we set an arbitrary point T and form pencil T(A B Γ Δ E...N - A' B' Γ' Δ' E' ...N'), the beams of which, we suppose intersect line (ε) at points α, β, γ, δ, ε,...ν, - α', β', γ', δ', ε',...,ν', respectively [If the Harmonic Open Cocyclic Series of Points of 2n Points: A, B, Γ, Δ, E,...,N, - A', B', Γ',Δ', E', ... ,N', is not given to us we can construct it in the same way as it is mentioned in construction 172 (§ 25.4)].

We say that the desired Harmonic Open Collinear Series of Points of 2n Points is the following: α, β, γ, δ, ε,...ν, - ,α', β', γ', δ', ε',...,ν'.

3/. Proof.

Since the Pair of the Open Cocyclic Series of Points of 2n Points: A, B, Γ, Δ, E,...,N,..-...A', B', Γ', Δ', E',...,N', is Harmonic then Pencil T(A B Γ Δ E...N - A' B' Γ' Δ' E' ...N'), ή T(αβγδε...ν,   α'β'γ'δ'ε'...ν'), will also be Harmonic (Criterion 174, direct).

Hence, according to the relevant definition (§ 24.2), since the Central Pencil T(αβγδε...ν, - α'β'γ'δ'ε'...ν'). is harmonic, then the following Open Collinear Series of Points of 2n Points will also be harmonic: α, β, γ, δ, ε,...ν, - α', β', γ', δ', ε',...,ν'.

Hence it is the desired one, as it is located on line (ε).

Remarks.

(a). The Problem above and its solution, we believe to be new and firstly appear here.

(b). Problem 176, the way it is given, it is undefined. If we want to define it we need to be given sufficient data.

(c). In an equivalent way we have constructed on a line harmonic groups of four, eight, ten, etc, or on a circle harmonic quadrilaterals, hexagons, octagons, etc (Constructions 99, 100, 101, 102, etc, respectively).

(d). Relevant Propositions are 35, 61-66, 96 - 103, 145, 171, 172, 174, 175, 179.

25.9. Converse of Diagonals of a Convex Pair of a Harmonic Open Cocyclic Series of Points of 2n Points (Significant Property. Extension to Proposition 106).

(Pair of a Convex Harmonic Cocyclic Polygonal Line).

Proposition 177 (Figure 109).

The diagonals, of each Pair of a Convex Harmonic Open Cocyclic Series of Points of 2n Points, are concurrent.

Proof.

We consider the Pair of a Convex Harmonic Open Cocyclic Series of Points of 2n Points: A, B, Γ, Δ, E,...,N - A', B', Γ', Δ', E',...,N'. (1).

We will prove that: $AA'\cap BB'\cap \Gamma\Gamma'\cap \Delta\Delta'\cap EE'\cap\ldots\cap NN'\equiv\Sigma$. (2).

1st Way [Based on the Auxiliary Proposition 7 (in Part E)].

Since the given Pair of a Harmonic Open Cocyclic Series of Points of 2n Points is Harmonic (implication), according to the definition (§ 23.2), it will have harmonic for example quadrilaterals ABΓB' and A'B'Γ'B, thus according to known Theorem, it will be:

$$\frac{AB}{B\Gamma}\cdot\frac{\Gamma B'}{B'A}=1 \text{ and } \frac{A'B'}{B'\Gamma'}\cdot\frac{\Gamma' B}{BA'}=1 \Rightarrow \frac{AB}{B\Gamma}\cdot\frac{\Gamma B'}{B'A}=\frac{A'B'}{B'\Gamma'}\cdot\frac{\Gamma' B}{BA'}. \quad (3).$$

Thus, from relation (3), according to Auxiliary Proposition 7 (converse) (in Part E) , it means that the following relation is true:

$$AA'\cap BB'\cap \Gamma\Gamma' \equiv\Sigma. \quad (4).$$

Moreover, since from the definition the given Pair, will have harmonic quad rilaterals ΒΓΔΓ' and Β'Γ'Δ'Γ as well, if we work in the exact same way as above, we easily find that : $BB'\cap\Gamma\Gamma'\cap\Delta\Delta' \equiv\Sigma$. (5).

Moreover, since from the definition the given Pair, will have harmonic quadrilaterals ΓΔΕΔ' and Γ'Δ'Ε'Δ, as well, if we work in the exact same way as above, we easily find that $\Gamma\Gamma'\cap\Delta\Delta'\cap EE'\equiv\Sigma$. (6).

We continue in the exact same way as above and finally reach the harmonic quadrilaterals ΛΜΝΜ' and Λ'Μ'Ν'Μ, from which we get:

$$\Lambda\Lambda'\cap MM'\cap NN'\equiv\Sigma. \qquad (\lambda).$$

Thus, from (4), (5), (6),…., (λ), relation (2) occurs.

2nd Way (Based on Proposition 51).

Since the given Pair of a Harmonic Open Cocyclic Series of Points of 2n Points is Harmonic (implication), according to the definition (§ 23.2), will have harmonic for example quadrilaterals ΑΒΓΒ' and Α'Β'Γ'Β, thus according to Proposition 51 (converse) (§ 9.12) , it will be $AA'\cap BB'\cap\Gamma\Gamma'\equiv\Sigma$. (4).

Moreover, since from the definition the given Pair, will have harmonic quadrilaterals ΒΓΔΓ' and Β'Γ'Δ'Γ as well, if we work in the exact same way as above, we easily find that : $BB'\cap\Gamma\Gamma'\cap\Delta\Delta' \equiv\Sigma$. (5).

Moreover, since from the definition the given Pair, will have harmonic quadrilaterals ΓΔΕΔ' and Γ'Δ'Ε'Δ, as well, if we work in the exact same way as above, we easily find that $\Gamma\Gamma'\cap\Delta\Delta'\cap EE'\equiv\Sigma$. (6).

We continue in the exact same way as above and finally reach the harmonic quadrilaterals ΛΜΝΜ' and Λ'Μ'Ν'Μ, from which we get:

$$\Lambda\Lambda'\cap MM'\cap NN'\equiv\Sigma. \qquad (\lambda).$$

Thus, from (4), (5), (6),…., (λ), relation (2) occurs.

3rd Way (Based on Construction 172).

If we take into consideration the way of construction of a Pair of a Harmonic Open Cocyclic Series of Points of 2n Points 172 (§ 25.4), it occurs that this pair, has concurrent diagonals. Hence relation (2) is true.

Remarks.

(a). Proposition 177, we believe to be new and firstly appear here.

(b). It is obvious that for Proposition 177, the converse is not true.

(c). Until now we have seen that, apart from the Pair of a Harmonic Open Cocyclic Series of Points of 2n Points, concurrent diagonals have the following Harmonic hexagons, octagons, decagons, etc (Propositions 104, 105, 106, respectively).
(d). Relevant Propositions are 51, 104-106, 172.

25.10. Criterion of Harmonic being of a Pair of a Convex Open Cocyclic Series of Points of 2n Points (Pair of a Convex Harmonic Cocyclic Polygonal Line).
Existence of a Lemoine Point on a Pair of a Harmonic Open Cocyclic Series of Points of 2n Points (Significant Property. Extension to Criterion 114).

Criterion 178 (Figure 109).
If a Pair of a Convex Open Cocyclic Series of Points of 2n Points, has concurrent diagonals and there exists a point on its plane for which its distance from each side of it, is analogous to the matching side, then and only then this Pair of a Convex Open Cocyclic Series of 2n Points, is Harmonic.

Proof.
(a). Direct.
We consider the inscribed in circle (o) Harmonic Pair of an Open Convex Series of Points of 2n Points:

$$A, B, \Gamma, \Delta, E,..,N - A', B', \Gamma', \Delta', E',...,N'. \qquad (1).$$

(Relevant terminology in § 23.2). We will prove that a point Σ' exists on its plane, for which it is:

$$\frac{\upsilon_\alpha}{\alpha}=\frac{\upsilon_\beta}{\beta}=\frac{\upsilon_\gamma}{\gamma}=\frac{\upsilon_\delta}{\delta}=\frac{\upsilon_\varepsilon}{\varepsilon}=....=\frac{\upsilon_\mu}{\mu}=\frac{\upsilon'_\alpha}{\alpha'}=\frac{\upsilon'_\beta}{\beta'}=\frac{\upsilon'_\gamma}{\gamma'}=\frac{\upsilon'_\delta}{\delta'}=\frac{\upsilon'_\varepsilon}{\varepsilon'}=....=\frac{\upsilon'_\mu}{\mu'}=\lambda. \qquad (2).$$

where $\upsilon_\alpha, \upsilon_\beta,...., \upsilon_\mu, - \upsilon'_\alpha, \upsilon'_\beta,...., \upsilon'_\mu$, is the distance of point Σ', from its side $AB=\alpha$, $B\Gamma=\beta$,..., $MN=\mu$, $A'B'=\alpha'$, $B'\Gamma'=\beta'$,..., $M'N'=\mu'$, respectively.
In fact, since the convex inscribed in circle (o) Pair of an Open Cocyclic Series of Points of 2n Points, in (1), is harmonic, according to Proposition 177

(§ 25.9), it will have AA'∩BB'∩ ΓΓ'∩ΔΔ'∩EE'∩...∩ NN'≡Σ, while according to the relevant definition (§ 23.2), its following quadrilaterals will be harmonic:

ABΓB', BΓΔΓ', ΓΔEΔ', ...,n-2 quadrilaterals
A'B'Γ'B, B'Γ'Δ'Γ, Γ'Δ'E'Δ,....,n-2 quadrilaterals } 2(n-2) quadrilaterals. (3).

This means that for example in quadrilateral ABΓB' its diagonals AΓ, BB', will be conjugate with respect to circle (o), thus according to auxiliary Proposition 6 (converse) in Part E, for triangle ABΓ BB' is its symmedian. Thus, according to known Theorem (Γεωμετρία [6] § 11-14.2), each point on BB' will be far from AB, BΓ distance analogous to each of these sides. This will obviously be true for the point Σ of convergence of the diagonals of the given inscribed in circle (o) Pair of Open Cocyclic Series of Points of 2n Points, which belongs to BB' as well, hence for point Σ, it will be:

$$\frac{u_\alpha}{\alpha}=\frac{u_\beta}{\beta}. \qquad (4).$$

If we work as above for the harmonic quadrilateral BΓΔΓ' and more specifically for triangle BΓΔ, we prove that for Σ it is as well:

$$\frac{u_\beta}{\beta}=\frac{u_\gamma}{\gamma}. \qquad (5).$$

Additionally, if we continue as exactly above for the rest triangles ΓΔE, ΔEZ,...,ΛMN, we find that for Σ it is as well:

$$\frac{u_\gamma}{\gamma}=\frac{u_\delta}{\delta},\ \frac{u_\delta}{\delta}=\frac{u_\varepsilon}{\varepsilon},\ldots,\ \frac{u_\lambda}{\lambda}=\frac{u_\mu}{\mu}. \qquad (6).$$

Moreover, from (3) it occurs that for example for quadrilateral A'B'Γ'B its diagonals A'Γ', B'B, will be conjugate with respect to circle (o), thus according to auxiliary Proposition 6 (converse) in Part E, for triangle A'B'Γ' B'B is its symmedian. Thus, according to known Theorem (Γεωμετρία [6] § 11-14.2), each point on BB' will be far from A′B′, B′Γ′ distance analogous to each of these sides. This will obviously be true for the point Σ of convergence of the diagonals of the given inscribed in circle (o) Pair of Open Cocyclic Series of Points of 2n Points, which belongs to BB' as well, hence for point Σ, it will be:

$$\frac{\upsilon'_\alpha}{\alpha'}=\frac{\upsilon'_\beta}{\beta'}. \qquad (7).$$

If we work as above for the harmonic quadrilateral Β'Γ'Δ'Γ and more specifically for triangle Β'Γ'Δ', we prove that for Σ it is as well:

$$\frac{\upsilon'_\beta}{\beta'}=\frac{\upsilon'_\gamma}{\gamma'}. \qquad (8).$$

Additionally, if we continue as exactly above for the rest triangles Γ'Δ'Ε', Δ'Ε'Ζ',...,Λ'Μ'Ν', we find that for Σ it is as well:

$$\frac{\upsilon'_\gamma}{\gamma'}=\frac{\upsilon'_\delta}{\delta'},\ \frac{\upsilon'_\delta}{\delta'}=\frac{\upsilon'_\varepsilon}{\varepsilon'},\ldots,\frac{\upsilon'_\lambda}{\lambda'}=\frac{\upsilon'_\mu}{\mu'}. \qquad (9).$$

After all, for example from the similar triangles ΑΣΒ, Β'ΣΑ', it occurs that:

$$\frac{\upsilon_\alpha}{\alpha}=\frac{\upsilon'_\alpha}{\alpha'}. \qquad (10).$$

Consequently, from relations (4) – (10) above, we easily get relation (2), which is true for intersection Σ. This means that the desired point Σ' is Σ or Σ'≡Σ.

This means that the desired point is point Σ where the diagonals of the given inscribed in circle (o) Pair of an Open Series of Points of 2n Points in (1).

(b). Converse.

I.e., if the inscribed in circle (o) Pair of an Open Convex Cocyclic Series of Points of 2n Points in (1), has concurrent diagonals at point Σ and for Σ relation (2) is true, then we will prove that this pair will be harmonic..

In fact, since (2) is true, for Σ it will be:

$$\frac{\upsilon_\alpha}{\alpha}=\frac{\upsilon_\beta}{\beta}. \qquad (4).$$

This, according to the converse of the Theorem in § 11-14.2 in Γεωμετρία [6], shows that Σ belongs to the symmedian of side ΑΓ in triangle ΑΒΓ and thus, according to the direct of the above auxiliary Proposition 6 (in Part E), ΒΒ' and ΑΓ are conjugate with respect to circle (o), or quadrilateral ΑΒΓΒ' is harmonic.

If we take into consideration, that the rest two-part equalities that occur from (2) are true, i.e. the following ones:

$$\frac{υ_β}{β}=\frac{υ_γ}{γ}, \quad \frac{υ_γ}{γ}=\frac{υ_δ}{δ}, \quad \frac{υ_δ}{δ}=\frac{υ_ε}{ε},\dots,\frac{υ_λ}{λ}=\frac{υ_μ}{μ}, \quad \frac{υ'_α}{α'}=\frac{υ'_β}{β'}, \quad \frac{υ'_β}{β'}=\frac{υ'_γ}{γ'},$$

$$\frac{υ'_γ}{γ'}=\frac{υ'_δ}{δ'}, \quad \frac{υ'_δ}{δ'}=\frac{υ'_ε}{ε'},\dots,\frac{υ'_λ}{λ'}=\frac{υ'_μ}{μ'} \qquad (11).$$

and if we work as exactly above, we find that the rest quadrilaterals in (3), ΒΓΔΓ', ΓΔΕΔ', ΔΕΖΕ', ΕΖΗΖ',....., Α'Β'Γ'Β, Β'Γ'Δ'Γ, Γ'Δ'Ε'Δ,..., are harmonic, hence, according to the definition of the Pair of an Open Cocyclic Series of Points of 2n Points (§ 23.2) then the given Pair of an Open Cocyclic Series of Points of 2n Points in (1), is in fact Harmonic too.

Remarks.

(a). Proposition 178 (criterion), we believe to be new and firstly appear here.

(b). We have proven earlier here that an equivalent Proposition (Criterion), is also true for Harmonic Hexagons, Octagons, Decagons, etc (Criteria 112, 113, 114, etc, respectively).

(c). Relevant Propositions are 112, 113 - 115, 177.

Comment.

Due to the above Criterion 178 of the Pair of a Harmonic Open Cocyclic Series of Points of 2n Points, we notice that:

(a). The Lemoine Point of every triangle, has the property, according to which, this point is far from each side of the triangle, distance analogous to the respective side (of the triangle).

(b). Equivalent Property is true for Harmonic quadrilaterals, Hexagons, Octagons, Decagons, etc (Criteria 112-114, etc).

This is the reason why the intersection of the diagonals of the Harmonic Polygonal Lines, is called Lemoine Point of these Harmonic Polygonal Lines, so we reach the following conclusion.

Due to what we have mentioned before and in equivalence to these, it occurs that:

The point of convergence of the diagonals of every Harmonic Pair of an Open Cocyclic Series of Points of 2n Points, is a LEMOINE POINT, for this Harmonic pair.

Due to the above Criterion 178, we reach another conclusion as well:

Every convex Harmonic Pair of an Open Cocyclic Series of Points of 2n Points, with concurrent diagonals, is harmonic, if and only if, it has a LEMOINE POINT, which coincides with the point of convergence of its diagonals.

Alternative Definition of a Harmonic Pair of an Open Cocyclic Series of Points of 2n Points.

Consequently, after the second conclusion above, from now on we can use for the Harmonic Pair of an Open Cocyclic Series of Points of 2n Points, the Alternative Definition below:

«Harmonic Pair of an Open Cocyclic Series of Points of 2n Points (Harmonic Cocyclic Line), is called every convex Pair of an Open Cocyclic Series of Points of 2n Points, then and only then, if it has a LEMOINE POINT.

25.11. Construction of a Pair of a Harmonic Open Cocyclic Series of Points of 2n Points, using our New Method of the «HARMONIC TRANSFORMATION». (Pair of a Convex Harmonic Cocyclic Polygonal Line).

How from a Regular Pair of an Open Cocyclic Series of Points of 2n Points Occurs a Harmonic one (Significant Construction).

In General.

We have always believed that a Harmonic Line (Collinear and Cocyclic), comes from a Regular Cocyclic Line. But how though?

The solution to the Construction of a Pair of a Harmonic Open Cocyclic Series of Points of 2n Points, based on a Pair of a Regular Open Cocyclic Series of Points of 2n Points, though we tried, we could not achieve until we came up with the Method of the Harmonic Transformation.

Another of construction of the Harmonic Cocyclic Line, is given in Construction 172 (§ 25.4), but not based on a Regular Cocyclic Line.

Finally this Construction, is easily achieved applying our New Method of the «Harmonic Transformation», (See relevantly § 21 and the Problem of the Vicennium 157, § 22.6), which we will also apply here given that this method can be easily applied to all relevant problems, but here based on a Regular Cocyclic Line.

Using this Method we firstly achieve the construction of the Collinear and consequently of the Cocyclic Harmonic Line.

A first taste of this application is given in paragraph 24.5.

The application of this method in the construction of a Collinear and Cocyclic Line is complex and achieved through the proper SYNTHESIS of Constructions 175 and 176 (Synthesis of figures 116 and 115).

For Problems 152-155 we have applied this method for quadrilaterals, hexagons, octagons, decagons and now below for a Pair of a Harmonic Open Collinear Series of Points of 2n Points.

General Construction 179 (Figure 111).

Based on proper projective transformations and intersections from a line and circles, construct a Pair of a Harmonic Convex Open Cocyclic Series of Points of 2n Points, based on a Regular respective pair, and conversely in a similar way and based on a Harmonic Cocyclic Pair construct the respective Regular one.

Solution.

1/. Direct.

a/. Analysis.

Here we are given the Pair of a Regular Open Cocyclic Series of Points of 2n Points: A, B, Γ, Δ, E,...,N - A', B', Γ', Δ', E',...,N', (1).

which is inscrcibed in circle (o) and we are asked, based on that and with proper projective transformations to construct a Pair of a Harmonic Open Cocyclic Series of Points of 2n Points:

A_1, B_1, $Γ_1$, $Δ_1$, E_1,...,N_1 – A'_1, B'_1, $Γ'_1$, $Δ'_1$, E'_1,...,N'_1. (2).

If we take into consideration Constructions 175 (§ 25.7) and 176 (§ 25.8) and we combine them properly, we are led to the construction below.

b/. Construction.

Initially, we work in the exact same way as in Construction 175 (converse) and after we define an arbitrary point Σ on circle (o), we form the regular pencil Σ(AΓBΔE ...N - A'B'Γ'Δ'E'...N'), the beams of which intersect an arbitrary line (ε), at points α, β, γ, δ, ε,...,ν - α', β', γ', δ', ε',...,ν', respectively.

Consequently, we work in the exact same way as in Construction 176 (direct) and after we define an arbitrary point T on a random circle (o'), we form Pencil T(αβγδε...ν - α'β'γ'δ'ε'...ν'), whose beams re-intersect circle (o'), at points A_1, B_1, $Γ_1$, $Δ_1$, E_1,...,N_1 – A'_1, B'_1, $Γ'_1$, $Δ'_1$, E'_1,...,N'_1, respectively.

We say that the Pair of an Open Cocyclic Series of Points of 2n Points A_1, B_1, $Γ_1$, $Δ_1$, E_1,...,N_1 – A'_1, B'_1, $Γ'_1$, $Δ'_1$, E'_1,...,N'_1, is the desired Harmonic one.

c/. Proof.

Initially we work in the exact same way as in the proof of Construction 175 (converse) and we prove, similarly to there, that the Pair of the Regular Open Collinear Series of Points of 2n Points α, β, γ, δ, ε,...,ν - α', β', γ', δ', ε',..., ν', is Harmonic.

Consequently we work, in the exact same way as in the proof of Construction 176 (direct) and we prove that the Pair of the Open Cocyclic Series of Points of 2n Points A_1, B_1, $Γ_1$, $Δ_1$, E_1,...,N_1 – A'_1, B'_1, $Γ'_1$, $Δ'_1$, E'_1,...,N'_1, is Harmonic.

2/. Converse.

For the converse we follow the exactly reverse route.

a/. Analysis.

I.e., here we are given the Pair of the Harmonic Open Cocyclic Series of Points of 2n Points: A_1, B_1, $Γ_1$, $Δ_1$, E_1,..., N_1 – A'_1, B'_1, $Γ'_1$, $Δ'_1$, E'_1,...,N'_1, which is inscribed in circle (o') and we ask, based on that and with proper projective transformations to construct a Pair of a Regular Open Cocyclic Series of Points of 2n Points: A, B, Γ, Δ, E,...,N - A', B', Γ', Δ', E',...,N'.

If we take into consideration here as well Constructions 176 and 175 and we combine them properly, we are led to the Construction below.

b/. Construction.

Initially, we work in the exact same way as in Construction 176 (converse) and after we define an arbitrary point T on circle (o'), we form Pencil $T(A_1B_1Γ_1Δ_1E_1...N_1 – A'_1B'_1Γ'_1Δ'_1E'_1...N'_1)$, the beams of which intersect an arbitrary line (ε), at points α, β, γ, δ, ε,...,ν - α', β', γ', δ', ε',...,ν', respectively.

Consequently, we work in the exact same way as in Construction 175 (direct) and after we define at least one «Orthooptical Point» Σ, (as the intersection of circles with diameters for example segments ββ', γγ'), of the Pair of the Harmonic Open Collinear Series of Points of 2n Points: α, β, γ, δ, ε,...,ν - α', β', γ', δ', ε',...,ν', we draw an arbitrary circle (o), that runs through Σ.

If A, B, Γ, Δ, E,...,N - A', B', Γ', Δ', E',...,N'. are the points where Pencil T(αβγδε...ν - α'β'γ'δ'ε'...ν') whose beams re-intersect circle (o), respectively, we say that the Pair of the Open Cocyclic Series of Points of 2n Points A, B, Γ, Δ, E,...,N - A', B', Γ', Δ', E',...,N'. is the desired Regular one.

[If circle (o) is given and does not run through Σ, then we work in the exact same way as mentioned in Construction 175 (direct, § (a)2/)].

c/. Proof.

Initially we work in the exact same way as in the proof of Construction 176 (converse) and we prove, similarly to there, that the Pair of an Open Collinear Series of Points of 2n Points: α, β, γ, δ, ε,...,ν - α', β', γ', δ', ε',...,ν', is Harmonic.

Consequently we work in the exact same way as in the proof of Construction 175 (direct) and we prove, similarly to there, that Pencil Σ(αβγδε...ν' - α'β'γ'δ'ε'...ν'), or Σ(ΑΒΓΔΕ...Ν - Α'Β'Γ'Δ'Ε'...Ν'), is Regular, thus the Pair of a Harmonic Open Cocyclic Series of Points of 2n Points, Α, Β, Γ, Δ, Ε,...,Ν - Α', Β', Γ', Δ', Ε',...,Ν', is Regular too and thus it is the desired one.

Remarks.

(a). Problem 179, the way it is given, it is undefined. If we want to define it we need to be given sufficient data.

(b). The Problem above and its solution, we believe to be new and firstly appear here.

(c). We notice that this way above, is applied to a large number of relevant construction. This means that it is a method, which we have named «Harmonic Transformation» (§ 21).

The detailed interpretation of this method, we tried to practically give in the proof of Problem 179 above.

(d). We notice that using this method, the Harmonic Collinear Group of 2n occurs and then the Cocyclic one.

(e). The application of the above Method, is seen rather thoroughly and clearly in the above exact figure 111, which obviously occurs through the synthesis of figures 115 and 116.

(f). Relevant Propositions are 35, 74, 98, 152-157, 173-176.

25.12. General Conclusion of the new Method of Harmonic Transfomation.

The Method of the Harmonic Transformation is applied to the solution of Problems regarding:

(a). Harmonic Involutions (Collinear and Cocyclic)

Simple, for which we use inscribed multilaterals, symmetrical with respect to one of their diagonals (main) (Corollaries 158, 163, 165) and

Double, for which we use inscribed multilaterals, symmetrical with respect to two perpendicular diagonals (main) or theirs (Corollary 167).

(b). Harmonic Multilaterals and Harmonic Lines(Collinear and Cocyclic)

for which we use Regular Multilaterals, that are obviously symmetrical with respect to all their diagonals (main) (Constructions 152-157) and Pairs of Regular inscribed Lines (Construction 179).

(In Part D, we give other relevant complementary features of the Harmonic Multilaterals and Harmonic Lines, mostly applying the Method of the Harmonic Transformation).

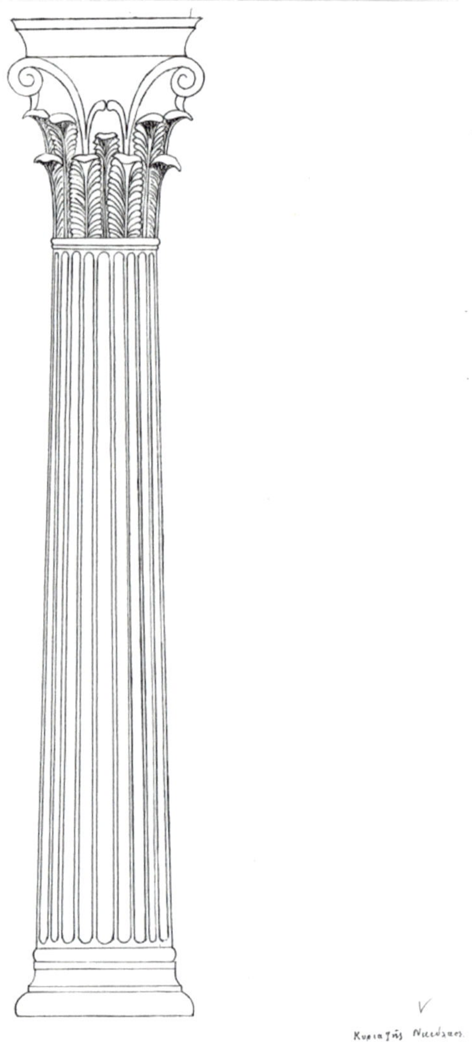

PART C

SUPPLEMENT OF PART A

Applications of Harmonic Involutions, Mostly using the New Method of Harmonic Transformation.

Carl Friedrich Gauss (1777-1855).

PART C

SUPPLEMENT OF PART A

Applications of Harmonic Involutions, Mostly using the New Method of Harmonic Transformation.

26. In General.

Part C is actually an extension of Part A′ and includes some additional information about Harmonic Involutions (Collinear and Cocyclic), that came up after having written Part B. These data will obviously be inserted in Part A in a new edition of the book.

Mostly though it includes applications of Harmonic Involutions using our New Method of Harmonic Transformation, which we have developed in paragraph 21 (although here it is adjusted to Harmonic Involutions) as while writing Part A, we had not invented this method yet.

Please note here that using this important Method of «Harmonic Transformation», we can methodically, easily and spherically solve the majority of the Problems in Harmonic Geometry, even the ones that have been impossible to solve so far.

Especially for the application of this method in Harmonic Involutions, we will use the Rules (Corollaries), we have given with Corollaries 158 till 168.

As we will later discover, while applying the Harmonic Transformation, on Harmonic Involutions:

a. Simple Harmonic Involutions (Collinear and Cocyclic), occur from inscribed 2n-laterals in circle, symmetrical respective to one of their diagonals, or from rhomboids inscribed in circle that have a common diagonal respective to which they are symmetrical.

b. Double Harmonic Involution (Collinear and Cocyclic), occur from 2n-laterals inscribed in circle, symmetrical respectively to two perpendicular diagonals, or from rectangles inscribed in circle that have their two axes of symmetry in common.

27. Introducing the Method of Harmonic Transformation in Harmonic Involutions.

27.1. Simple Harmonic Involutions (Collinear and Cocylic).

Just like we have mentioned earlier, Simple Harmonic Involutions (Collinear and Cocylic), occur from 2n-laterals inscribed in circle symmetrical respective to one of their main diagonals, or from rhomboids inscribed in circle which share their diagonal, the one that is their axis of symmetry.

27.2. Applications of Simple Harmonic Involutions.

27.21. Construction of a Simple Harmonic Involution with One Pair of Conjugate Points (See § 3.2, 8.11, 8.61 and Construction 135).

(How a Simple Harmonic Cocyclic Involution occurs from an Inscribed Rhomboid).

Construction 180 (Figure 100).

Based on appropriate projective transformations and intersections with line and circles, construct a Simple Cocyclic Harmonic Involution with One Pair of Conjugate Points, based on an Inscribed Rhomboid, given to us, and conversely, similarly based on a Simple Cocylic Harmonic Involution of One Pair of Conjugate Points, given to us, construct an Inscribed Rhomboid.

Solution.

1/. Direct.

(a). Analysis.

Here we have as given the rhomboid ΑΒΓΔ inscribed in circle (o) and, based on that and appropriate projective transformations, we are asked to find a Simple Cocyclic Harmonic Involution with double points Β', Δ' and a pair of Conjugate Points Α', Γ'.

Taking into consideration Corollary 158 (§ 22.61 A), Criteria 35 and 75, as well as paragraphs 3.2, 8.11, 8.61, are led to the construction below:

(b). Construction.

We work in the same way as in the implication (first case) of Corollary 158, we firstly construct the Symmetrical Pencil Σ(ΑΒΓΔ) or Σ(αβγδ) and then its intersection with line (ε), thus the Collinear Harmonic Involution of Double Points β, δ and of One Pair of Conjugate Points α, γ appears.

Consequently, we construct the Harmonic Pencil Τ(αγβδ) or Τ(Α'Γ'Β'Δ') [where Τ∈(o′)], or the Simple Cocyclic on circle (o′) Harmonic Involution with Β', Δ' double points and of One Pair of Conjugate Points Α', Γ', which we say is the demanded one.

(c). Proof.

According to the relevant definition (§ 3.1), we can easily find that pencil Σ(ΑΒΓΔ) or Σ(αβγδ) is symmetrical respectively to ΣΒ and ΣΔ. Thus the points of its intersection with line (ε) α, γ, will be in Harmonic Involution, with β, δ its Double Points, or that the Group of Four Points α, β, γ, δ, is harmonic, therefore pencil Τ(αγβδ) or Τ(Α'Γ'Β'Δ'), will be harmonic (Criterion 35). This means that Α' and Γ', harmonically divide Β', Δ', thus according to the definition in paragraph 8.11, points Α', Γ', are in Harmonic Involution with Double Points Β', Δ'.

2/. Converse.

For the converse we have to follow the opposite direction.

(a). Analysis.

So, we are given the Harmonic Involution of One Pair of Conjugate Points Α', Γ' and of Double Points Β', Δ', in a circle (o′), and we are asked to inscribe in circle (o) a rhomboid ΑΒΓΔ, with ΒΔ its axis of symmetry.

Taking into consideration Corollary 158, Criteria 35 and 75, as well as paragraphs 8.11, 8.61, 3.2, we are led to the construction below.

(b). Construction.

We work similarly to the converse (first case) of Corollary 158, we firstly construct pencil Τ(Α'Γ'Β'Δ') or Τ(αγβδ) which intersects with line (ε) at points α, β, γ, δ.

Because we want the quadrilateral ΑΒΓΔ, that will occur to be symmetrical with respect to ΒΔ, this is why we set a point Σ on the circle of diameter βδ and on circle (o). This means that we set as Σ the point of intersection of these two circles.

Consequently, we construct pencil Σ(αγβδ), which also intersects circle (o) at points A, B, Γ, Δ and we say that quadrilateral ΑΒΓΔ, is the wanted rhomboid, symmetrical with respect to ΒΔ.

[In the case that the circle of diameter βδ does not intersect with circle (o), then we use an auxiliary circle, equal to circle (o), with which it intersects and after we construct ΑΒΓΔ, we move it to circle (o)].

(c). Proof.

In fact, because points A', B', Γ', Δ' are in Harmonic Involution, from the implication, with Double Points B', Δ' and Pair of Conjugate Points A', Γ', quadrilateral A'Γ'B'Δ' is harmonic, thus according to the converse of Criterion 35, pencil Σ(A'Γ'B'Δ') or Σ(αγβδ) would be harmonic. Therefore the group of four points α, γ, β, δ, would be harmonic, or these points will be in Harmonic Involution, with Double Points β, δ and Pair of Conjugate Points α, γ.

Moreover, because the group of four points α, γ, β, δ is harmonic and angle βΣδ is right, in triangle Σαγ, Σβ will be the bisector of angle αΣγ.

Thus ΣΒ is the bisector of angle ΑΣΓ, therefore ΑΒ=ΒΓ and because angle ΒΣΔ is right, ΒΔ is a diameter of circle (o). As a result, we can easily find that quadrilateral ΑΒΓΔ is in fact symmetrical with respect to ΒΔ, this means that it is the wanted rhomboid.

(d). Investigation.

In the special case that for the rhomboid ΑΒΓΔ, ΑΓ is also a diameter, then this will be a square, so for the implication we work similarly to paragraph 1/ above. Concerning the converse to find square ΑΒΓΔ we have to set the intersection of circles of diameters βδ and αγ as Σ (Orthooptical Point) (See Construction 152).

Remarks.

(a). Problem 180, the way it is given is undefined. If we want to define it, we need sufficient data.

(b). The above Problem and its solution, we believe to be new and firstly appear here.

(c). We see that for the solution of the above Construction, where we used the Method of Harmonic Transformation, we did not use the Theory about Poles and Polar Coordinates (§ 8.61).

(d). We notice that with this method, firstly a Harmonic Collinear Involution occurs and then the wanted Harmonic Involution. This means that with the New Method of Harmonic Transformation, we can simultaneously achieve the Construction of the Harmonic Collinear Involution.

(e). Two other solutions of Problem 180, can be found in Construction 135).

(f). Other relevant Propositions are 35, 61, 74, 75, 79, 99, 135, 158.

27.22. Construction of the Simple Cocyclic Harmonic Involution of two Pairs of Conjugate Points (Relevant paragraphs 3.4 and 8.62).

(How a Simple Harmonic Cocylic Involution occurs from a Convex Inscribed Symmetrical Hexagon).

Construction 181 (Figure 103).

Based on appropriate projective transformations and intersections with line and circles, construct a Simple Cocyclic Harmonic Involution of two Pairs of Conjugate Points, based on a Inscribed Convex Symmetrical with respect to one of its Diagonals Hexagon, which is given to us, and conversely, similarly and based on a Simple Cocyclic Involution of two Pairs of Conjugate Points, which is given to us, construct an Inscribed Convex Symmetrical with respect to one of its Diagonals Hexagon.

Solution.

1st Way (Based on Construction 180).

1/. Direct.

(a). Analysis.

Here a hexagon ΑΒΓΔΕΖ inscribed in circle (o) convex symmetrical with respect to ΒΕ is given and we are asked, based on that and with appropriate projective transformations to find a Simple Cocyclic Harmonic Involution with double points Β', Ε' and two pairs Α', Γ' and Δ', Ζ' of Conjugate Points.

Taking into consideration Construction 180 (§ 27.21), Criteria 35 and 75, just like in paragraphs 3.2, 8.11, 8.62, we are led to the construction below:

(b). Construction.

We notice that the symmetrical with respect to BE hexagon ΑΒΓΔΕΖ, is analyzed in two rhomboids ΑΒΓΕ and ΔΕΖΒ, with common diagonal BE.

This way, based on these rhomboids and Construction 180, we create on line (ε) the Collinear Harmonic Involution of one pair of Conjugate Points α, γ and Double points β, ε and afterwards in circle (o′) we construct the Cocyclic Harmonic Involution of one pair of Conjugate Points A′, Γ′ and Double points B′, E′.

Consequently, working just like above, we construct line (ε) the Collinear Harmonic Involution of one pair of Conjugate Points, δ, ζ and Double Points β, ε and afterwards on circle (o′) we construct the Cocyclic Harmonic Involution of one pair of Conjugate Points Δ′, Z′ and Double Points B′, E′.

We say that the desired involution, consists of two Pairs of Conjugate Points A′, Γ′ and Δ′, Z′, and Double Points B′, E′.

(c). Proof.

In fact, if we combine together the above constructed involutions of one Pair of Conjugate Points each one of line (ε), we will have the Harmonic Involution of two Pairs of Conjugate Points α, γ and δ, ζ, with Double Points β, ε.

In addition, if we combine together the above constructed involutions of one Pair of Conjugate Points each one of circle (o′), we will have the Harmonic Involution of two Pairs of Conjugate Points A′, Γ′ and Δ′, Z′ and with Double Points B′, E′, which is in fact the desired one.

2/. Converse.

For the converse we follow the opposite direction.

(a). Analysis.

Here we are given the Harmonic Involution, with two Pairs of Conjugate Points A′, Γ′ and Δ′, Z′ and with Double Points B′, E′ and we are asked, based on that and with the appropriate projective transformations to find a hexagon ΑΒΓΔΕΖ inscribed in circle (o) convex and symmetrical with respect to BE.

Taking into consideration Construction 180 (§ 27.2), Criteria 35 and 75, as well as paragraphs 3.2, 8.11, 8.62, we are led to the construction below:

(b). Construction.

Based on Construction 180, we construct figure 103. In which we notice that the Harmonic Involution of two Pairs of Conjugate Points A′, Γ′ and Δ′, Z′ and with Double Points B′, E′, is analyzed in two Harmonic Involutions with common Double Points B′, E′ and each one of one Pair of Conjugate Points, which are A′, Γ′ for one and Δ′, Z′ for other, thus based on Construction 180 (converse), the two Involutions of one Pair of Conjugate Points each occur on line (ε) respectively, from which with the appropriate combination a collinear involution of two Pairs of Conjugate Points occurs.

Thus, from the two involutions above and based on Construction 180 (converse), we realize that the two quadrilaterals ABΓE and ΔEZB are rhomboids that have BE as a common axis of symmetry and from which the desired symmetrical hexagon ABΓΔEZ occurs.

(c). Proof.

In fact, if we combine together the two involutions of line (ε) above the Pair of Conjugate Points of the first is, α, γ and of the other δ, ζ, and with Pair of Double Points β, ε, then the Harmonic Involution of two Pairs of Conjugate Points α, γ and δ, ζ and with Double Points β, ε occur.

Moreover, we easily realize that if we combine the two rhomboids ABΓE and ΔBZE, that share their axis of symmetry BE, hexagon ABΓΔEZ occurs, which we easily realize that it is symmetrical with respect to BE.

2nd Way (Based on Corollary 159).

1/. Direct.

(a). Analysis.

Taking into consideration Corollary 159, we are led to the construction below:

(b). Construction.

Based on Construction 180, we construct figure 103, in which we say that the two pairs of points A′, Γ′ and Δ′, Z′, are in Harmonic Involution, with its Double Points B′, E′ and that it is the desired involution.

(c). Proof.

Because hexagon ΑΒΓΔΕΖ is symmetrical with respect to ΒΕ, ΑΓ and ΔΖ will be perpendicular to ΒΕ, thus, according to Corollary 159 (direct) these, transform to the conjugate lines Α′Γ′ and Δ′Ζ′, with Β′Ε′ both.

Therefore, according to known facts, the two pairs of points Α′, Γ′ and Δ′, Ζ′, are in Harmonic Involution, with Double Points Β′, Ε′ and therefore this is in fact the one desired.

2/. Converse.

For the converse we follow the opposite direction.

(a). Analysis.

Taking into consideration Corollary 159, we are led to following construction:

(b). Construction.

Based on Construction 180, we construct figure 103. in which we say that hexagon ΑΒΓΔΕΖ, is the desired symmetrical with respect to ΒΕ.

(c). Proof.

Because pairs of points Α′, Γ′ and Δ′, Ζ′, are in Harmonic Involution, with Double Points Β′, Ε′, Α′Γ′ and Δ′Ζ′, are conjugate with respect to Β′Ε′.

Thus, according to the converse of Corollary 159, Α′Γ′ and Δ′Ζ′, transform into the lines ΑΓ and ΔΖ perpendicular to diameter ΒΕ, therefore it easily occurs that hexagon ΑΒΓΔΕΖ, is in fact the desired symmetric with respect to ΒΕ one.

Remarks.

(a). Problem 181, the way it is given, it is undefined. If we want to define it, then we need sufficient data.

(b). The above Problem and its solutions, we believe are new and firstly appear here.

(c). We notice that with this method, the Harmonic Collinear Involution occurs first and then the desired Harmonic Cocyclic Involution. <u>This means that with the New Method of Harmonic Transformation, we simultaneously achieve the Construction of the Harmonic Collinear Involution.</u>

(d). In figure 103, the involution of line (ε) is a two-branch one, as point Σ was set on the arc AZ which is placed between the two parallel lines ΑΓ and ΔZ.
Would we want it to be a one-branch one, we should place point Σ on arcs ΑΒΓ or ΔEZ.
(e). Construction 181 and its proof, can be based on Corollary 162, or Corollary 163, or paragraphs 3.3, 3.4 and 8.62 (Simple Harmonic Involute Pencil of three pairs of beams), etc.
(f). Other relevant Propositions are 35, 42, 45, 75, 135, 158, 159, 162, 163, 180.

27.23. Construction of a Simple Cocyclic Harmonic Involution of three Pairs of Conjugate Points (Relevant § 8.64).

(How a Simple Harmonic Cocyclic Involution occurs from a Convex Inscribed Symmetrical Octagon).

Construction 182 (Figure 104).

Based on appropriate projective transformations and intersections with line and circles, construct a Simple Cocyclic Harmonic Involution of three Pairs of Conjugate Points, based on an Inscribed Convex Symmetrical with respect to its Diagonal Octagon, which is given to us, and conversely, similarly and based on a Simple Cocyclic Harmonic Involution of three Pairs of Conjugate Points, which is given to us, construct an Inscribed Convex Symmetrical with respect to its Diagonal Octagon.

Solution.

Based on Corollary 164 (§22.61G), we construct figure 104.
Hence, based on this figure and through analyzing it to and combining it with figures equivalent to figures 100 and 103, working similarly to Constructions 180 and 181, we can easily solve this Problem as well, taking into consideration paragraph 8.64).

Remarks.

(a). Problem 182, the way it is given, it is undefined. Would we want to define it, then we need sufficient data.

(b). The above Problem and its solution, we believe to be new and firstly appear here.

(c). We notice that with this method, the Harmonic Collinear Involution occurs first and then the desired Harmonic Cocyclic Involution. This means that, with the New Method of Harmonic Transformation, we simultaneously achieve the Construction of the Harmonic Collinear Involution.

(d). In figure 104, the involution of line (ε) is a two-branch one, as point Σ is set on arc ΑΘ which is placed between the two parallel lines ΑΓ and ΕΗ. Would we want it to be an one-branch one, point Σ needs to be set on arcs ΑΒΓ or ΕΖΗ, which are placed outside of the two parallel lines ΑΓ and ΕΗ.

(e). Construction 182 and its proof, is possible to be based on Corollary 163, or on Corollary 162, or on paragraphs 3.5, 3.6 and 8.64 (Simple Harmonic Involute Pencil of three beams), etc.

(f). Other relevant Propositions are 35, 75, 135, 158, 159, 162-164, 180, 181.

27.24. Generalization of Constructions 180-182.

Construction of a Simple Cocyclic Harmonic Involution of n Pairs of Conjugate Points.

(How a Simple Harmonic Cocyclic Involution occurs from a Convex Inscribed Symmetrical 2(n+1)-lateral).

Let us continue in the same way as in Constructions 180-182, easily and similarly, it is possible to construct every other Simple Cocyclic Harmonic Involution of n Pairs of Conjugate Points, applying the New Method of Harmonic Transformation.

General Construction 183.

Based on proper projective transformations and intersections with line and circles, construct a Simple Cocyclic Harmonic Involution of n Pairs of Conjugate Points, based on an Inscribed Convex Symmetrical with respect to its Diagonal 2(n+1)-lateral, which is given to us, and conversely, similarly and based on Simple Cocyclic Harmonic Involution of n Pairs of Conjugate Points, which is given to us, construct an Inscribed Convex Symmetrical with respect to its Diagonal 2(n+1)-lateral.

Solution.

The solution to this Problem, for every Simple Cocyclic Harmonic Involution of n Pairs of Conjugate Points, is carried out just like in Constructions 180- 182, with proper combinations of them.

Conclusions.

From Constructions 180-182 above and the Universal Problem 183, the conclusions below occur, concerning the value of the Method of Harmonic Transformation:

(a). While applying the Harmonic Transformation, occurs at the same time, a Cocyclic Simple Harmonic Involution of n Pairs of Conjugate Points and a Collinear one occur at the same time.

(b). The Construction of a Simple Harmonic Involution of n Pairs of Conjugate Points (Collinear or Cocyclic), becomes easy when applying the Method of Harmonic Transformation.

General Remarks.

(a). Problem 183, the way it is given, it is undefined. Would we want to define it, then we need sufficient data.

(b). The above Universal Problem and its solution, we believe to be new and firstly appear here.

(c). We tried to present the detailed way of applying this method in involutions, through the proof of Constructions 180-182 above.

(d). The way of applying the Method above, is presented more descriptively and unambiguously in figures 97-106, 111, 117-120, 122-126, 128 above.

(e). Other relevant Propositions are, 28, 29, 35, 74, 75, 79, 99, 102, 135, 145, 149, 158, 159, 162-164, 180-182.

27.3. Double Harmonic Involutions (Collinear and Cocyclic).

As we have mentioned elsewhere, Double Harmonic Involutions (Collinear and Cocyclic), occur from 2n-laterals inscribed in circle, symmetrical with respect to two of their diagonals, perpendicular to each other, or from rectangles inscribed in circle and their two axis of symmetry in common.

27.4. Applications of Double Harmonic Involutions.

27.41. Construction of a Double Cocyclic Harmonic Involution of two Pairs of Conjugate Points.

(How a Double Harmonic Cocyclic Involution occurs from a Rectangle).

Construction 184 (Figure 117).

Based on proper projective transformations and intersections with line and circles, construct a Double Cocyclic Harmonic Involution of two Pairs of Conjugate Points, based on a Rectangle, which is given to us, and conversely, similarly and based on a Double Cocyclic Harmonic Involution of two Pairs of Conjugate Points, which is given to us, construct a Rectangle.

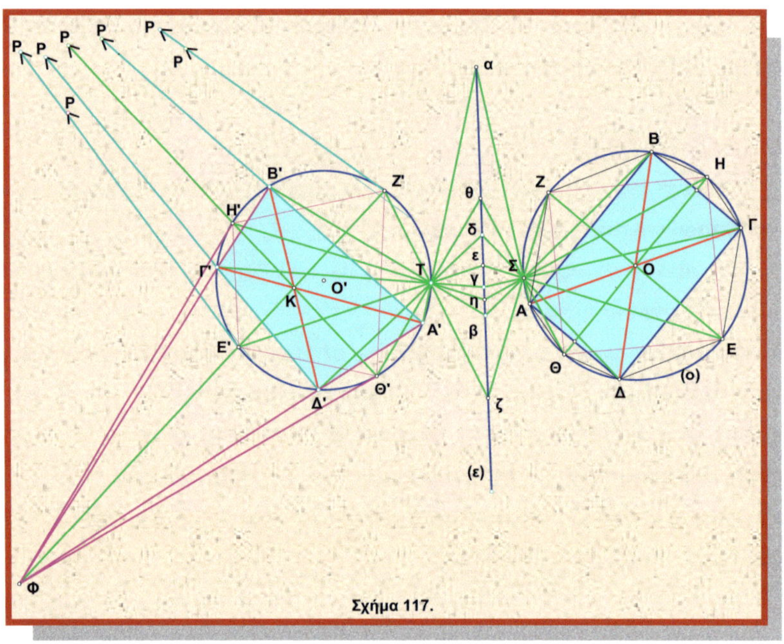

Σχήμα 117.

Solution.

1/. Direct.

(a). Analysis.

Let there be a rectangle ΑΒΓΔ, which we are asked to transform, into a Double Harmonic Involution of two Pairs of Conjugate Points (Definitions

§ 8.22 and 8.63), inscribed in circle (o′), using the Method of Harmonic Transformation.

If the two axes of symmetry of the rectangle ΑΒΓΔ intersect with the circle it is inscribed in (o) at points Ε, Ζ, Η, Θ (as in figure 117) and if we take into consideration Construction 181, for every hexagon ΑΘΔΓΗΒ and ΑΖΒΓΕΔ symmetrical with respect to its diagonal , we are led to the following construction:

(b). Construction.

We set a random point Σ on circle (o) and create pencils Σ(ΑΘΔΓΗΒ) and Σ(ΑΖΒΓΕΔ) or Σ(αθδγηβ) and Σ(αζβγεδ), which intersect a random line (ε) at these two groups of six points α, θ, δ, γ, η, β and α, ζ, β, γ, ε, δ.

Consequently on circle (o) we set a random point Τ and create pencils Τ(αθδγηβ) and Τ(αζβγεδ), or Τ(Α′Θ′Δ′Γ′Η′Β′) and Τ(Α′Ζ′Β′Γ′Ε′Δ′), which intersect circle (o′) at the following groups of six points Α′, Θ′, Δ′, Γ′, Η′, Β′ and Α′, Ζ′, Β′, Γ′, Ε′, Δ′, which should we combine in a proper way, then we say that they constitute together the desired Double Harmonic Involution of two Pairs of Conjugate Points.

(c). Proof.

We prove, as in the proof of Construction 181, that each one of the following groups of six points α, θ, δ, γ, η, β and α, ζ, β, γ, ε, δ, constitutes a Simple Harmonic Involution of two Pairs of Conjugate Points.

The first one with Double Points θ, η and of two Pairs of Conjugate Points α, δ and β, γ, and the second one with Double Points ε, ζ and of two Pairs of Conjugate Points α, β and γ, δ, hence these two involutions, combined, would constitute a Double Harmonic Involution of two Pairs of Conjugate Points (definitions § 2.24, 3.81, 7.14 and 8.63) (See also Constructions 11, 30).

According to Construction 181, each one from the following groups of six points Α′, Θ′, Δ′, Γ′, Η′, Β′ and Α′, Ζ′, Β′, Γ′, Ε′, Δ′, constitutes a Simple Harmonic Involution of two Pairs of Conjugate Points.

The first one with Double Points Θ′, Η′ and two Pairs of Conjugate Points Α′, Δ′ and Β′, Γ′, and the second with Double Points Ε′, Ζ′ and two Pairs of

Conjugate Points A′, B′ and Γ′, Δ′. Hence, combined together these two involutions make up the desired Double Harmonic Involution of two Pairs of Conjugate Points (definitions § 8.22 and 8.63) (See also relevant Construction 43, which constitutes another form of Construction 184).

1/. Converse.

(a). Analysis.

Let there be a Double Harmonic Involution with two Pairs of Double Points E′, Z′ and H′, Θ′ and two Pairs of Conjugate Points A′, Δ′ and B′, Γ′, based on which and applying the method of Harmonic Transformation we are asked to transform quadrilateral A′B′Γ′Δ′ into a rectangle ABΓΔ, inscribed in a given circle (o).

If we analyze the given Double Harmonic Involution into two simple ones and take into consideration Construction 181, we are led to the following construction:

(b). Construction.

On circle (o′) we set a random point T and create pencils T(A′Θ′Δ′Γ′H′B′) and T(A′Z′B′Γ′E′Δ′) or T(αθδγηβ) and T(αζβγεδ) which a random line (ε) intersects at the following groups of six points α, θ, δ, γ, η, β and α, ζ, β, γ, ε, δ.

Consequently we set as point Σ the intersection of the two circles of diameters εζ and ηθ, hence, if the given circle (o) runs through Σ and pencils Σ(αθδγηβ) and Σ(αζβγεδ) or Σ(ΑΘΔΓΗΒ) and Σ(ΑΖΒΓΕΔ), intersect circle (o), at the following groups of six points A, Θ, Δ, Γ, H, B and A, Z, B, Γ, E, Δ, of which we say that the quadrilateral ABΓΔ, is the desired rectangle. [If the given circle (o) does not run through Σ, then we use a supplementary circle, equal to the given one (o), that runs through Σ and after we construct rectangle ABΓΔ, we move it on circle (o)].

(c). Proof.

Because it is given that the following eight points A′, Θ′, Δ′, E′, Γ′, H′, B′, Z′, constitute a Double Harmonic Involution of two Pairs of Conjugate Points, which due to its properties (§ 8.63), is analyzed into two Simple Harmonic Involutions of two Pairs of Conjugate Points, from which the first's Double Points are H′, Θ′ and the two Pairs of Conjugate Points are A′, Δ′ and B′, Γ′

and the second with Double Points E′, Z′ and two Pairs of Conjugate Points A′, B′ and Γ′, Δ′, thus, according to Construction 181, let T be a random point on (o′), pencils T(A′Θ′Δ′Γ′H′B′) and T(A′Z′B′Γ′E′Δ′) or T(αθδγηβ) and T(αζβγεδ), would intersect a random line (ε), at points which constitute two Simple Collinear Harmonic Involutions of two Pairs of Conjugate Points, which if we combine in a proper way would constitute a Double Harmonic Involution of two Pairs of Conjugate Points, of line (ε).

Hence, according to Construction 181, from each one of the Simple Harmonic Involutions of two Pairs of Conjugate Points on line (ε), from the two Simple Involutions above, a symmetrical hexagon will occur, respectively. This means that, the symmetrical, with respect to HΘ, hexagon A-ΘΔΓHB will occur and the symmetrical, with respect to EZ, hexagon AZBΓEΔ, as well as Σ belongs to both circles of diameters ηθ and εζ (Orthooptical Point, definition § 7.14).

From the above it appears that HΘ is the perpendicular bisector of AΔ and BΓ, thus diameter of the circle (o) and that AΔ//BΓ.

Similarly we find that EZ is the perpendicular bisector of AB and ΓΔ, thus diameter of circle (o) and that AB//ΓΔ.

As a result, quadrilateral ABΓΔ, is the desired rectangle, as it is inscribed in circle (o).

Remarks.

(a). The two Double Involutions (collinear and cocyclic), that above were constructed to be harmonic and of course have all the properties, that have been mentioned earlier in relevant paragraphs. All these properties are easily proved here as well, if we depend on Corollaries 158-168.

(b). Problem 184, the way it is given, it is undefined. Would we want to define it, then we need sufficient data.

(c). The solution to the Problem above, we believe to be new and firstly appears here (Alternative solution in Construction 43).

(d). We notice that with this method, firstly a Double Harmonic Collinear Involution occurs, and then the desired Harmonic Cocyclic Involution with the New Method of Harmonic Transformation, hence we simultaneously achieve the Construction of the Harmonic Collinear Involution.

(e). In figure 117, the involution of line (ε) is of one and of two branches, as Σ was set on arc AZ, which is outside the parallel lines AB and ΓΔ and between the parallel lines AΔ and BΓ, respectively.

(f). The proof of Construction 184, can be based on Corollaries of paragraph 22.61.

(g). Other relevant propositions are 11, 31, 35, 43, 75, 180-183, 185.

27.42. Construction of a Double Cocyclic Harmonic Involution of four Pairs of Conjugate Points (Solution to Problem 44, with Harmonic Transformation).

(How a Double Harmonic Cocyclic Involution occurs from a Convex Inscribed Symmetrical Dodecagon).

Construction 185 (Figure 106).

Based on proper projective transformations and intersections with line and circles, construct a Double Cocyclic Harmonic Involution of Four Pairs of Conjugate Points, based on a Pair of Rectangles Inscribed in the same circle with two axes of symmetry in common, which is given to us, and conversely, similarly and based on a Double Cocyclic Harmonic Involution of Four Pairs of Conjugate Points, which is given to us, construct a Pair of Rectangles Inscribed in the same circle with two axes of symmetry in common.

Solution.

Based on Corollary 167 (§ 22.61 I), we construct figure 106.

Hence, based on this figure and by analyzing and combining it, to two figures equivalent to figure 117, working in the same way as in Construction 184, we easily solve this Problem as well.

Remarks.

(a). Problem 185, the way it is given, it is undefined. Should we want to define it, then we need sufficient data.

(b). The Problem above and its solution, we believe to be new and firstly appear here.

(c). We notice that with this method, firstly the Harmonic Collinear Involution occurs and then the desired Harmonic Cocyclic Involution. Hence, with the New Method of Harmonic Transformation, we simultaneously achieve the Construction of the Harmonic Collinear Involution as well.

(d). The proof of Construction 185, can be based on Corollaries of paragraph 22.61.

(e). Construction 185 is a solution to Problem 44, but here with the Method of Harmonic Transformation.

(f). Other relevant Propositions are 30, 31, 35, 43, 44, 75, 180-184, 186.

27.43. Generalization of Constructions 184, 185.

Construction of a Double Cocyclic Harmonic Involution of 2n Pairs of Conjugate Points.

(How a Double Harmonic Cocyclic Involution occurs from a Convex Inscribed Symmetrical Multilateral).

Should we continue in a way similar to Constructions 184, 185, easily and similarly, it is possible to construct every other Double Cocyclic Harmonic Involution of 2n Pairs of Conjugate Points, applying the New Method of Harmonic Transformation.

General Construction 186.

Based on proper projective transformations and intersections with line and circles, construct a Double Cocyclic Harmonic Involution of 2n Pairs of Conjugate Points, based on n Rectangles Insribed in the Same Circle, with their two Axes of Symmetry in common, which is given to us, and conversely, similarly and based on a Double Cocyclic Harmonic Involution of 2n Pairs of Conjugate Points, which is given to us, construct n Rectangles Insribed in the Same Circle, with their two Axes of Symmetry in common.

Solution.

The solution to this General Problem, for every Double Cocyclic Harmonic

Involution of 2n Pairs of Conjugate Points, is implemented just as in Constructions 184, 185, with the proper combinations of these.

Conclusions.

From the above Constructions 184, 185 and the Universal Problem 186, the following conclusions occur, regarding the Value of the Method of Harmonic Transformation:

(a). When applying the Harmonic Transformation, simultaneously with the Cocyclic Double Harmonic Involution of 2n Pairs of Conjugate Points, the Collinear one also occurs.

(b). The Construction of the Double Harmonic Involution of 2n Pairs of Conjugate Points (Collinear or Cocyclic), becomes easier with the Method of Harmonic Transformation.

General Remarks.

(a). Problem 186, the way it is given, it is undefined. Should we want to define it, we need sufficient data.

(b). The above General Problem and its solution, we believe to be new and firstly appear here.

(c). We have tried to provide the detailed way of applying this method in involutions, with the proof of Constructions 184, 185 above.

(d). The execution of the Method above, is descriptively and unambiguously developed in figures 106, 117 above.

(e). Other relevant Propositions are 30, 31, 158, 159, 162-164, 167, 180-185.

27.5 Various other Constructions of Collinear and Simultaneously Cocyclic Involutions, using the Method of Harmonic Transformation.

It is possible, if we continue working, in a way similar to Constructions 180 to 188, to have other Constructions of Involutions Collinear and at the same time Cocyclic, applying the Method of Harmonic Transformation, as you can see in the following Constructions 187 and 188.

This, though, will be left for another edition of the book, and as an exercise to help friends of Geometry understand all the above better. Here we will mention only a few of such examples, that are also mentioned in Part A.

Hence, it is possible, using the Method of Harmonic Transformation, to solve the following pairs of Problems about Collinear and Cocyclic Involutions, with the following equivalence in Problems:

12 → 46, 13 and 27 → 47, 14 → 48, 20 → 52, 21 → 53, etc.

27.51. Construction of a Simple Cocyclic Harmonic Involution of two Pairs of Conjugate Points.

Solution to the Problems 9, 10 and 42 (§ 9.3), using the Method of Harmonic Transformation.

Construction 187 (Figure 118).

On a circle (o′), there is an ordered group of points A′, Γ′, Δ′, Z′ given. If pairs of points A′, Γ′ and Δ′, Z′ are the two Pairs of Conjugate Points of a Simple Cocyclic Harmonic Involution, we are asked to define the pair of Double Points B′, E′ of the involution using the Method of Harmonic Transformation, or to state it otherwise:

Let there be a circle (o′) and the group of four ordered points A′, Γ′, Δ′, Z′ on it. Prove, using the Method of Harmonic Transformation, that only one pair of points B′, E′ on the same circle (o′) exists, which harmonically divides both pairs of points A′-Γ′ and Δ′-Z′.

Solution.

(a). Analysis.

Let us assume that the desired Double Points of the involution are set and are B′, E′.

If we take into consideration Construction 181 (§ 27.22) (converse) and locus 16 (§ 6.10), we can easily construct figure 118, as shown below:

We set a random point T on circle (o′) and create pencil T(A′B′Γ′Δ′E′Z′) or T(αβγδεζ).

If we intersect this pencil with a random line (ε), according to construction 181 (converse), the Simple Harmonic Involution with Double Points β, ε and two Pairs of Conjugate Points α, γ and δ, ζ will occur.

Let us draw arcs of circles in correspondence to chords αδ and γζ and equal acute (preferably) angles, towards the same side of (ε) and if these arcs intersect at point Σ, then the pencil Σ(αβγδεζ) or Σ(ΑΒΓΔΕΖ) is formed.

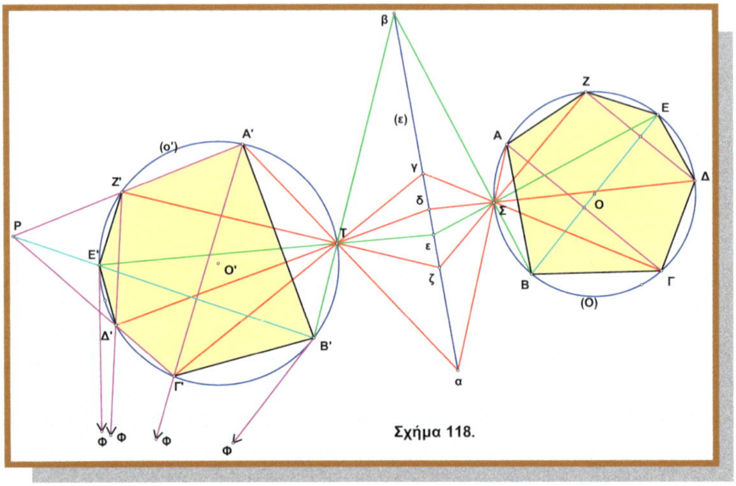

Σχήμα 118.

Because it is true that ∠αΣδ=∠γΣζ⇒∠αΣζ=∠γΣδ, hence pencil Σ(αζδγ) is symmetrical with respect to Σε and thus, according to locus 16, point Σ will belong to the Apollonian circle of diameter βε.

Let (o) be a random circle that runs through point Σ, then pencil Σ(αβγδεζ) re-intersects circle (o) at points A, B, Γ, Δ, E, Z.

Since as well ∠ΑΣΔ=∠ΓΣΖ quadrilateral ΑΓΔΖ is a trapezium with ΑΓ // ΔΖ.

But, we can easily find that: ∠ΑΓΒ=∠ΒΣα=∠βΣγ=∠ΒΣΓ=∠ΒΑΓ or

∠ΑΓΒ=∠ΒΑΓ⇒ΑΒ=ΒΓ.

In a similar way we also find that ΔΕ=ΕΖ, thus hexagon ΑΒΓΔΕΖ, είναι is symmetrical with respect to BE, which will also be a diameter of circle (o), as it is also true that ∠βΣε=∠ΒΣΕ=1L.

From all the above, we realize that hexagon ΑΒΓΔΕΖ is constructed, so we are led to the construction below.

(b). Construction.

On circle (o′), we set a random point T.

Pencil T(Α′Γ′Δ′Ζ′) intersects a random line (ε) at points α, γ, δ, ζ. We write arcs with chords αδ and γζ and equal acute angles, that intersect at point Σ.

Because ∠αΣζ=∠γΣδ⇒∠ΑΣΖ=∠ΓΣΔ, we will have ΑΖ=ΓΔ and therefore ΑΓ // ΔΖ.

Let us draw the diameter that is perpendicular to ΑΓ and ΔΖ, that intersect circle (o) at points Β and Ε.

Based on Β and Ε we can easily define intersections β and ε on (ε) and ultimately intersections Β′ and Ε′ on circle (o′), which we say are the desired Double Points of the involution, with Pairs of Conjugate Points Α′, Γ′ and Δ′, Ζ.′

(c). Proof.

Because hexagon ΑΒΓΔΕΖ is symmetrical with respect to ΒΕ, if we work in a way similar to the implication of Construction 181, we prove that points α, β, γ, δ, ε, ζ and Α′, Β′, Γ′, Δ′, Ε′, Ζ′, constitute Simple Harmonic Involutions of two Pairs of Conjugate Points each and Double Points β, ε for the first and Β′, Ε′ for the second. This means we have defined the desired Double Points Β′, Ε′, of the involution.

Remarks.

(a). From Construction 187 it occurs that the Double Points Β′, Ε′, of the involution were defined, without using poles and polar coordinates, that we had used in Construction 42. Here, as we already saw, we used the method of Harmonic Transformation.

(b). The Problem above and its solution, we believe to be new and firstly appear here.

(c). We notice that using this method, firstly the Double Points β, ε of the Harmonic Collinear Involution occur and then the desired Double Points Β′, Ε′, of the Harmonic Cocyclic Involution (Construction 42). This means that with this New Method of Harmonic Transformation, we simultaneously achieve the Construction of Double Points and of the Harmonic Collinear Involution (Constructions 9 and 10).

(d). The proof of Construction 187, can be based on Corollaries of paragraph 22.61.

(e). If angles αΣδ and γΣζ are drawn to be right, then it easily occurs that ΑΔ and ΓΖ that will appear, would be diameters of circle (o).

(f). In the above proof of Constructions 180-186 we tried to show the detailed way of applying this method in Simple Involutions.

(g). The execution of the Method above, is descriptively and unambiguously developed in figures 118 above.

(h). Taking into consideration what was mentioned in paragraph 8.62 and in some of Corollaries 158-168, or in Construction 42, then we can easily solve Problem 9 (§ 6.3), based on the Theory of Poles and Polar Coordinates.

In fact, in figure 118, if on line (ε), points α, γ, δ, ζ are given and Double Points β, ε of the involution are demanded, then we set a random point T on a random circle (o′) and intersections A′, Γ′, Δ′, Z′, we work similarly to Constructions 42 and after we define intersections P, Φ, $K \equiv A'\Delta' \cap \Gamma'Z'$, we can easily define intersections B′, E′ and through these the desired points β, ε.

(i). Other relevant Propositions are 1, 9. 10, 16, 42, 180-186, 188.

27.52. Constructions of a Double Cocyclic Harmonic Involution of two Pairs of Conjugate Points.

Solution to Problems 11 (§ 6.5) and 43 (§ 9.4), using the Method of Harmonic Transformation.

Construction 188 (Figure 117).

On circle (o′), we are given the ordered set of points A′, B′, Γ′, Δ′, which constitute a Double Cocyclic Harmonic Involution, with two Pairs of Conjugate Points A′, Δ′ and B′, Γ′ of one simple involution and three pairs of conjugate points A′, B′ - H′, Θ′ and Γ′, Δ′ of the other simple involution and we are asked to define their two pairs of Double Points H′, Θ′ and E′, Z′, respectively, using the Method of Harmonic Transformation, or to state it differently:

On circle (o′) we are given an ordered set of its points A′, B′, Γ′, Δ′. Prove that only one pair of points H′, Θ′ on the same circle (o′) exists, which harmonically define the two pairs of points A′, Δ′ και B′, Γ′ and only one pair of points E′, Z′ on the same circle (o′), which harmonically divide the three pairs of points A′, B′ - H′, Θ′ and Γ′, Δ′.

Solution.

(a). Analysis.

We assume that the two Pairs of Double Points of the Double Involution, we constructed and are H′, Θ′ and E′, Z′.

According to Construction 184 (converse), we can easily construct firstly, on line (ε), the Double Harmonic Involution of two Pairs of Conjugate Points α, β and γ, δ with their Double Points the following pairs η, θ and ε, ζ, and consequently, after we define the Orthooptical Point Σ (§ 7.14), we draw in circle (o) that runs through Σ, the Rectangle ΑΒΓΔ, as well as pairs of points Η, Θ and Ε, Ζ. For the rectangle ΗΘ and ΕΖ are its two axes of symmetry, which are obviously perpendicular and diameters of circle (o).

We notice that points Α, Β, Γ, Δ are fixed on circle (o) and constructed as points Α′, Β′, Γ′, Δ′, are given.

In this way, rectangle ΑΒΓΔ is constructed, thus its axes of symmetry ΗΘ and ΕΖ are constructed as well, so we are led to Construction below.

(b). Construction.

On circle (o′) we set a random point T and let pencil T(Α′Β′Γ′Δ′) intersect line (ε) at points α, β, γ, δ, respectively.

Using diameters αγ and βδ we draw circles that intersect each other at point Σ. If pencil Σ(αβγδ) re-intersects circle (o) at points Α, Β, Γ, Δ, we say that ΑΒΓΔ is a rectangle, whose axes of symmetry ΗΘ and ΕΖ are easily constructed.

Consequently, based on rectangle ΑΒΓΔ and points of intersection Η, Θ and Ε, Ζ of the axes of symmetry and the circle (o), easily, as in the direct of Construction 184 we set on line (ε) points α, β, γ, δ, ε, ζ, η, θ and afterwards on circle (o′) points Α′, Β′, Γ′, Δ′, Ε′, Ζ′, Η′, Θ′, which we say constitute a Double Harmonic Involution, of which the two desired Pairs of Double Points are Η′, Θ′ and Ε′, Ζ′.

(c). Proof.

Because angles αΣγ and βΣδ or ΑΣΓ and ΒΣΔ are right, then ΑΓ and ΒΔ are diameters of the circle (o). Hence quadrilateral ΑΒΓΔ is actually a rectangle and its two axes of symmetry ΗΘ and ΕΖ are diameters of the circle (o).

Thus, as points A′, B′, Γ′, Δ′, E′, Z′, H′, Θ′, were drawn based on points A, B, Γ, Δ, E, Z, H, Θ, according to the proof of the implication in Construction 184 and points A′, B′, Γ′, Δ′, E′, Z′, H′, Θ′, would constitute a Double Harmonic Involution of two Pairs of Conjugate Points A, B and Γ, Δ and two Pairs of Double Points H′, Θ′ and E′, Z′, this means that H′, Θ′ and E′, Z′, are the desired Double Points of the Involution, as they also belong in circle (o′), in addition H′Θ′ and E′Z′ are conjugate, as they stem from HΘ, EZ which are perpendicular to each other [Corollary 159 (implication)].
[It is possible for us to have another simple solution to Problem 188, if we apply twice Construction 187. Once to define Double Points H′, Θ′ and another one to define Double Points E′, Z′. Attention is needed though when constructing point Σ, which should be Orthooptical (§ 7.14), and is defined from the intersection of circles with diameters εζ and ηθ].

Remarks.

(a). From Construction 188 it occurs that Double Points H′, Θ′ και E′, Z′, of the involution, were defined without using poles and polar coordinates, that we used in Construction 43. Here, as seen, we achieved the proof using the method of Harmonic Transformation.

(b). The Problem above and its solution, we believe to be new and firstly appear here.

(c). We notice that using the method of Harmonic Transformation, firstly the Double Points η, θ and ε, ζ of the Harmonic Collinear Involution occur and then the desired Double Points H′, Θ′ and E′, Z′, of the Harmonic Cocyclic Involution. <u>This means that with this New Method of Harmonic Transformation, we simultaneously achieve the Construction of Double Points and of the Harmonic Collinear Involution.</u>

(d). The proof of Construction 188, can be based on Corollaries of paragraph 22.61.

(e). We tried to provide the detailed way of applying this method on Double Involutions, in the proofs of Constructions 184-188 above.

(f). Execution of the above Method, is shown step by step, descriptively and unambiguously in figure 117 above.

(g). Taking into consideration what was mentioned in paragraph 8.63 and in some of Corollaries 158-168, or in Construction 43, then we can easily solve Problem 11 (§ 6.5), based on the Theory of Poles and Polar Coordinates.

In fact, in figure 117, if on line (ε), we are given points α, β, γ, δ and are asked for Double Points η, θ and ε, ζ of the involution, then we set a random point T on a random circle (o′) and intersections A′, B′, Γ′, Δ′, we work similarly to Construction 43 and after we define intersections P, Φ, K ≡ A′Γ′ ∩B′Δ′, then intersections H′, Θ′ and E′, Z′ are easily defined, and through these the desired points η, θ, and ε, ζ.

(h). If we work in the same way as in Propositions 180-188, we can obviously prove or solve many other Propositions and Problems mentioned in Part A.

(i). Other relevant Propositions are 1, 11, 16, 43, 180-187.

PART D

SUPPLEMENT OF PART B

Applications of Harmonic Multilaterals, Mainly using the New Method of Harmonic Transformation.

Euclid of Alexandria.

PART D

COMPLEMENT OF PART B

Applications of Harmonic Multilaterals, Mainly using the New Method of Harmonic Transformation.

28. Generally.

Part D is actually an extension to Part B and includes some additional data about Harmonic Multilaterals (Collinear and Cocyclic), that came up after having written Part C. These data will obviously be included in Part B in a new edition of the book. This did not happen here, so as to not change the sequence of their conception, mostly for historic purposes, something we did not want to happen.

Mainly though, it includes applications of Harmonic Multilaterals, using our New Method of Harmonic Transformation, which we have developed in paragraph 21.

We note that using this significant method of «Harmonic Transformation», we can solve methodically, easily and spherically most of the Problems of Harmonic Geometry, even those that were impossible to solve until now.

29. Applications of Harmonic Multilaterals, mainly using the Mathod of Harmonic Transformation.

29.1._Application of the Method of Harmonic Transformation, Significant Property of the Harmonic Decagon. Another proof to Proposition 106. Convergence of Diagonals in Harmonic Decagons.

Proposition 200 (Figure 120).

The diagonals (main), of Harmonic Decagons, are concurrent.

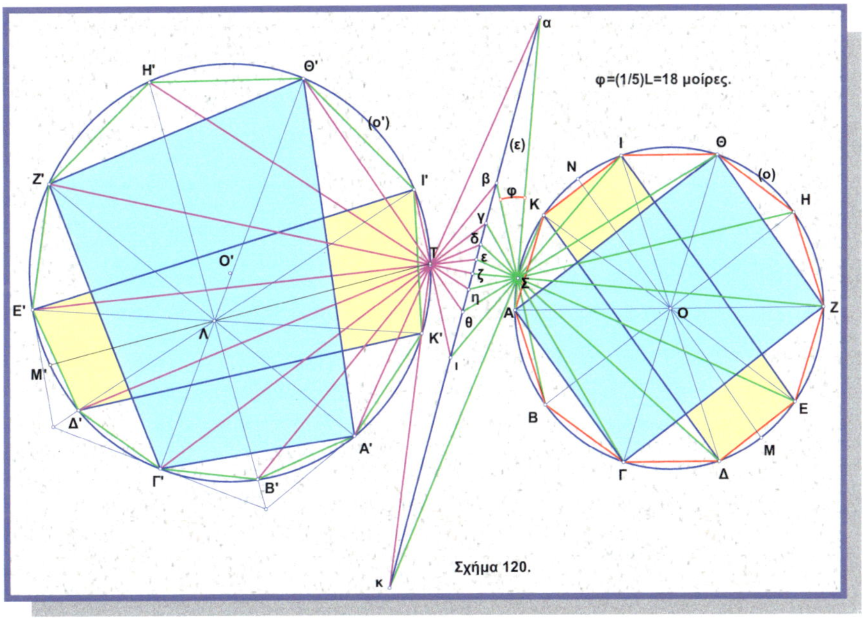

Σχήμα 120.

Proof.

Let the Α'Β'Γ'Δ'Ε'Ζ'Η'Θ'Ι'Κ' be a Harmonic Decagon. We will prove that it is:

$$Α'Ζ' \cap Β'Η' \cap Γ'Θ' \cap Δ'Ι' \cap Ε'Κ' \equiv Λ. \qquad (1).$$

Since this Decagon is Harmonic, according to Construction 155 (§ 22.4), it will have occurred from a Regular Decagon ΑΒΓΔΕΖΗΘΙΚ, after applying the New Method of Harmonic Transformations (§ 21).

Hence, if we receive as an axis of symmetry of the Decagon ΑΒΓΔΕΖΗΘΙΚ for example its diagonal ΒΗ and draw the perpendicular to it at point O, then we will have a second axis MN (where M, N are the points of intersection of this axis and circle (o)].

It easily occurs that, quadrilaterals ΑΓΖΘ and ΔΕΙΚ are rectangles and share their axes of symmetry, diameters ΒΗ and MN. This, according to Corollary 167 (§ 22.61 I), means that these rectangles are Harmonically Transformed to quadrilaterals Α'Γ'Ζ'Θ' and Δ'Ε'Ι'Κ' respectively, whose diagonals Α'Ζ', Γ'Θ', Δ'Ι', Ε'Κ', are concurrent at the intersections of chords Β'Η' and M'N', that have occurred from the axes ΒΗ and MN respectively, through the above Harmonic Transformation.

Thus we see that the above relation (1) is actually true.

Remarks.

(a). We see that using the Method of Harmonic Transformation, we achieved an analytical proof of Proposition 106 that until now was thought difficult to happen.

(b). The other two ways of proving Proposition 200, have been given in Proposition 106 (paragraph 19.13).

(c). Axis ΒΗ was randomly chosen. We could have received any other diagonal of ΑΒΓΔΕΖΗΘΙΚ.

(d). In an equivalent way we work for the Harmonic Hexagon, Octagon, Dodecagon, etc. This process is suggested as an exercise for those who are truly friends of Geometry.

(e). According to Corollary 167, Κ'Ι', Α'Θ', Β'Η', Γ'Ζ', Δ'Ε' are also concurrent at the intersection P of the tangents at M', N', while Α'Γ', Κ'Δ', N'M', Ι'Ε', Θ'Ζ', are concurrent at intersection Φ of the tangents at Β', Η'.

(f). Proposition 200, we believe, to be new and firstly appear here.

(g). Up until now we have seen that, besides Harmonic Decagon, concurrent diagonals exist in Harmonic Hexagons and Octagons as well, which can also be proven in a way equivalent to the one mentioned above.

(h). Other relevant Propositions are 106, 155, 167, 201.

29.2. Application of the Method of Harmonic Transformation.

Astonishing Property of Harmonic Decagons.

Series of Points of Eighteen Pairs of Lines, Simple Harmonic Involution.

Proposition 201 (Figures 99 and 121).

In every Simple Harmonic Decagon ΑΒΓΔΕΖΗΘΙΚ (figure 121), the sixteen intersections below belong to its diagonal ΒΗ:

ΑΖ ∩ΓΘ≡Λ, ΔΙ ∩ΕΚ≡Λ', ΔΘ ∩ΖΚ≡α, ΕΘ ∩ΖΙ≡β, intersection γ of tangents at Ζ and Θ, ΕΖ ∩ΘΙ≡δ, ΔΖ ∩ΘΚ≡ε, intersection ζ of tangents at Ε and Ι, ΑΕ ∩ΓΙ≡α', ΑΔ ∩ΓΚ≡β', intersection γ' of tangents at Α and Γ, ΑΚ ∩ΓΔ≡δ', ΑΙ ∩ΓΕ≡ε', intersection ζ' of tangents at Δ and Κ, ΑΘ ∩ΓΖ≡η', ΔΕ ∩ΙΚ≡θ',

or to state it otherwise:

In every Simple Harmonic Decagon ΑΒΓΔΕΖΗΘΙΚ, eighteen (18) intersections and points, Λ, α, β, Η, γ, δ, ε, ζ, α', β', Β, γ', δ', ε', ζ', η', Λ', θ', of figure 121, are collinear.

Additionally, if ΒΗ ∩ΕΙ≡ι, ΒΗ ∩ΖΘ≡κ, ΒΗ ∩ΔΚ≡λ, ΒΗ ∩ΑΓ≡μ, the above Collinear Series of Points, constitutes a Simple Collinear Harmonic Involution, with Double Points Β, Η and these Pairs of Conjugate Points:

ι, ζ - α, ε - β, δ – κ, γ – Λ, η' -Λ', θ' - λ, ζ' - α', ε' - β', δ' - μ, γ'. (1).

Proof (Figures 99 and 121).

(a). Analysis (Figure 99).

Here we will remember our new Method of «Harmonic Transformation» and more specifically, the construction way of a Harmonic Decagon through a Regular Decagon and conversely the way of constructing a Regular Decagon using a Harmonic Decagon, which we have already mentioned here in Construction 155 (§ 22.4) and mainly the converse, according to which, from the given Harmonic Decagon Α'Β'Γ'Δ'Ε'Ζ'Η'Θ'Ι'Κ', we get a Regular Decagon ΑΒΓΔΕΖΗΘΙΚ, through the intervention of a Harmonic Collinear Group of Ten Points α, β, γ, δ, ε, ζ, η, θ, ι, κ and its Orthooptical Point Σ, as seen in figure 99.

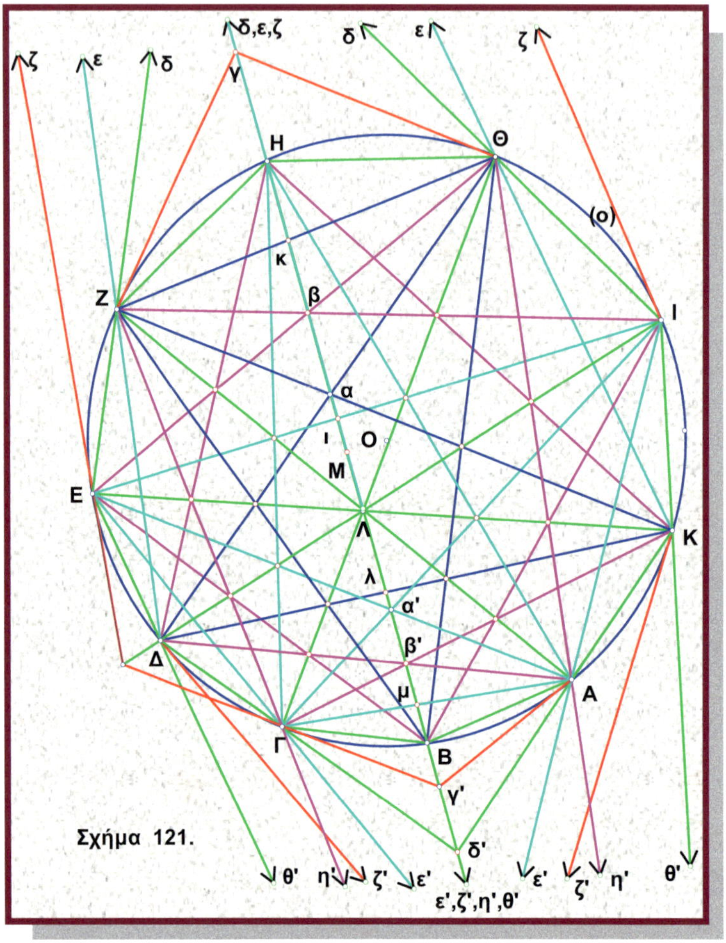

Σχήμα 121.

More specifically, if we work in figure 99 and based on the given Harmonic Decagon Α'Β'Γ'Δ'Ε'Ζ'Η'Θ'Ι'Κ' and an arbitrary point T of its circumscribed circle (o'), we construct on an arbitrary line (ε) the Harmonic Group of Ten Points α, β, γ, δ, ε, ζ, η, θ, ι, κ and after we define its one «Orthooptical Point» Σ, we draw a random circle (o), that runs through Σ and in this way we get the Regular Pencil Σ(αβγδεζηθικ) of ten beams with $\varphi=L/5=18^0$, whose beams intersect circle (o), at the apexes of the Regular Decagon ΑΒΓΔΕΖΗΘΙΚ.

After this construction, we notice that in triangle Σει, Ση is the bisector of angle εΣι (∠εΣη=∠ηΣι=2.18⁰=36⁰) and that angle ηΣβ is right (∠ηΣβ=5.18⁰=90⁰). This means that Pencil Σ(βηει) is Harmonic, hence Pencil T(βηει) or T(B'H'E'I') is also Harmonic or that the quadrilateral B'E'H'I' is harmonic and thus B'H' and E'I' are conjugate with respect to circle (o').

If we work in the exact same way, we also find that B'H' and Δ'K' are conjugate with respect to circle (o'), as in triangle Σδκ, Ση is the bisector of angle δΣκ (∠δΣη=∠ηΣκ=3.18⁰=54⁰) and angle βΣη is right (∠βΣη=5.18⁰=90⁰), thus Pencil Σ(βηδκ) is Harmonic too, and so on and so forth.

(b). Proof (Figure 121).

Based on what is mentioned above and if we now move to figure 121, we see that for the Harmonic Decagon ΑΒΓΔΕΖΗΘΙΚ too, ΔΚ and ΕΙ are conjugate to ΒΗ.

After all, from the definition of the Simple Harmonic Decagon (§ 16.1d), quadrilaterals ΑΒΓΗ, ΖΗΘΒ are Harmonic too and ΑΓ, ΖΘ are conjugate, to ΒΗ.

Consequently all the following ΑΓ, ΔΚ, ΕΙ, ΖΘ are conjugate to ΒΗ, with respect to circle (o).

Hence, because ΑΓ and ΒΗ are conjugate with respect to circle (o), ΒΗ will run through pole γ' of ΑΓ or that γ'∈ΒΗ.

Similarly, because ΘΖ and ΒΗ are also conjugate with respect to circle (o), ΒΗ will run through pole γ of ΖΘ or that γ∈ΒΗ.

Additionally, because ΔΚ, ΕΙ are conjugate to ΒΗ too, with respect to circle (o), ΒΗ will run through poles ζ', ζ of ΔΚ, ΕΙ, respectively, or that ζ', ζ ∈ΒΗ.

Additionally, because ΔΚ, ΖΘ are also conjugate to ΒΗ, pair of points Β, Η will harmonically divide pairs of points Δ, Κ and Ζ, Θ. Thus, according to Proposition 51 (§ 9.12), intersections ΔΘ ∩ΖΚ≡α and ΔΖ ∩ΘΚ≡ε, will belong to ΒΗ.

Similarly, because ΑΓ, ΕΙ are also conjugate to ΒΗ, pair of points Β, Η will harmonically divide pairs of points A, Γ and E, I. Thus, according to Proposition 51 (§ 9.12), intersections AE ∩ΓI≡α' and ΓE ∩AI≡ε', will belong to BH. Moreover, because EI and ZΘ are conjugate to BH, intersections β and δ∈BH.

Additionally, because AΓ and ΔK are conjugate BH, intersections β' and δ'∈BH.

Similarly, because AΓ and ZΘ are conjugate to BH, intersections AZ ∩ΓΘ≡Λ and AΘ ∩ΓZ≡η' will belong to BH.

Finally, because ΔK and EI are, too, conjugate to BH, intersections ΔI ∩EK≡Λ', ΔE ∩KI≡θ' will belong to BH.

From the above we see that in fact the following sixteen points γ', γ, ζ, ζ', α, ε, α', ε', β, δ, β', δ', Λ, η', Λ', θ', belong to diagonal BH.

After all, because diagonals of the Simple Harmonic Decagon are concurrent (Proposition 106, paragraph 19.13), it will be Λ'≡Λ, hence it will be θ'≡η', such as (HBΛθ')=1, (HBΛη')=1⇒ θ'≡η' too. This means that, intersections and points, that are collinear, while in reality are 18 in number, because it is Λ'≡Λ, θ'≡η', in fact these points are 16.

From the above it occurs that the series points above, is a Simple Harmonic Involution of Two-Branches with Double Points B, H and Pairs of Conjugate Points these of relation (1).

Hence, if M is the midpoint of BH, then it will also be:

$MB^2=MH^2 = M\iota.M\zeta = M\alpha.M\varepsilon = M\beta.M\delta = M\kappa.M\gamma = M\Lambda.M\eta' = M\Lambda'.M\theta' = M\lambda.M\zeta'=M\alpha'.M\varepsilon' = M\beta'.M\delta' = M\mu.M\gamma'$. (2).

(c). Investigation.

Diagonal BH, because it was received as arbitrary, a Proposition equivalent to 201, will be true for each one of the rest diagonals AZ, ΓΘ, ΔI, EK, of Decagon ΑΒΓΔΕΖΗΘΙΚ and thus we actually have the properties of a complex of a complete Harmonic Decagon, the tangential of the complete Decagon, the complete Decagon of the first, the second and the third diagonals of the Harmonic Decagon (terminology § 5). This complex that will occur, will be equivalent to figures 3.2.6.(4).(b).1/ and 4.2.5.(5).(b).5/, in our book [1], which refer to a Complex of a Harmonic Hexagon and a Harmonic Octagon, respectively (See also General Remarks, § 29.10 below).

Remarks.

(a). The above Proposition and its proof, we believe to be new and firstly appear here.

(b). Analogous Propositions are true for all other Harmonic Multilaterals as well.

(c). Relevant Propositions are 202-204.

29.3. Another Discussion on Proposition 201, Mainly with the Simple Application of Corollaries 158 to 168.

Remarkable Properties of Harmonic Decagons.

Proposition 202 (Figures 122).

Every Regular Decagon ΑΒΓΔΕΖΗΘΙΚ, is Harmonically transformed, to a Harmonic Decagon Α'Β'Γ'Δ'Ε'Ζ'Η'Θ'Ι'Κ', which, concerning its arbitrary diagonal Β'Η' (main), has the following properties:

Α'Γ', Κ'Δ', Ι'Ε', Θ'Ζ', are concurrent at the intersection Φ of the tangents at points Β', Η'.

Intersections Α'Ζ'∩Γ'Θ'≡Λ, Δ'Ι'∩Ε'Κ'≡Λ', Δ'Θ'∩Κ'Ζ'≡α, Ε'Θ'∩Ι'Ζ'≡β, Ε'Ζ'∩Ι'Θ'≡δ, Δ'Ζ'∩Κ'Θ'≡ε, Κ'Ι'∩Δ'Ε'≡θ, Α'Ε'∩Γ'Ι'≡α', Α'Δ'∩Γ'Κ'≡β', Α'Κ'∩Γ'Δ'≡δ', Α'Ι'∩Γ'Ε'≡ε', Α'Θ'∩Γ'Ζ'≡η', Δ'Ε'∩Κ'Ι'≡θ' and intersections γ, ζ, γ' ζ', of the tangents at the following pairs of points Θ'- Ζ', Ι'- Ε', Α'- Γ', Κ'- Δ', respectively, lie on diagonal Β'Η', while Λ≡Λ', θ'≡ η'≡Ρ ∈ Β'Η'.

Quadrilaterals Α'Β'Γ'Η', Ζ'Η'Θ'Β', Β'Δ'Η'Κ', Β'Ε'Η'Ι', are Harmonic.

Moreover, if Β'Η' ∩Ε'Ι'≡ι, Β'Η' ∩Ζ'Θ'≡κ, Β'Η' ∩Δ'Κ'≡λ, Β'Η' ∩Α'Γ'≡μ, the above Collinear Series of Points of line Β'Η', constitutes a Simple Collinear Harmonic Involution, with Double Points Β', Η' and Pairs of Conjugate Points:

ι, ζ - α, ε - β, δ – κ, γ – Λ, η' -Λ', θ' - λ, ζ' - α', ε' - β', δ' - μ, γ'. (1).

Additionally, the Cocyclic Series of Points Α', Β', Γ', Δ', Ε', Ζ', Η', Θ', Ι', Κ', constitutes a Simple Cocyclic Harmonic Involution, with Double Points Β', Η' and Pairs of Conjugate Points: Α', Γ' - Δ', Κ' - Ε', Ι' - Ζ', Θ', (2).

If M is the midpoint of B′H′, then it will also be: $MB'^2=MH'^2 = M\iota.M\zeta = M\alpha.M\varepsilon = M\beta.M\delta = M\kappa.M\gamma = M\Lambda.M\eta' = M\Lambda'.M\theta' = M\lambda.M\zeta'=M\alpha'.M\varepsilon' = M\beta'.M\delta' = M\mu.M\gamma'= MA'.M\Gamma'= M\Delta'.MK'= ME'.MI'= MZ'.M\Theta'$. (3).

Properties analogous to the above, are true for every other diagonal of the Harmonic Decagon A'B'Γ'Δ'E'Z'H'Θ'I'K',

and conversely,

every Harmonic Decagon A'B'Γ'Δ'E'Z'H'Θ'I'K', is possible to be Harmonically transformed, to a Regular Decagon ΑΒΓΔΕΖΗΘΙΚ, which has the following properties:

ΑΓ, ΚΔ, ΙΕ, ΘΖ, are concurrent at infinity (are parallel).

It is symmetric with respect to BH.

Intersections of these pairs of chords AZ -ΓΘ, ΔΙ-ΕΚ, ΔΘ-ΚΖ, ΕΘ-ΙΖ, ΕΖ-ΙΘ, ΔΖ-ΚΘ, ΚΙ-ΔΕ, ΑΕ-ΓΙ, ΑΔ-ΓΚ, ΑΚ-ΓΔ, ΑΙ-ΓΕ, ΑΘ-ΓΖ, ΔΕ-ΙΚ, such as intersections of tangents at these pairs of symmetric points Θ- Ζ, Ι- Ε, Α- Γ, Κ- Δ respectively, will lie on BH, while chords AZ, ΓΘ, ΔΙ, EK, BH (diagonals), would be concurrent at center O of circle (o) and ΑΘ, BH, ΓΖ, ΔΕ, IK, will be concurrent at infinity (they are parallel).

The two above involutions would be true [Collinear (1) and cocyclic (2)], such as relation (3), properly adjusted.

Quadrilaterals ΑΒΓΗ, ΖΗΘΒ, ΒΔΗΚ, ΒΕΗΙ, would be symmetric, with respect to BH.

Analogous properties, to the above, are true for every other diagonal of the Regular Decagon ΑΒΓΔΕΖΗΘΙΚ.

Proof (Figure 122).

(a). Direct.

Since Decagon A'B'Γ'Δ'E'Z'H'Θ'I'K' is Harmonic, according to Construction 155 (direct) (§ 22.4), this will have occurred from a Regular Decagon ΑΒΓΔΕΖΗΘΙΚ, with the application of the New Method of Harmonic Transformations (§ 21 and Construction 155).

Hence, if we receive as axis for example diagonal BH of ΑΒΓΔΕΖΗΘΙΚ and to this draw a perpendicular to its center O, then we have a second axis MN [where M, N are the points of intersection of this axis and circle (o)].

. Supplement of Harmonic Multilaterals.

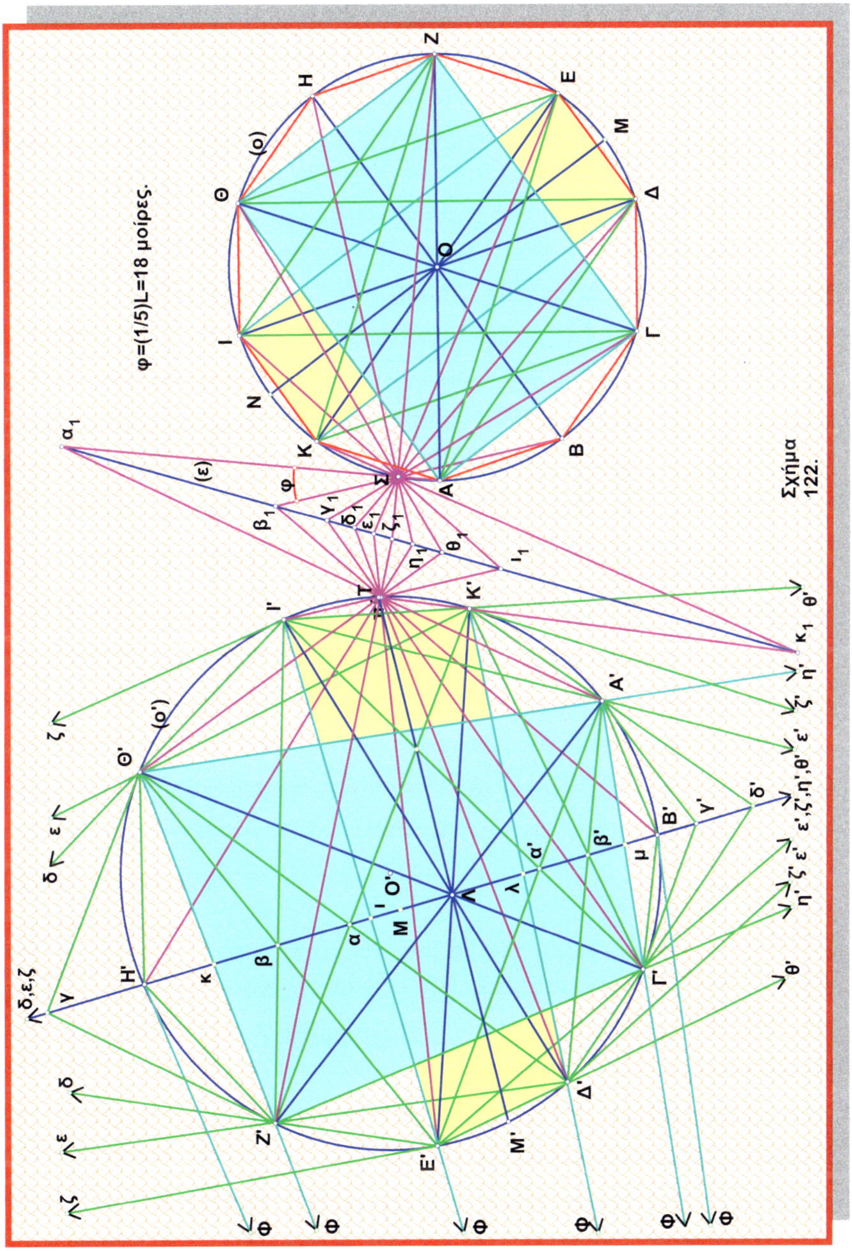

Σχήμα 122.

It easily occurs that, quadrilaterals ΑΓΖΘ and ΔΕΙΚ are rectangles and share their axes of symmetry, diameters ΒΗ and ΜΝ. This, according to Corollary 167 (direct) (§ 22.61 I), means that these rectangles are Harmonically Transformed to Harmonic Quadrilaterals Α'Γ'Ζ'Θ' and Δ'Ε'Ι'Κ' whose diagonals Α'Ζ', Γ'Θ', Δ'Ι' Ε'Κ', are concurrent at the intersection of chords Β'Η' and Μ'Ν', that have occurred from axes ΒΗ and ΜΝ respectively, according to the above Harmonic Transformation. Hence it will be:

Β'Η'∩Μ'Ν'∩Α'Ζ'∩Γ'Θ'≡Λ≡Ι'Δ'∩Ε'Κ'≡Λ' ή Λ'≡Λ≡Β'Η'∩Μ'Ν'. (4).

Moreover, according to the same Corollary 167 (direct) (§ 22.61 I),

It will also be: Α'Γ'∩Κ'Δ'∩Μ'Ν'∩Ι'Ε'∩Θ'Ζ'≡Φ. (5).

Additionally, according to the above Corollary 167 (direct), it will be:

Β'Η'∩Α'Θ'∩Γ'Ζ'≡η'≡Κ'Ι'∩Δ'Ε'≡θ'≡Ρ ∈Β'Η' ή θ'≡η'≡Ρ ∈Β'Η'. (6).

After all, since Decagon Α'Β'Γ'Δ'Ε'Ζ'Η'Θ'Ι'Κ', comes from Regular Decagon ΑΒΓΔΕΖΗΘΙΚ, which is symmetrical with respect to ΒΗ, according to Corollary 165 (direct) (§ 22.61 H), intersections:

Λ≡ Β'Η'∩Μ'Ν', Ρ≡θ'≡η', α, β, γ, δ, ε, ζ, α', β', γ', δ', ε', ζ', all in fact lie on diagonal Β'Η'.

Similarly, according to the same Corollary 165 (direct), quadrilaterals Α'Β'Γ'Η', Ζ'Η'Θ'Β', Β'Δ'Η'Κ', Β'Ε'Η'Ι', are Harmonic.

Analogous properties, to the above, Harmonic Decagon Α'Β'Γ'Δ'Ε'Ζ'Η'Θ'Ι'Κ' has concerning every other diagonal of it too, as this Harmonic Decagon, occurs from Regular Decagon ΑΒΓΔΕΖΗΘΙΚ, which is obviously symmetric with respect to every other of its diagonals, since ΒΗ, was randomly received.

From the above, we easily realize that the above series of points of line Β'Η', constitutes a Simple Collinear Harmonic Involution of two Branches, with Double Points Β', Η' and Pairs of Conjugate Points those in relation (1), (definition § 2.11).

In fact, because for example Ζ'Θ' is the polar of γ, with respect to circle (ο'), it will be (γκ Β'Η')=1. Additionally it is (δβ Β'Η')=1, and so on and so forth.

Moreover, because chords Α'Γ', Δ'Κ', Ε'Ι', Ζ'Θ' are conjugate to line Β'Η', the Cocyclic Series of Points Α', Β', Γ', Δ', Ε', Ζ', Η', Θ', Ι', Κ', will constitute a

Simple Cocyclic Involution, with Double Points also B', H' and Pairs of Conjugate Points those in relation (2), (definition § 8.11), as it will be for example (Δ'K'B'H')=1, (A'Γ'B'H')=1, and so on and so forth.
Therefore, from the two Involutions above, it occurs that the above relation (3) is true (§ 8,11).

(b). Converse.

If Decagon ΑΒΓΔΕΖΗΘΙΚ has been Harmonically transformed, from the Harmonic Decagon A'B'Γ'Δ'E'Z'H'Θ'I'K' in the known way [Construction 155 (converse) (§ 22.4)], in such way that it is Regular, then in this all what is desired in Proposition 202 will be true (Converses of Corollaries 165 and 167).
These, besides, are rather easy to prove, if we depend on Theorem 253 in volume A' of book [28] and take into consideration that ever other diagonal of the regular Decagon is an axis of symmetry for it.
Moreover because Decagon ΑΒΓΔΕΖΗΘΙΚ is Regular and because of its symmetry with respect to BH, we easily realize that here as well, analogous Involutions occur (collinear and cocyclic), to those that are mentioned in the direct, consequently for those too, a relation analogous to (3) will be true, properly adjusted, (without the marks on the letters).
For example, from the Cocyclic Involution of points A, B, Γ, Δ, E, Z, H, Θ, I, K, easily and directly it occurs that:

$$OB^2=OH^2=R^2=OA.O\Gamma=O\Delta.OK=OI.OE=OZ.O\Theta.$$

(c). Investigation.

We easily prove that, the five intersections of the opposite sides (of their carriers) of the Harmonic Decagon (figure 122) are collinear, καθώς as these belong to the polar of Λ (point of convergence of the main diagonals of the Harmonic Decagon). This means that, intersections A'B'∩Z'H', B'Γ'∩H'Θ', Γ'Δ'∩Θ'I', Δ'E'∩I'K', E'Z'∩K'A', are collinear (equivalent to Pascal's Theorem).
From these five intersections, some of its diagonals run through too, main and not, uniformly distributed (three from every intersection), in a way that

they are completely in order and constitute a Harmonic Complex of lines and Central Pencils of five beams each (this figure can count for an exercise for those interested).

Moreover, the carriers of the first (terminology §5) diagonals of the Harmonic Decagon, constitute two Complete Harmonic Pentagons, while the second and third diagonals of this Decagon, respectively constitute two Complete Harmonic Asteroid Decagons.

After all, except for the above mentioned five collinear intersections, that belong to the polar of intersection Λ, we also have intersections of sides and diagonals that intersect in pairs on the main diagonals {See above (figure 122), which though we have approached in alternative ways and extensively, in our book [1]}.

Such complexes, of Complete Harmonic Quadrilaterals, Hexagons, Octagons and not only that, are mentioned in our book of five volumes [1] 1990 (See also General Annotations § 29.10 below).

There, we have approached these matters from another point of view, while here we use the respective circumscribed Complete Multilaterals (polar converses).

After making up the Method of Harmonic Transformation (§ 21), Corollaries from 158 to 168 and its applications, in Harmonic Multilaterals, the General Conclusion below occurs:

General Conclusion.

All Harmonic Multilaterals occur from the respective Regular Multilaterals (with an equal amount of sides), applying the Method of Harmonic Transformation and conversely.

Every apex, side, diagonal of a Regular Multilateral matches to an apex, side, diagonal of its respective Harmonic Multilateral.

In general, every center of a pencil of a Regular Multilateral, matches to a center of a pencil, of an equal amount of beams, of its respective Harmonic Multilateral with an equal amount of sides and every line of a Regular Multilateral matches to a line of its respective Harmonic Multilateral.

Moreover, to each diagonal (main) of the Regular Multilateral, belongs an involutory series of points that consists of the convergence of equal number of its sides or its diagonals. The exact same thing happens in the respective Harmonic Multilateral.

Remarks

(a). Proposition 202 above is actually Proposition 201 (paragraph 29.2), in another point of view, extended and with more data desired, while its converse is also given.

(b). This Proposition, is significant as it provides us the whole image of every Complete Harmonic Decagon, as with it various and impressive properties of the Harmonic Decagon are revealed, which no other multilateral has, some of which will be mentioned below in Construction 203.

(c). Axis BH was randomly chosen. We could have chosen any other diagonal (main) of ΑΒΓΔΕΖΗΘΙΚ.

(d). We notice that in Harmonic Transformations, multilaterals (Regular and its respective Harmonic), have an equal number of sides and angles, while every point on one multilateral matches a point on the other.

The same happens for rectilinear segments, and by extension for lines.

(e). The Proposition above and its proof, we believe to be new and firstly appear here.

(f). Analogous Propositions are true in the rest Harmonic Multilaterals.

(g). Relevant Propositions are 152, 155, 165, 167, 201, 203.

29.4. Gigantic Problem (in Size and in Difficulty). Big Challenge.

The Big, Rather Difficult and Impresive Problem.

Construction of a Special Convex Inscribed in Circle Decagon, with Various and Impressive Properties.

Construction 203 (Figures 123-126).

In a given circle (o), we inscribe a special convex Decagon ΑΒΓΔΕΖΗΘΙΚ (not regular), which has the main properties below:

(a). Its main diagonals (terminology § 4), are concurrent at point Λ. I.e. it is:

$$AZ \cap BH \cap \Gamma\Theta \cap \Delta I \cap EK \equiv \Lambda. \quad (1).$$

(b). Additionally, it is: $\frac{AB}{B\Gamma}\cdot\frac{\Gamma\Delta}{\Delta E}\cdot\frac{EZ}{ZH}\cdot\frac{H\Theta}{\Theta I}\cdot\frac{IK}{KA}=1.$ (2).

(c). If α, β, γ, ...κ, are the distances of Λ from sides AB, ΒΓ, ΓΔ,..., KA respectively, it would be:

$$\frac{\alpha}{AB}=\frac{\beta}{B\Gamma}=\frac{\gamma}{\Gamma\Delta}=\frac{\delta}{\Delta E}=\frac{\varepsilon}{EZ}=\frac{\zeta}{ZH}=\frac{\eta}{H\Theta}=\frac{\theta}{\Theta I}=\frac{\iota}{IK}=\frac{\kappa}{KA}. \quad (3).$$

(d). Moreover, it will be: $\frac{\alpha}{\beta}\cdot\frac{\gamma}{\delta}\cdot\frac{\varepsilon}{\zeta}\cdot\frac{\theta}{\eta}\cdot\frac{\iota}{\kappa}=1.$ (4).

(e). The five intersections of its opposite sides, are collinear. Hence intersections:

ΓΔ∩ΘΙ≡Π, ΔΕ∩ΙΚ≡Ρ, ΕΖ∩ΚΑ≡Σ, ΖΗ∩ΑΒ ≡Τ, ΗΘ∩ΒΓ≡Υ, (5).

belong on a line (ε), while every other of its main diagonals, runs through an intersection from these. I.e. it is:

Π ∈ΑΖ, Ρ ∈ΒΗ, Σ ∈ΓΘ, Τ ∈ΔΙ, Υ ∈ΕΚ. (6).

(f). Each one from the five intersections of its second opposite diagonals (terminology § 5), coincides to one of the above collinear intersections. Hence it is:

ΒΕ∩ΗΚ≡Π, ΓΖ∩ΑΘ≡Ρ, ΔΗ∩ΒΙ≡Σ, ΕΘ∩ΓΚ≡Τ, ΖΙ∩ΔΑ≡Υ. (7).

(g). The ten intersections

ΚΓ∩ΒΙ≡α′, ΑΔ∩ΓΚ≡β′, ΒΕ∩ΔΑ≡γ′,...,ΙΒ∩ΑΘ≡κ′, (8)

of its second diagonals, lie on its main diagonals: AZ, BH, ΓΘ,...,KE, respectively (two intersections on each diagonal) .

(h). The ten intersections

ΚΔ∩ΒΘ≡α″, ΑΕ∩ΓΙ≡β″, ΒΖ∩ΔΚ≡γ″,...,ΙΓ∩ΑΗ≡κ″, (9).

of its third diagonals, lie on its main diagonals: AZ, BH, ΓΘ,...,KE, respectively (two intersections on each diagonal).

(i). The ten intersections

ΙΚ∩ΒΓ≡α, ΚΑ∩ΓΔ≡β, ΑΒ∩ΔΕ≡γ,...,ΘΙ∩ΑΒ≡κ, (10).

of its one by two sides, lie on its main diagonals: AZ, BH, ΓΘ,..., KE, respectively (two intersections on each diagonal).

(j). The ten intersections

ΘΚ∩ΒΔ≡α_1, ΙΑ∩ΓΕ≡β_1, ΚΒ∩ΔΖ≡γ_1,...,ΗΙ∩ΑΓ≡κ_1, (11).

of its first diagonals, lie on its main diagonals: AZ, BH, ΓΘ,...,KE, respectively (two intersections on each diagonal).

(k). If the tangents of circle (o), at pairs of its opposite apexes Δ and I, E and K, Z and A, H and B, Θ and Γ, intersect at points Π′, P′, Σ′, T′, Y′, respectively, then these intersections lie on line (ε) above.

(l). If the tangents of circle (o), at pairs of its apexes K and B, A and Γ, B and Δ,….,I and A, intersect at points A′, B′, Γ′,…, K′, respectively, then these intersections lie on its main diagonals AZ, BH, ΓΘ, ….,KE, respectively (two intersections on each diagonal).

(m). If the tangents of circle (o), at pairs of points Γ and I, Δ and K, E and A,….,Θ and B, intersect at points A_1, B_1, $Γ_1$,…, K_1, respectively, then these intersections lie on its main diagonals AZ, BH, ΓΘ, ….,KE, respectively (two intersections on each diagonal).

(n). Each one of the five intersections of its opposite first diagonals, coincides to one of the above collinear intersections Π′, P′, Σ′, T′, Y′. Hence it is:

$$ΓE∩KΘ≡Π′,\ ΔZ∩IA≡P′,\ EH∩BK≡Σ′,\ ZΘ∩ΓA≡T′,\ BΔ∩HI≡Y′. \qquad (12).$$

(o). Each one of the five intersections of its opposite third diagonals, coincides to one of the above collinear intersections Π′, P′, Σ′, T′, Y′. Hence it is:

$$BZ∩HA≡Π′,\ ΓH∩BΘ≡P′,\ ΔΘ∩ΓI≡Σ′,\ EI∩ΔK≡T′,\ ZK∩AE≡Y′. \qquad (13).$$

(p). The eighteen intersections above that lie on each of its main diagonals, constitute a Simple Collinear Harmonic Involution, with Double Points the pairs of apexes of the decagon that lie on this same diagonal of it.

Moreover, every eight apexes of the decagon, each time, constitute a Simple Cocyclic Harmonic Involution with Double Points the two apexes remaining.

For example, on diagonal AZ, as we saw above, intersections Π, A_1, $α_1$, α, A′, λ≡BK∩AZ, α′, α″, v≡ΓI∩AZ, Λ, π≡ΔΘ∩AZ, ζ″, ζ′, μ≡EH∩AZ, Z′, ζ, $ζ_1$, Z_1 exist, which constitute a Simple Collinear Harmonic Involution, with Double Points the pair of its apexes A and Z. Additionally, apexes B, Γ, Δ, E, H, Θ, I, K, constitute a Simple Cocyclic Harmonic Involution, with Double Points the pair of apexes A and Z aswell.

This means that, let M be the midpoint of AZ, it will be:

$$MA^2= MZ^2= Mλ.MA′= Mα′.Mα= Mα″.Mα_1= Mv.MA_1= MΛ.MΠ= Mπ.MZ_1= Mζ″.Mζ_1= Mζ′.Mζ= Mμ.MZ′= MB.MK= MΓ.MI= MΔ.MΘ= ME.MH. \qquad (14).$$

(q). The ten intersections Π, Π′, P, P′, Σ, Σ′, T, T′, Y, Y′, that belong to line (ε), constitute a Collinear Harmonic Group of Ten Points.

(Its properties are endless. But the above are enough for us here).

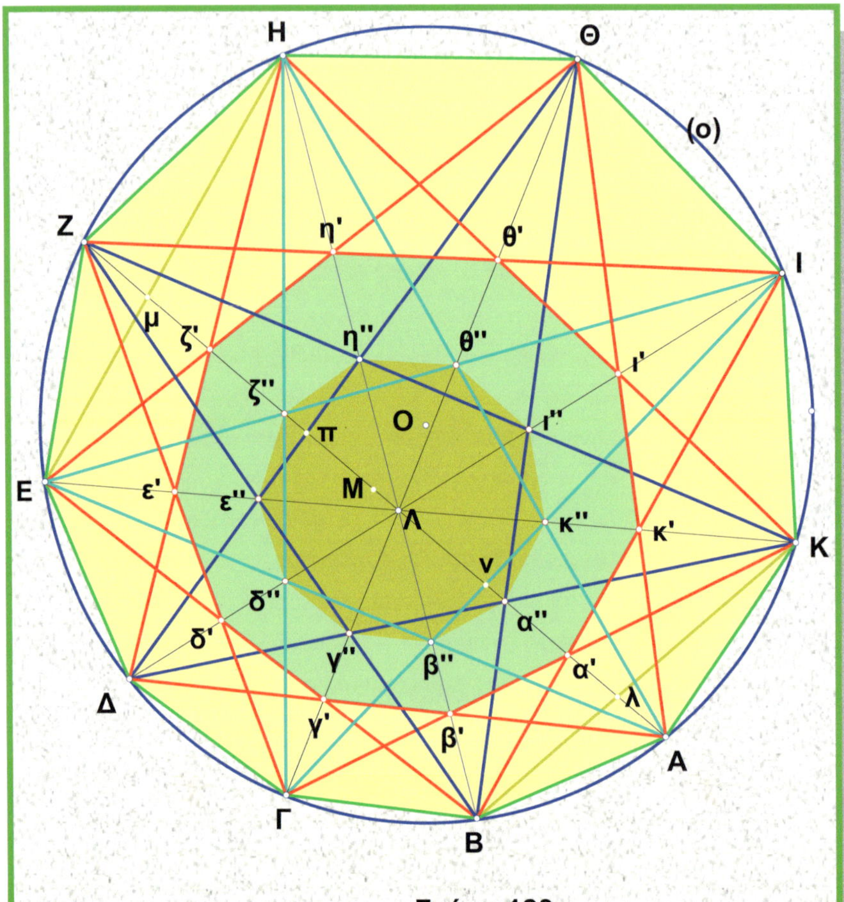

Σχήμα 123.

Solution (Figures 123-125).

[We highlight the fact that figures 123-125, regularly should be one uniform

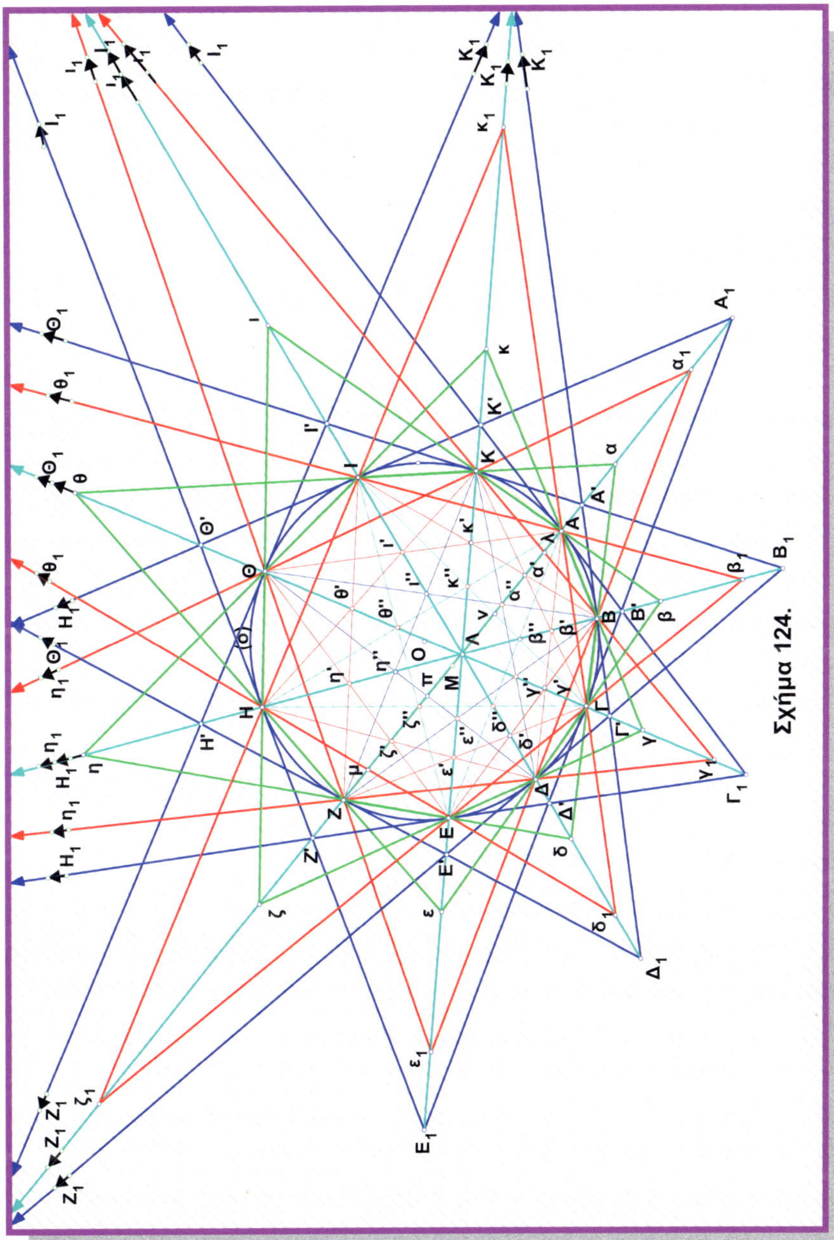

Σχήμα 124.

figure and circle (o), be of the same size, in all figures. In that case, though, the design paper would need to be huge. For this reason we separated it in three parts and with circle (o), in various sizes, depending on our needs. Thus the reader should each time go to the figure that includes the geometrical figures that interest him or her].

(a). Analysis.

We suppose that the desired Decagon is constructed and is the ΑΒΓΔΕΖΗΘΙΚ (Figure 123).

Thus its main diagonals would be concurrent at point Λ [relation (1)], for which, according to the data of the Problem, relation (3) will also be true.

This, according to the converse of Criterion 114 (§ 19.23), means that the desired Decagon needs to be Harmonic, whose Construction can easily be made with the application of our new Method of Harmonic Transformation [This is possible to depend on the converse of Criterion 98 (§19.3), as for example pencil Ι(Π′ΡΡ′ΣΣ′ΤΤ′ΥΥ′Π) or Ι(ΙΚΑΒΓΔΕΖΗΘ), would be harmonic].

(b). Construction.

For the Constructions of the Harmonic Decagon, we work in the same way as in Construction 155 (§ 22.4) (implication).

Hence, if the Decagon that occurs is ΑΒΓΔΕΖΗΘΙΚ, then we say that this is the desired one.

(c). Proof.

For the constructed, as above, to be the desired Harmonic Decagon, we need to prove that for it all properties mentioned in Construction 203 above are true, paragraphs (a) to (q).

In fact, since Decagon ΑΒΓΔΕΖΗΘΙΚ, is Harmonic:

1/. According to Proposition 106 (§ 19.13), relation (1) will be true.

2/. According to Proposition 121 (§ 19.34), relation (2) will be true.

3/. According to Criterion 114 (§ 19.23) (ευθύ), relation (3) will be true.

4/. According to Proposition 125 (§ 19.39), relation (4) will also be true.

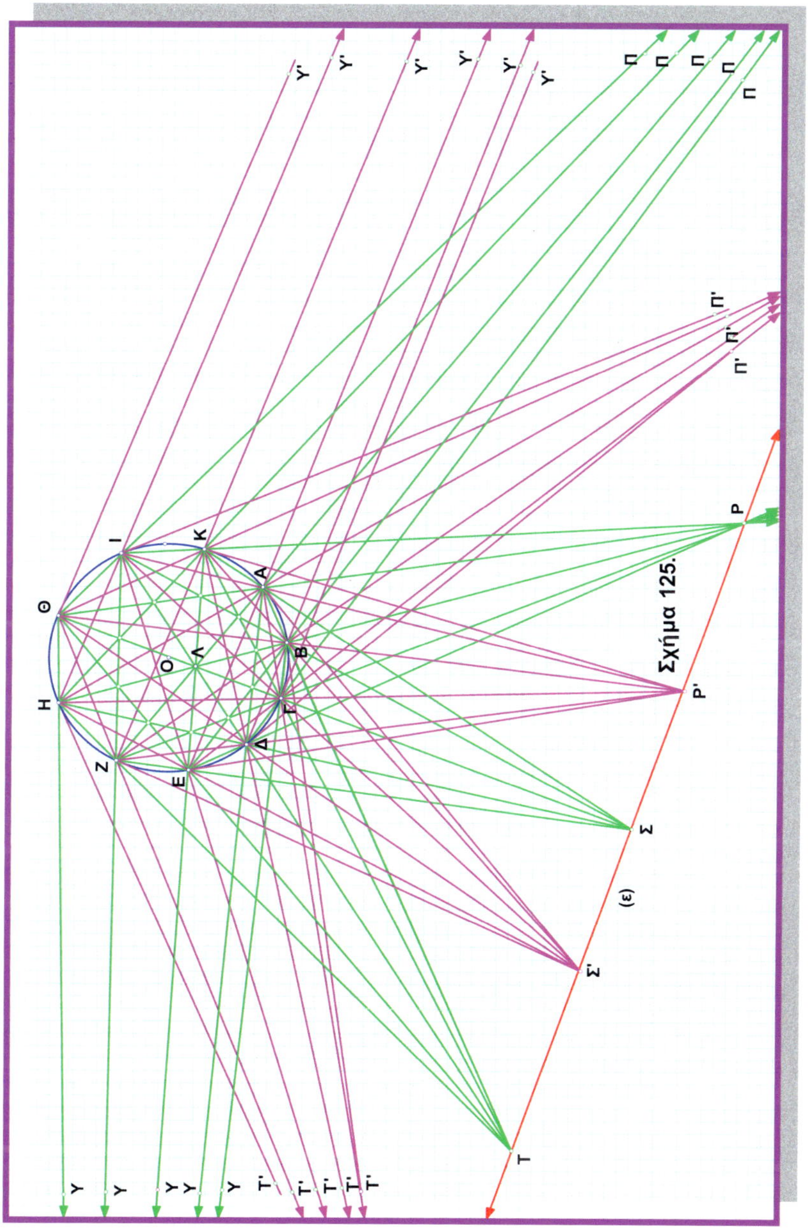

Σχήμα 125.

5/. We easily realize that intersections Π, Ρ, Σ, Τ, Υ are conjugate to point Λ, with respect to circle (o), thus according to what is known, all these will belong to the polar (ε) of point Λ.
After all, according to Proposition 201 (§ 29.2) or Proposition 202 (§ 29.3), relation (6) will also be true.
6/. Additionally, according to Proposition 201 (§ 29.2) or Proposition 202 (§ 29.3), relation (7) will also be true.
7/. According to Proposition 201 (§ 29.2) or Proposition 202 (§ 29.3), all relations from (8) to (11) will be true.
8/. We easily realize that intersections Π', Ρ', Σ', Τ', Υ', are conjugate to point Λ, with respect to circle (o), thus according to what we know, all these would belong on the polar (ε) of point Λ.
Hence all ten points Π, Ρ, Σ, Τ, Υ, Π', Ρ', Σ', Τ', Υ', lie on line (ε).
9/. According to Proposition 201 (§ 29.2) or Proposition 202 (§ 29.3), all the desired in paragraph (l) would be true.
10/ According to Proposition 201 (§ 29.2) or Proposition 202 (§ 29.3), the desired of paragraph (m) would be true.
11/. Because for example ΓΕ is the polar of Δ' and ΚΘ is the polar of Ι', Δ'Ι' is the polar of the intersection of ΓΕ and ΚΘ. Hence, because Δ'Ι', and ΔΙ coincide as we saw, their poles will coincide as well. This means that the intersection of ΓΕ and ΚΘ will coincide with Π'. Similarly we find that the rest of what is desire in (12) is also true.
12/. Because for example ΒΖ is the polar of $Δ_1$ and ΑΗ is the polar of $Ι_1$, $Δ_1Ι_1$ will be the polar of the intersection of ΒΖ and ΑΗ. Hence, because $Δ_1Ι_1$ and ΔΙ, coincide as we saw, their poles will also coincide. This means that the intersection of ΒΖ and ΑΗ will coincide with Π'. Similarly, we find that the rest of what is desired in (13) are true.
13/. According to Proposition 201 (§ 29.2) or Proposition 202 (§ 29.3), all of what is desired in paragraph (p) are true.
14/. We notice that for example the central pencil Θ(ΑΒΓΔΕΖΗΘΙΚ) of ten beams (with the tangent at Θ included), will be harmonic (§ 12.8), as it is

inscribed in circle (o) and Decagon ΑΒΓΔΕΖΗΘΙΚ will be Harmonic because of the way it is constructed (Criterion 98, § 19.3, direct). This means that this pencil is intersected from line (ε) at the following ten points Π, Π′, Ρ, Ρ′, Σ, Σ′, Τ, Τ′, Υ, Υ′, these points will in fact constitute a Harmonic Group of Ten (§ 11.1d and § 12.8).

(d). Description – Interpretation of Figures 123-126.

Figures 123-125, constitute a uniform figure, as each one completes the other two or is part of the other two. We made them into three figures though, in order to avoid using big size paper.

Figure 123.

In this figure intersections of diagonals of the Harmonic Decagon that lie inside circle (o) are included.

Hence, we notice that with these intersections the two following decagons are formed α′β′...κ′ and α″β″...κ″, from which, the first one is formed from intersections of the second diagonals of the Harmonic Decagon and the second one from intersections of its third diagonals. Figure 124.

In this figure intersections of sides and diagonals of the Harmonic Decagon and its tangential are included, which are located outside of circle (o), but are relatively close to it.

Hence, we notice that with these intersections the following decagons are formed:

1/ From the intersections of the sides and the first diagonals the following decagons are formed αβ...κ and $\alpha_1\beta_1...\kappa_1$, respectively.

2/. From the intersections of the sides of the tangential (polar converse) of the Harmonic Decagon, decagons Α′Β′...Κ′ and $A_1B_1...K_1$ are formed.

Figure 125.

In this figure intersections of the sides and the diagonals of the Harmonic Decagon and its respective tangential are included, that are located outside of circle (o) and actually are far away from circle (o).

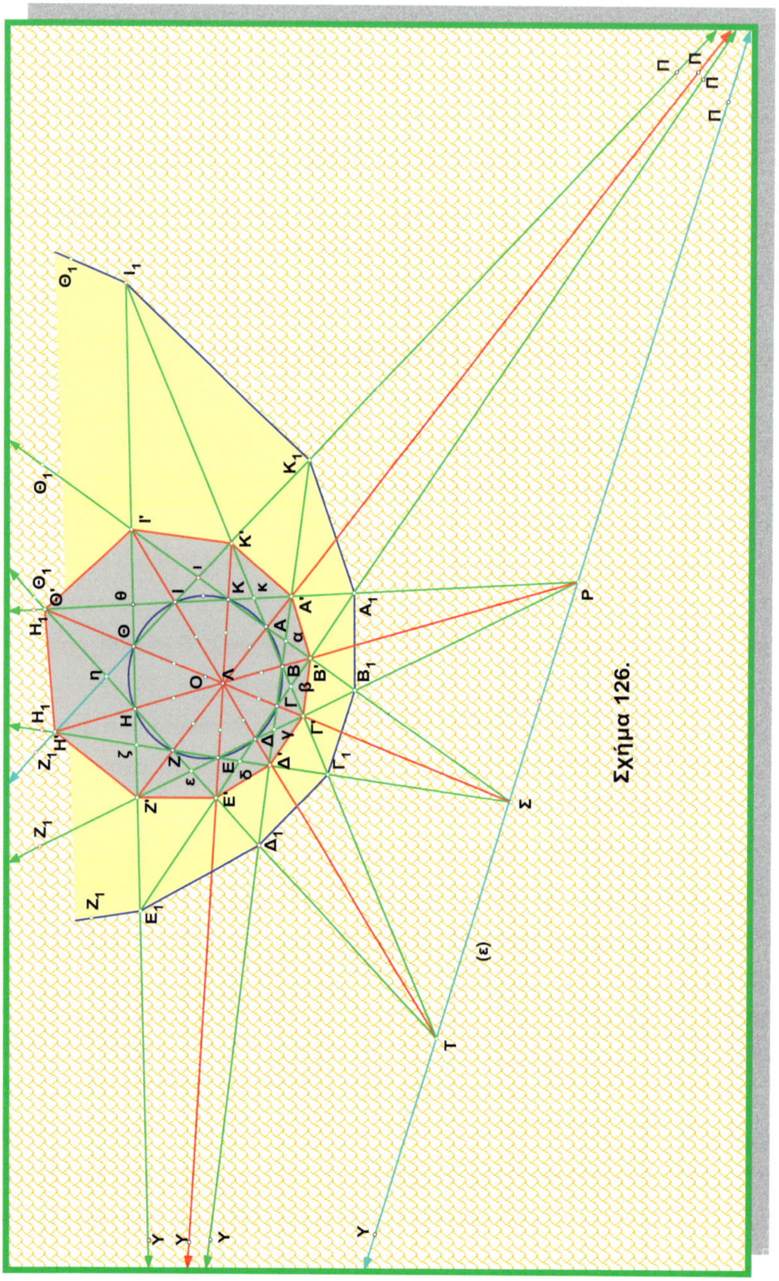
Π
Π
Π
Π
Ι1
Θ1
Κ1
Θ1
Γ'
Κ'
Ι
Θ1
Θ'
θ
Ι
Κ
κ
Α'
Α1
Ρ
Η1
Θ
Α
α
Β
η
Ο
Λ
Β'
Β1
β
Η1
Η
Γ
Γ'
Η'
Ζ1
ζ
Δ
γ
Ζ
Ε
Δ'
Γ1
ε
δ
Σ
Ζ'
Ε'
Ζ1
Δ1
Ζ1
Ε1
(ε)
Τ
Υ
Υ
Υ
Υ
Σχήμα 126.

In this figure we notice that ten central pencils are formed, whose centers Π, Π′, P, P′, Σ, Σ′, T, T′, Y, Y′, are collinear and constitute a Harmonic Group of Ten. From these:

1/. Five with centers Π, P, Σ, T, Y, are formed from the opposite sides, main diagonals and the opposite second diagonals of the Harmonic Decagon.

2/. Five with centers Π′, P′, Σ′, T′, Y′, are formed from the opposite sides of the tangential of the Harmonic Decagon, from the opposite first and third diagonals of the Harmonic Decagon.

It is obvious that, these three figures above constitute a Uniform Harmonic Set, that has analogous characteristics, to the Regular Decagon, which is (regular) and Harmonic (Proposition 151) and from which it is after all originated, through a Harmonic Transformation (Construction 155).

Figure 126.

This figure is independent from figures 123-125 and its purpose is to present the form of a finished Complete Harmonic Decagon ΑΒΓ…Κ αβγ…κ Α′Β′Γ′…Κ′ $Α_1Β_1Γ_1…Κ_1$ ΠΡΣΤΥ, which consists of the initial Simple Harmonic Decagon ΑΒΓ…Κ, the Primary Decagon αβγ…κ, the Secondary Decagon Α′Β′Γ′…Κ′, the Tertiary Decagon $Α_1Β_1Γ_1…Κ_1$ and the quaternary collinear Group of Five Points Π, Ρ, Σ, Τ, Υ, which constitutes a Harmonic Pentagon, degenerate to a line (ε).

Remarks.

(a). Problem 203, the way it is given, is undefined (does not constitute a specific Harmonic Decagon). If we want to define it, then we need to be given sufficient data.

(b). The above Problem and its solution, we believe to be new and firstly appear here.

(c). Regular Decagon, as all Regular Multilaterals are Harmonic too (Proposition 151, § 19.66), but constitutes a special case of the harmonic. Therefore the regular, has itself all properties of the harmonic, but in a special Regular form (for example the line of the intersections of the opposite sides, is found at infinity).

(d). Constructions equivalent to 203, are possible to be manufactured for all other Harmonic Multilaterals.

(e). As we will realize from the above, Problem 203 is gigantic, as its formulation is rather long and mostly because this decagon has various properties and because for its solution we need to depend on many Propositions and Constructions, that cover almost all material in this book. it is an application covering almost the whole outline of Harmonic Geometry.

We believe that there is no other way of solving it and for this reason it is a great challenge for Geometry fans.

(f). If we include, in a uniform figure, the Complete Decagons that are formed from the sides of the Simple Harmonic Decagon (as in figure 126), from the Tangential Complete Decagon and the three Complete Decagons that each one is formed from the first, second and third diagonals of the Harmonic Decagon, then a complex with various properties will occur. Such complexes are mentioned in our book [1], concerning Harmonic Quadrilaterals [Figures 2.2.2.(1).(α).3/ and 4/ pages 696 and 697], Hexagons [Figure 3.2.6.(4).(b).1/, page 1285] and Octagons [Figure 4.2.5.(5).(β).3/ page 1746], which we reached though in another way, different from the one we used here (See also General Annotations § 29.10 below).

(g). Some more properties of the decagon in Construction 203, are the following:

The main diagonals of its tangential decagon are concurrent at Λ (This is an extension to Newton's Theorem, and Brianhon's Theorem as well. See Proposition 206).

Many intersections of the sides and diagonals of the Harmonic Decagon, and its tangential, belong to every main diagonal of its tangential. Hence something equivalent to Proposition 201 or 202 is happening.

Each one of the main diagonals of the tangential, runs through one of the intersections Π′, Ρ′, Σ′, Τ′,Υ′.

(h). For more features of Harmonic Complexes, Quadrilaterals, Hexagons and Octagons, we cite book [1].

(i). Relevant Propositions are 98, 106, 114, 121, 125, 151, 155, 201, 202.

29.5. Notable Property of Harmonic Groups of Four.

Extension to Constructions 11, 32a and 63.

Proposition 204 (Figure 53).

On a line (ε) a group of eight points A, B, Γ, Δ, E, Z, H, Θ is given, for which the group of four points A, Γ, E, H is harmonic and for which group of four, the two pairs of points A, Γ and E, H, are in Harmonic Involution with Double Points B, Z, while the two pairs of points A, H and Γ, E are in Harmonic Involution with Double Points Δ, Θ [Hence A, Γ, E, H. Are in Double Harmonic Involution (definition § 2.24)].
Prove that the following group of eight points A, B, Γ, Δ, E, Z, H, Θ, will be harmonic.

Proof.

1st.Way (Based on Construction 32 a).

According to Construction 32a (§ 7.22), there will be a point P from which segments AB, BΓ, ΓΔ, ΔE, EZ, ZH, HΘ, will be seen in angles equal and of measure L/4 ή $\varphi=22.5^0$. This means that pencil P(ABΓΔEZHΘ) will be Closed Regular of ¼ of a right (§ 12.5), such as also $\angle AP\Theta'=22{,}5^0$, thus the group of eight points A, B, Γ, Δ, E, Z, H, Θ, will also be harmonic (§ 12.6 and general conclusion of § 15.6).

2nd.Way (Based on Construction 63).

From what is given in Proposition 204, it occurs that the four below «one by one» groups of four points in (1), are harmonic:

A, B, Γ, Z - Γ, Δ, E, Θ - E, Z, H, B - H, Θ, A, Δ. (1).

In addition, from the hypothesis, the group of points A, Γ, E, H, is also harmonic, thus, if we work in the same way as in the proof of way 3 in Construction 63, it will easily occur that the group of eight points A, B, Γ, Δ, E, Z, H, Θ, will in fact be harmonic.

Conclusion.

Regarding Proposition 204 and if we take into consideration Constructions 11, 32a and 63, the conclusion below occurs:

If, regarding Proposition 204, the group of four points A, Γ, E, H, is not harmonic then the group of eight points A, B, Γ, Δ, E, Z, H, Θ, will also not be harmonic, while if the group of four points A, Γ, E, H, is harmonic then the group of eight points A, B, Γ, Δ, E, Z, H, Θ, will also be harmonic.

Remarks.

(a). The Proposition above and its solution, we believe to be new and firstly appear here.

(b). We reach an equivalent result also if it is given that the group of four points B, Δ, Z, Θ, is harmonic and the pairs of points B, Δ and Z, Θ, are in Harmonic Involution with Double Points Γ, H and pairs of points B, Θ and Δ, Z, are in Harmonic Involution with Double Points A, E. Extension to a circle in Proposition 205.

(c). Relevant Propositions are 11, 32a, 63, 149, 205.

29.6. Notable Property of Harmonic Quadrilaterals.

Extension of Proposition 204, from a line to a circle.

Proposition 205 (Figure 96).

In circle (o), its inscribed convex octagon ΑΒΓΔΕΖΗΘ is given whose diagonals are concurrent and for which quadrilateral ΑΓΕΗ is harmonic, the two pairs of points A, Γ and E, H, are in Harmonic Involution with Double Points B, Z, while the two pairs of points A, H and Γ, E are in Harmonic Involution with Double Points Δ, Θ [Hence points A, Γ, E, H. Are in Double Harmonic Involution (definition § 8.22)].

Prove that octagon ΑΒΓΔΕΖΗΘ, is harmonic.

Proof.

(a). Analysis.

If we take into consideration Construction 149 (§ 29.64), we easily reach the proof below.

(b). Proof.

In figure 96, using Construction 149 [§ (b)], we have constructed:

The random Simple Harmonic Quadrilateral ΑΓΕΗ. The two pairs of points Δ, Θ and Β, Ζ which harmonically divide pairs of points Α, Η - Γ, Ε, and Α, Γ – Ε, Η, respectively and which are constructed in such way that :

$$ΑΕ∩ΒΖ∩ΓΗ∩ΔΘ≡Κ.$$

We will prove that octagon ΑΒΓΔΕΖΗΘ, is harmonic.

In fact, if we work in the exact same way, as in the proof of Construction 149 [§ (c)], we prove, based on Criterion 110 (§ 19.18) too, that octagon Α-ΒΓΔΕΖΗΘ, is harmonic.

Conclusion.

Regarding Proposition 205, if we take into consideration Construction 149 and Proposition 204, the conclusion below occurs:

If, regarding Proposition 205, quadrilateral ΑΓΕΗ, is not harmonic then octagon ΑΒΓΔΕΖΗΘ, will not be harmonic as well, while if quadrilateral ΑΓΕΗ, is harmonic then octagon ΑΒΓΔΕΖΗΘ, will be harmonic.

Remarks.

(a). The Proposition above and its solution, we believe to be new and firstly appear here.

(b). We would reach an equivalent result also if it was given that quadrilateral ΒΔΖΘ is harmonic and pairs of points Β, Δ and Ζ, Θ, are in Harmonic Involution with Double Points Γ, Η and pairs of points Β, Θ and , Δ, Ζ, are in Harmonic Involution with Double Points Α, Ε.

(c). Relevant Propositions are 63, 110, 149, 204.

29.7. Convergence of Diagonals of the Tangential Decagon of each Simple Harmonic Decagon.(Equivalent to Point Brianchon and Newton's Theorem, for a Harmonic Decagon).

Extension of Proposition 146 to a Harmonic Decagon.

Proposition 206 (Figure 127).

The diagonals (main), of the tangential or Polar Converse decagon, of every Simple Harmonic Decagon, are concurrent or to state it otherwise:
The tangential or Polar Converse decagon, of every Simple Harmonic Decagon, has an equivalent Brianchon point.

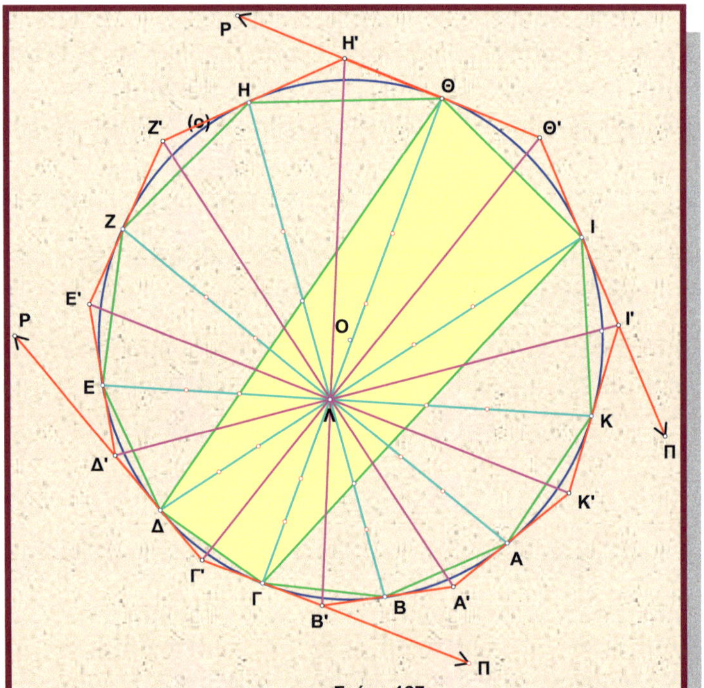

Σχήμα 127.

Proof.

Let there be a Harmonic Decagon ΑΒΓΔΕΖΗΘΙΚ and Α'Β'Γ'Δ'Ε'Ζ'Η'Θ'Ι'Κ' (where Α' is the intersection of the tangents at points Α, Β - Β' is the intersection of the tangents at points Β, Γ, and so on and so forth) its respective tangential or the polar converse of this Harmonic Decagon, with its operator circle (o) (Geometry [26], paragraph 246a).
We will prove that it is: Α'Ζ'∩Β'Η'∩Γ'Θ'∩Δ'Ι'∩Ε'Κ'≡Ν. (1).

1st Way (Analytical).

Let the tangents at points Γ, I intersect at point Π and the tangents at points Δ, Θ, intersect at point P.

In this way, the circumscribed quadrilateral Γ'PΘ'Π occurs in quadrilateral ΓΔΘΙ, for which, according to known Newton's Theorem (Geometry [26], § 236α), they have ΓΘ∩ΔΙ∩Γ'Θ'∩ΠΡ≡Λ, where Λ≡ ΑΖ∩ΒΗ∩ΓΘ∩ΔΙ∩ΕΚ [Criterion 106 (§ 19.15)].

Hence diagonal Γ'Θ' runs through Λ, which is the point of convergence of the diagonals of the Harmonic Decagon ΑΒΓΔΕΖΗΘΙΚ.

In the exact same way we work to find that A'Z', B'H', Δ'I', E'K', run through Λ.

Hence it finally is: ΑΖ∩ΒΗ∩ΓΘ∩ΔΙ∩ΕΚ∩Α'Ζ'∩Β'Η'∩Γ'Θ'∩Δ'Ι'∩Ε'Κ'≡Λ≡Ν.

2nd Way [Based on Proposition 1δ(1)].

There is a known Newton's Theorem according to which, two polar converse quadrilaterals have concurrent diagonals.

We have generalized this Theorem for polar converse 2n-laterals, in Proposition 1δ(1) in our book «Geometry [31] (volume 1), which Proposition is always true for 2n-laterals, as long as the diagonals of their inscribed 2n-lateral are concurrent.

Therefore, if we consider known and depend on Proposition 1δ(1), then because the Harmonic Decagon ΑΒΓΔΕΖΗΘΙΚ has always diagonals concurrent at point Λ [Criterion 106 (§ 19.15)], at the same point Λ diagonals of the decagon Α'Β'Γ'Δ'Ε'Ζ'Η'Θ'Ι'Κ' will be concurrent.

Hence it will be: ΑΖ∩ΒΗ∩ΓΘ∩ΔΙ∩ΕΚ≡Α'Ζ'∩Β'Η'∩Γ'Θ'∩Δ'Ι'∩Ε'Κ'≡Λ≡Ν.

Conclusion.

Through the above and because the Lemoine point of the Harmonic Decagon coincides with the convergence point of its diagonals Λ [Criterion 114 (§ 19.23)], this conclusion is reached:

«The diagonals (main), of the tangential or Polar Converse decagon, of every Simple Harmonic Decagon, are concurrent at the convergence point of the diagonals of this Harmonic Decagon, which is not only the Lemoine point of this Harmonic Decagon, but also the equivalent to the Brianchon and Newton point, of the tangential decagon, or their Lemoine, Newton and Brianchon points coincide at the convergence point of the Harmonic Decagon's diagonals ».

Remarks.

(a). In analogous ways we can prove that equivalent Propositions and Conclusions, are true for the Quadrilateral, the Harmonic Hexagon (Proposition 146), for the Harmonic Octagon, etc.

(b). This Proposition, we believe to be new and firstly appear here.

(c). It is obvious that Proposition 206 is an extension to Newton's Theorem (Geometry [26], § 236a), which is known to be true only for quadrilateral Polar Converses.

(d). A Proposition equivalent to Proposition 201, is proven to be true for every diagonal of the tangential decagon, that matches to a Harmonic Decagon.

(e). Proof of Proposition 206, is possible to be achieved using the Method of Harmonic Transformation, as done in Propositions 200, 202, based on Corollaries 158-168.

(f). Relevant Propositions are 106, 114, 146.

29.8. Auxiliary Proposition (Lemma),

A New Discussion on Criterion 34 (§ 8.71).

Criterion 207 (Figure 128).

In every convex inscribed in circle (o) hexagon ΑΒΓΔΕΖ with ΑΔ∩ΒΕ∩ΓΖ≡Κ, if one of its quadrilaterals:

ΑΒΓΕ, ΒΓΔΖ, ΓΔΕΑ, ΔΕΖΒ, ΕΖΑΓ, ΖΑΒΔ, (1).

Is harmonic, then every other of the quadrilaterals in (1) will also be harmonic, following the equivalences below:

ΑΒΓΕ↔ΔΕΖΒ, ΓΔΕΑ↔ΖΑΒΔ, ΕΖΑΓ↔ΒΓΔΖ, (2).

And conversely, if one pair, of the pairs of quadrilaterals in relation (2) are harmoniv, then it will also be ΑΔ∩ΒΕ∩ΓΖ≡Κ (and not only that).

Proof.

1st Way (Based on Lemma 5).

(a). Direct.

We suppose for example that the harmonic quadrilateral is ΑΒΓΕ. We will prove that its respective quadrilateral ΔΕΖΒ is harmonic, if ΑΔ∩ΒΕ∩ΓΖ≡Κ.

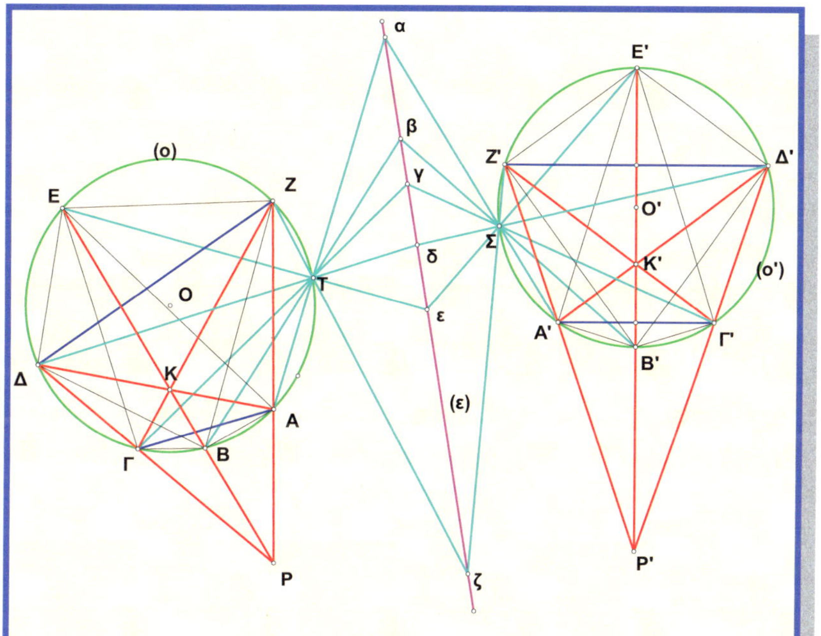

Σχήμα 128.

In fact, according to Lemma 5 (5η Auxiliary Proposition, in Part E), regarding hexagon ΑΒΓΔΕΖ, for which it is ΑΔ∩ΒΕ∩ΓΖ≡Κ (hypothesis), it will also be :

$$\frac{AB}{B\Gamma}\cdot\frac{\Gamma\Delta}{\Delta E}\cdot\frac{EZ}{ZA}=1. \qquad (3).$$

Additionally, according to Lemma 5 and for the concave hexagon ΖΑΕΓΔΒ, because it shares its diagonals (main) with hexagon ΑΒΓΔΕΖ it will have: ΑΔ∩ΒΕ∩ΓΖ≡Κ, thus it will be:

$$\frac{ZA}{AE}\cdot\frac{E\Gamma}{\Gamma\Delta}\cdot\frac{\Delta B}{BZ}=1. \qquad (4).$$

From relations (3), (4) $\Rightarrow \frac{AB}{B\Gamma} \cdot \frac{EZ}{\Delta E} \cdot \frac{E\Gamma}{AE} \cdot \frac{\Delta B}{BZ} = 1.$ (5).

Moreover, because quadrilateral ABΓE is harmonic (hypothesis), according to known Theorem {[6] § 8-4.10}, it will be: AB.ΓE=BΓ.EA. (6).

Hence, relation (5), because of (6), will be: ΔE.ZB=BΔ.EZ. (7).

The latter relation (7), according to known Theorem [6], means that the inscribed quadrilateral ΔEZB is harmonic.

(b). Converse.

Hence, if the given convex hexagon ABΓΔEZ, has harmonic for example the two quadrilaterals ABΓE and ΔEZB, then it will also be AΔ∩BE∩ΓZ≡K.

In fact, because quadrilaterals ABΓE and ΔEZB are harmonic from the hypothesis, AΓ and ΔZ are conjugate to BE and thus if Π is the pole of BE, AΓ and ΔZ will run through Π.

Hence, intersection Π′ of AΓ and ΔZ coincides with pole Π of BE.

But intersection Π′ of AΓ and BZ, as it is known {[4], § 81}, has KP as its polar (where P≡AZ∩ΓΔ).

So we see that the poles of BE and KP coincide, thus their polars will coincide too. Hence it is BE≡KP or BE runs through K and P, therefore it is in fact AΔ∩BE∩ΓZ≡K (but also AZ∩BE∩ΓΔ≡P).

For the proof of the rest pairs of quadrilaterals in relation (2) we work in a similar way.

2nd Way (Based on Lemma 7).

(a). Direct.

We suppose again that ABΓE is the harmonic quadrilateral. We will prove that its respective quadrilateral ΔEZB is harmonic, if AΔ∩∩BE∩ΓZ≡K.

In fact, according to Lemma 7 (7th Auxiliary Proposition, Part E), concerning Central Pencil K(ABΓE) or K(ΔEZB), it is:

$$\frac{AB}{B\Gamma} \cdot \frac{\Gamma E}{EA} = \frac{\Delta E}{EZ} \cdot \frac{ZB}{B\Delta}. \quad (8).$$

After all, because from hypothesis quadrilateral ABΓE, is harmonic, it will also be: AB.ΓE=BΓ.EA or $\frac{AB}{B\Gamma} \cdot \frac{\Gamma E}{EA} = 1.$ (9).

Therefore, relation (8), because of (9), becomes: $\frac{\Delta E}{EZ}\cdot\frac{ZB}{B\Delta}=1$ or

$$\Delta E.ZB=EZ.B\Delta.$$

The latter relation means that quadrilateral ΔEZB is harmonic etc.

(b). Converse.

I.e., if the given convex hexagon ABΓΔEZ, has harmonic for example two quadrilaterals ABΓE and ΔEZB, then it will also be $A\Delta\cap BE\cap \Gamma Z\equiv K$.

In fact, because quadrilaterals ABΓE and ΔEZB are harmonic from the hypothesis, they will have:

$$\frac{AB}{B\Gamma}\cdot\frac{\Gamma E}{EA}=1,\ \frac{\Delta E}{EZ}\cdot\frac{ZB}{B\Delta}=1 \Rightarrow \frac{AB}{B\Gamma}\cdot\frac{\Gamma E}{EA}=\frac{\Delta E}{EZ}\cdot\frac{ZB}{B\Delta}.$$

The last relation, according to the converse of the above Auxiliary Proposition 7 (Part E), means that it is: $A\Delta\cap BE\cap \Gamma Z\equiv K$.

Similarly we work for the proof of the rest pairs of quadrilaterals in relation (2).

3rd Way (Using the Method of Harmonic Transformation).

(a). Direct.

We suppose for example that the harmonic quadrilateral is ABΓE. We will prove that its respective quadrilateral ΔEZB is also harmonic if:

$$A\Delta\cap BE\cap \Gamma Z\equiv K. \qquad (10).$$

In fact, since quadrilateral ABΓE is harmonic, AΓ and BE will be conjugate with respect to circle (o). This means that during the Harmonic Transformation, AΓ, BE will have come from two perpendicular chords A′Γ′, B′E′ of a circle (o′), from which B′E′ is the diameter of circle (o′) (Converse of Corollary 159), thus B′E′ is the perpendicular bisector to A′Γ′.

Additionally, during the above Harmonic Transformation, because relation (10) is true, it will be $A'\Delta'\cap\Gamma'Z'\equiv P\in B'E'$ and A′Δ′, Γ′Z′, will be symmetrical chords, with respect to diameter B′E′ of circle (o′) (Converse of Corollary 161).

Therefore A′K′, Γ′K′ will also be symmetrical, with respect to B′K′ and because B′E′ is the perpendicular bisector to A′Γ′, triangle K′A′ Γ′ is isosceles thus ∠Δ′A′Γ′=∠Z′Γ′A′, but it also is ∠A′Z′Γ′=∠A′Δ′Γ′.

Hence, triangles A′Z′Γ′, Γ′Δ′A′ are similar and because they share side A′Γ′, they will be equal, thus A′Δ′=Γ′Z′. Therefore A′Γ′Δ′Z′ is an isosceles trapezium and because B′E′ ⊥A′Γ′, it will also be B′E′ ⊥Δ′Z′.

Thus, from the Harmonic Transformation and because B′E′ ⊥Δ′Z′, B′E′, Δ′Z′, are transformed to the conjugate chords BE, ΔZ with respect to circle (o) (Direct of Corollary 159).

Therefore, since BE, ΔZ are conjugate then quadrilateral ΔEZB, will be harmonic (definition).

(b). Converse.

I.e., if the given convex hexagon ΑΒΓΔΕΖ, has harmonic for example these two quadrilaterals ΑΒΓΕ and ΔEZB, then it will also be AΔ∩BE∩ΓZ≡K.

In fact, then AΓ and ΔZ will be conjugate, with respect to circle (o), to BE (definition). This means that during the Harmonic Transformation, AΓ, ΔZ came the chords A′Γ′, Δ′Z′ perpendicular to diameter B′E′ of circle (o′) (converse of the Corollary 159), Hence, A′Γ′Δ′Z′ is an isosceles trapezium, as B′E′ is the perpendicular bisector to A′Γ′, Δ′Z′.

Therefore chords A′Z′ and Γ′Δ′ are symmetrical to diameter B′E′, thus intersection A′Δ′∩Γ′Z′≡K′∈B′E′.

Through all the above, during the Harmonic Transformation (Implication of Corollary 161), A′Δ′, Γ′Z′ are transformed to chords AΔ, ΓZ of circle (o), which intersect at chord BE, thus relation (10) will be true (but also AZ∩BE∩ΓΔ≡P).

We work similarly for the proof of the rest pairs of quadrilaterals in relation (2).

Remarks.

(a). Proposition 207, is stated otherwise and proven in a diffirent way in Criterion 34 (§ 8.71).

(b). We notice that quadrilaterals ΑΒΓΕ, ΔEZB, share diagonal BE.

(c). It is obvious that Criterion 207 is true and if instead of Harmonic Quadrilateral ΑΒΓΕ, quadrilateral ΔΕΖΒ is given as the harmonic one. This is proved in the same way as for ΑΒΓΕ above.
(d). Relevant propositions are 34, 159, 161.

29.9. Another Proof of Criterion 51, using the method of Harmonic Transformation.

Criterion 208 (Figure 128).

Every convex hexagon ΑΒΓΔΕΖ inscribed in circle (o) with concurrent diagonals (main) and whose diagonal for example ΒΕ runs through intersection P of its two opposite sides ΑΖ and ΓΔ, has its two quadrilaterals ΑΒΓΕ and ΔΕΖΒ harmonic, and conversely.

Proof.

(a). Direct.

In fact, because from hypothesis we have ΑΔ∩ΒΕ∩ΓΖ≡Κ, (1).

and ΑΖ∩ΒΕ∩ΓΔ≡P, (2).

hexagon ΑΒΓΔΕΖ, during the Harmonic Transformation [Converse of Corollary 163 (Rule 6)], has come from a hexagon Α′Β′Γ′Δ′Ε′Ζ′, inscribed in circle (o′), which besides everything else, is symmetrical with respect to its diagonal Β′Ε′, with respect to which its quadrilaterals Α′Β′Γ′Ε′ and Δ′Ε′Ζ′Β′ are obviously symmetrical too.

Therefore, symmetrical hexagon Α′Β′Γ′Δ′Ε′Ζ′ with respect to its diagonal Β′Ε′, during the Harmonic Transformation (Direct of Corollary 163), this will be transformed to a hexagon ΑΒΓΔΕΖ, inscribed in circle (o), which besides everything else, has harmonic its quadrilaterals ΑΒΓΕ, ΔΕΖΒ.

(b). Converse.

I.e., if the given convex hexagon ΑΒΓΔΕΖ, has harmonic its two quadrilaterals from example ΑΒΓΕ and ΔΕΖΒ, then it will be ΑΔ∩ΒΕ∩ΓΖ≡Κ and ΑΖ∩ΒΕ∩ΓΔ≡P.

To prove this we will depend on Corollaries 159 and 161.

In fact, because quadrilaterals ABΓE andı ΔEZB are harmonic from the hypothesis, AΓ and ΔZ will be conjugate to BE (definition), with respect to circle (o). This means that during the Harmonic Transformation, AΓ, ΔZ came from perpendicular chords A′Γ′, Δ′Z′ to diameter B′E′ of circle (o′) (converse of Corollary 159), Thus, A′Γ′Δ′Z′ is an isosceles trapezium , since B′E′ is the perpendicular bisector to A′Γ′, Δ′Z′.

Therefore chords A′Z′ and Γ′Δ′ are symmetrical to diameter B′E′, hence intersections A′Δ′∩Γ′Z′≡K′ and A′Z′∩Γ′Δ′≡P′, will belong to B′E′.

Due to the above, during the Harmonic Transformation (Implication of Corollary 161), A′Z′, Γ′Δ′ are transformed to chords AZ, ΓΔ of circle (o), which will intersect on the chord BE. This means that relation (2) is true. Moreover, for the same reason, relation (1).

We work similarly to prove the other two pairs of quadrilaterals BΓΔZ- EZAΓ and ΓΔEA- ZABΔ.

Remarks.

(a). The proof above is obviously another new proof of Criterion 51, using the method of Harmonic Transformation.

(b). Relevant Propositions are 34, 51, 159, 161, 163, 207.

29.10. GENERAL ANNOTATIONS REGARDING HARMONIC MULTILATERALS.

(a). Complexes of Harmonic Multilaterals.

The whole Theory above (B and D Parts) regarding Simple Harmonic Multilaterals (Definition, Theories, Methods, Lemmas, Propositions, Corollaries, etc.), in the way they are presented here, we believe that firstly appear in our paper above, since so far we have not stumbled upon them in our known bibliography or online, except for our books [1] and [2], there though their presentation is thorough {mainly in [1]}, and there they are reached through a different way, while Complexes of Complete Harmonic Multilaterals are mentioned as well.

Specifically:

1/. In book [1] (volume 2) at least 12 Propositions-Criteria about Simple Harmonic Quadrilaterals are mentioned (pages 734 to 751) and at least 33 Propositions-Criteria regarding Complexes of Complete Harmonic Quadrilaterals pages 611 to 692. {An example of the complexity, of a Complex of Complete Harmonic Quadrilatelars, can be seen in figure 2.2.2.(1).(a).3/ and 4/, pages 696 and 697, in book [1]). In this same book of ours [1] regarding these figures, we prove that all the cases of line (central pencils) convergence are true, as well as all cases of collinear centers of pencils (series of points) and more than that}.

2/. In book [1] (volume 3) at least 10 Propositions-Criteria about Simple Harmonic Hexagons are mentioned (pages 1321 to 1336) and at least 37 Propositions-Criteria about Complete Harmonic Hexagons (pages 1174 to 1286). {An example of the complexity, of a Complex of Complete Harmonic Hexagons, can be seen in figure 3.2.6.(4).(b).1/ (attached), page 1285, in book [1]. In this same book of ours [1] regarding these figures, we prove that all the cases of line (central pencils) convergence are true, as well as all cases of collinear centers of pencils (series of points) and more than that}.

3/. In book [1] (volume 4) at least 10 Propositions-Criteria about Simple Harmonic Octagons are mentioned (pages 1769 to 1778) and at least 76 Propositions-Criteria about Complete Harmonic Octagons (pages 1663 to 1749). {A sample of the complexity, of a Complex of Complete Harmonic Octagons, can be seen in figures 4.2.5.(5).(b).1/ to 6/, in book [1] page 1746. In this same book of ours [1] regarding these figures, we prove that all the cases of line (central pencils) convergence are true, as well as all cases of collinear points (series of points) and more than that}.

4/. If we consider, in a uniform figure, the Complete Decagons that are formed from the sides of a Simple Harmonic Decagon (figures 123-126), from its Tangential Complete Decagon and the three Complete Decagons

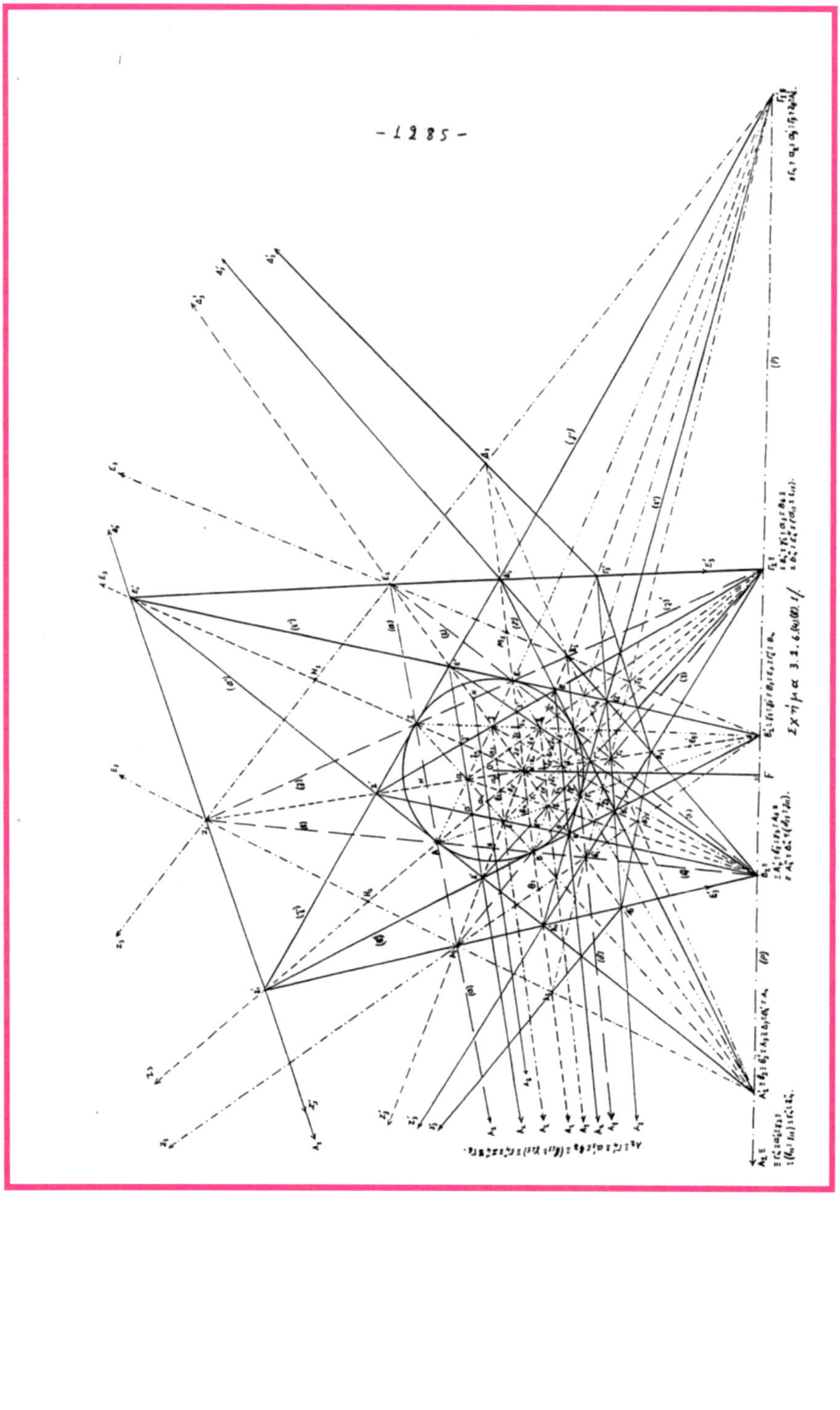

that each is formed from the first, second and third diagonals of this Harmonic Decagon, then a Complex of Harmonic Decagon will occur analogous to the ones mentioned above, which has various properties, some of which refer to Construction 203 above.

5/. Analogous Harmonic Complexes, can obviously be formed for every other Harmonic Multilateral.

6/. Relevant Propositions are 201-203.

(b). It is obvious that from our present paper we only meet the properties of Simple Harmonic Multilaterals, and if we were able to prolong it, drawing data (Criteria, Propositions, etc), with many other properties of them, from book [1]. But we wrapped it up here, as this work would be huge and these data are given in book [1].

(c). From Criterion 148 (Figure 95), it occurs that: «Twin triangles ΑΓΕ, ΒΔΖ, in every harmonic hexagon ΑΒΓΔΕΖ are four times homological, with homology centers $K \equiv A\Delta \cap BE \cap \Gamma Z$, $A_2 \equiv ZA \cap \Gamma\Delta \cap BE$, $B_2 \equiv AB \cap \Delta E \cap \Gamma Z$, $\Gamma_2 \equiv B\Gamma \cap EZ \cap A\Delta$), that is an important Criterion, as up until now only three times homological triangles were known to us.

30. GENERAL CONCLUSIONS OF HARMONIC GEOMETRY

Through meticulous reading of this book, the following general conclusions occur, regarding Harmonic Geometry:

1. Intersections, from lines, of Central Pencils of Pencils Symmetric with respect to pair of perpendicular axes, give us (constitute) Collinear Simple Harmonic Involutions.
2. Intersections, from lines, of Central Pencils of Pencils Symmetric with respect to two pairs of perpendicular axes simultaneously, which (axes) constitute a Central Regular Pencil 45^0, give us (constitute) Collinear Double Harmonic Involutions.
3. Intersections, from lines, of Closed Regular Central Pencils, give us (constitute) Collinear Closed Harmonic Series of Points (Harmonic Multilaterals degenerate to a line).
4. Intersections, from lines, of Pairs of Open Regular Central Pencils, give us (constitute) Pairs of Collinear Open Harmonic Series of Points (Pairs of Harmonic Lines degenerate to a straight line).
5. Pairs of apexes of multilaterals inscribed in circles, conjugate with respect to one of their main diagonals, give us (constitute) Cocyclic Simple Harmonic Involutions.
6. Pairs of apexes of multilaterals inscribed in circles, conjugate with respect to two of their main diagonals simultaneously, give us (constitute) Cocyclic Double Harmonic Involutions.
7. Pairs of apexes of of Pairs of Open Lines inscribed in circles, conjugate with respect to one of their main diagonals, give us (constitute) Cocyclic Pairs of Open Harmonic Lines.
8. Apexes of Regular Multilaterals inscribed in circle, give us (constitute) Cocyclic Simple Harmonic Series of Points (Harmonic Multilaterals) (Proposition 151).
9. Symmetrical Multilaterals inscribed in circle, with respect to one of their main diagonals, using the New Method of Harmonic Transformation, give us Symmetrical Central Pencils with respect to a pair of perpendicular axes

and consequently Collinear Simple Harmonic Involutions and finally Cocyclic Simple Harmonic Involutions and conversely.

10. Symmetrical Multilaterals inscribed in circle, with respect to two perpendicular main diagonals of their, using the New Method of Harmonic Transformation, give us Symmetrical Central Pencils with respect to two pairs of perpendicular axes that constitute a Central Regular Pencil of four beams 45^0 and consequently Collinear Double Harmonic Involutions and finally Cocyclic Double Harmonic Involutions and conversely.

11. Regular 2n-laterals, using the Method of «Harmonic Transformation», give us Regular Central Pencils of 2n beams and consequently Collinear Closed Harmonic Series of Points with 2n points and finally Harmonic Cocyclic Series of Points with 2n points (Harmonic 2n-laterals) and conversely.

12. All the features (about Harmonic Involutions and Harmonic Multilaterals) given in Harmonic Geometry, refer to features, concerning elements that belong to circles (cocyclic elements), as well as elements that belong to lines (collinear elements), but then, the lines they belong to, can be considered degenerate circles with radius of infinite length (extreme case, then these circles are of zero curvature).

13. Harmonic Multilaterals (collinear and cocyclic), can be considered as special cases of involutions, as in Harmonic Multilaterals with 2n apexes, each pair of opposite apexes, harmonically divides the rest pairs of apexes.

14. Regular Central Pencils, can be considered Symmetrical Pencils, with repsect to one of their beams (special case of symmetrical Central Pencils).

15. Every Regular Multilateral is Harmonic too (Proposition 151). Hence Regular Multilaterals are a special case of Harmonic ones, in which case all their sides are equal.

16. All Harmonic Multilaterals (Collinear and Cocyclic, mostly Cocyclic), show a rather big number of significant properties that other multilaterals do not (See Construction 203), except for Regular ones though that are also Harmonic.

17. The New Method of Harmonic Transformation, is the main tool (key), for for the prove-solution of Propositions-Problems in Harmonic Geometry.

18. After all what is mentioned in the book and not only, the conclusion is reached, that there are many issues of Geometry, that are not mentioned In bibliography, and that Geometry is not over here, but we believe it is endless and can continue its development. Consequently, we can say that even though various enemies of Geometry tried to abolish it, they did not make it. Geometry is here, magica, strong and evergrowing.

PART E.

AUXILIARY PROPOSITIONS.

(Lemmas).

Constantin Carathéodory (1873-1950).

PART E.

AUXILIARY PROPOSITIONS.

(Lemmas).

1. In General.

Part E, as we have mentioned in the foreword, includes some Auxiliary Propositions (AP) (Lemmas) of Euclidean Geometry, which we believe firstly appear here, together with their proofs, which are needed for this paper, but which do not include Harmonic Elements, as to be included in the core of this book.

In addition, this Part includes known Propositions, whose converses are not included in bibliography, or there are not detailed proofs of them, mainly in their special cases, which are necessary for this paper though.

Moreover, here there are also given new proofs appropriate for this paper, regarding some known Propositions, as their proofs found in bibliography, do not help understand some parts of this book.

2. Geometrical Auxiliary Propositions (AP).

We present these auxiliary Propositions just below, in the same sequence as they are used in our paper, as here new definitions and theories are not needed, because we use what is known from Euclidean Geometry.

In addition, Theorems and Propositions, that already exist in known bibliography with proofs sufficient for the needs of our paper, are not repeated here.

2.1. Isogonal Property in Triangles.

This is a New Auxiliary Proposition. For this Proposition we try to perform an analytical proof and cover all the possible cases, as the needs of this book demand so.

Proposition 1. (Figures 1-3).

Let there be a pair of isogonal lines with respect to angle A of triangle ΑΒΓ, that intersect triangle's side ΒΓ, at points Δ, Δ' and let the internal bisector of angle A of the same triangle, intersect ΒΓ at E, while the external bisector intersects ΒΓ at E', then we will have:

$$\frac{(ABE)}{(A\Gamma E)}\cdot\frac{(A\Gamma\Delta')}{(A\Delta' E)}\cdot\frac{(A\Delta E)}{(AB\Delta)}=1, \qquad (1).$$

$$\frac{(ABE')}{(A\Gamma E')}\cdot\frac{(A\Gamma\Delta')}{(A\Delta' E')}\cdot\frac{(A\Delta E')}{(AB\Delta)}=1, \qquad (2).$$

and conversely: Let AE be the internal bisector of angle ΒΑΓ of triangle ΑΒΓ and AE' the external bisector of the same angle and if one of the above relationships is true, then AE is the bisector of angle ΔΑΔ' of triangle ΑΔΔ' too, or if AE is the internal bisector of angle ΔΑΔ' of triangle ΑΔΔ' and AE' its external one and one of the relationships above is true, then AE is the bisector of angle ΒΑΓ of triangle ΑΒΓ too.

Proof.

A′. If we work in triangle ΑΒΓ, with ΑΔ, ΑΔ′ isogonal:

(a). Implication.

1/. Proof of relation (1).

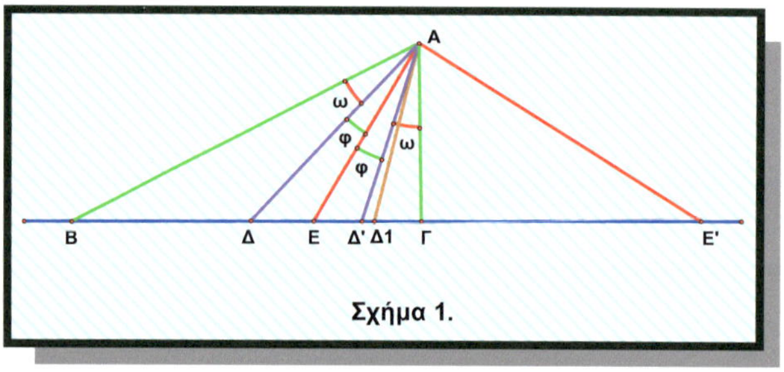

Σχήμα 1.

1st Case (Figure 1). Let points Δ, Δ' be on side ΒΓ.

Since triangles ΑΒΕ and ΑΓΕ have equal angles ΒΑΕ, ΓΑΕ, we will have:

$$\frac{(ΑΒΕ)}{(ΑΓΕ)}=\frac{ΑΒ\bullet ΑΕ}{ΑΕ\bullet ΑΓ}. \qquad (3).$$

Also since, ∠ΓΑΔ'=∠ΒΑΔ=ω, ∠ΔΑΕ=∠Δ'ΑΕ=φ, similarly we find that:

$$\frac{(ΑΓΔ')}{(ΑΒΔ)}=\frac{ΑΓ\bullet ΑΔ'}{ΑΒ\bullet ΑΔ},\quad \frac{(ΑΔΕ)}{(ΑΔ'Ε)}=\frac{ΑΔ\bullet ΑΕ}{ΑΕ\bullet ΑΔ'}. \qquad (4).$$

Thus, from relations (3), (4)⇒(1).

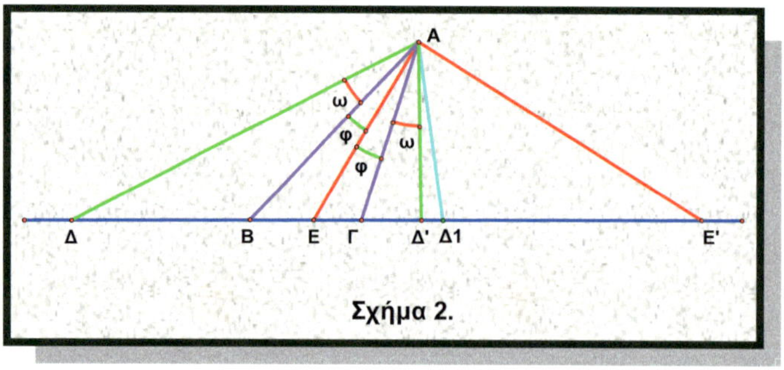

Σχήμα 2.

2nd Case (Figure 2). **If points Δ, Δ' are not on side ΒΓ and lie on both its sides.**

Since in triangle ΑΔΔ' ΑΒ, ΑΓ are isogonal, if we apply relation (1) from above, then: $\frac{(A\Delta E)}{(A\Delta' E)}\cdot\frac{(A\Gamma\Delta')}{(A\Gamma E)}\cdot\frac{(ABE)}{(AB\Delta)}=1 \Rightarrow$ (1).

This means that in this case as well, relation (1) is true.

3rd Case (Figure 3). **If points Δ, Δ' are not on side ΒΓ and towards the same side of midpoint Μ of segment ΕΕ'.**

This case obviously occurs, if φ=∠δΑΕ=∠ΕΑΔ'>∠ΑΕΕ' (where δ a point on the extension of ΑΔ and towards Α). Then we easily find that ∠ΔΑΕ' =∠Ε'ΑΔ'=φ', φ+φ'=∠ΕΑΕ'=1L.

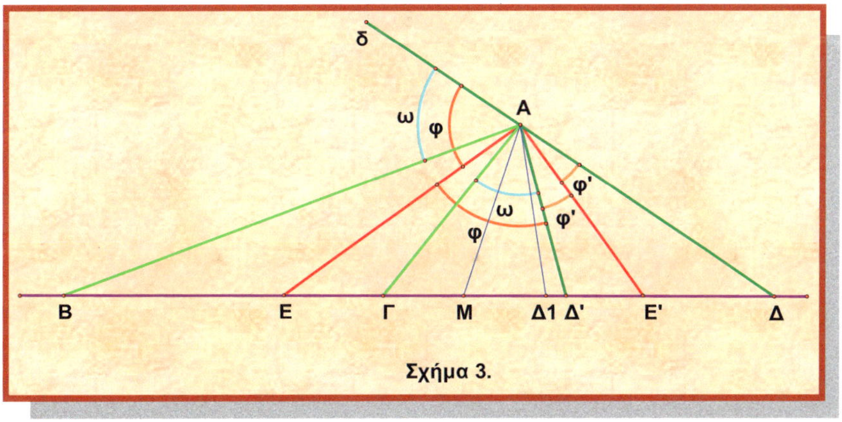

Σχήμα 3.

Because triangles ΑΒΕ, ΑΕΓ have equal angles ∠ΒΑΕ, ∠ΕΑΓ it will hold true that:

$$\frac{(ABE)}{(A\Gamma E)}=\frac{AB\cdot AE}{AE\cdot A\Gamma}. \qquad (5).$$

Moreover, if ∠Δ'ΑΓ=ω, because ∠Δ'ΑΓ+∠ΒΑΔ =ω+∠ΒΑΔ=2L, ∠ΔΑΕ

$+\angle E A\Delta' = \angle \Delta AE + \varphi = 2L$, regarding the two pairs of triangles $A\Gamma\Delta'$, $AB\Delta$ and $A\Delta E$, $A\Delta' E$, we similarly find that:

$$\frac{(A\Gamma\Delta')}{(AB\Delta)} = \frac{A\Gamma \bullet A\Delta'}{AB \bullet A\Delta}, \quad \frac{(A\Delta E)}{(A\Delta' E)} = \frac{A\Delta \bullet AE}{AE \bullet A\Delta'}, \qquad (6).$$

respectively.

Thus, from relations (5), (6)⇒(1).

Thus in this case also relation (1) is true.

Therefore in all three cases above (Figures 1-3), relation (1) is true.

2/. Proof of relation (2).

1st Case (Figure 1). If points Δ, Δ' lie on side ΒΓ.

Because for triangles ABE' and AΓE' it is true that:

$\angle E'A\Gamma + \angle E'AB = \angle E'AE - \omega - \varphi + \angle E'AE + \omega + \varphi = 2\angle E'AE = 2L$, it will be:

$$\frac{(ABE')}{(A\Gamma E')} = \frac{AB \bullet AE'}{AE' \bullet A\Gamma}. \qquad (7).$$

Additionally because, $\angle \Gamma A\Delta' = \angle BA\Delta = \omega$, $\angle \Delta AE' + \angle \Delta' AE' = 2L$, we similarly find that:

$$\frac{(A\Gamma\Delta')}{(AB\Delta)} = \frac{A\Gamma \bullet A\Delta'}{AB \bullet A\Delta}, \quad \frac{(A\Delta E')}{(A\Delta' E')} = \frac{A\Delta \bullet AE'}{AE' \bullet A\Delta'}. \qquad (8).$$

Thus, from relations (7), (8)⇒(2).

2nd Case (Figure 2). If points Δ, Δ' are outside ΒΓ and lie on both its sides.

Because in triangle AΔΔ' AB, AΓ are isogonal and if we apply relationship (2) from above, then for triangle AΔΔ', it will be:

$$\frac{(A\Delta E')}{(A\Delta' E')}\cdot\frac{(A\Gamma\Delta')}{(A\Gamma E')}\cdot\frac{(ABE')}{(AB\Delta)}=1 \Rightarrow (2). \qquad (9).$$

Thus, in this case as well, relation (2) is true.

3rd Case (Figure 3). <u>If points Δ, Δ' are outside BΓ and towards the same side of M midpoint of segment EE'.</u>

This case will obviously occur, if $\varphi=\angle\delta AE=\angle EA\Delta'>\angle AEE'$ (where δ is a point on the extension of AΔ and towards A). Then we easily find that $\angle\Delta AE'=\angle E'A\Delta'=\varphi'$, $\varphi+\varphi'=\angle EAE'=1L$.

Because for triangles ABE', AΓE': $\angle BAE'+\angle E'A\Gamma=\angle BAE'+\omega+\varphi'=2L$,

It will be: $$\frac{(ABE')}{(A\Gamma E')}=\frac{AB\bullet AE'}{AE'\bullet A\Gamma}. \qquad (10).$$

Additionally, because $\angle\Delta' A\Gamma+\angle BA\Delta=\omega+\angle BA\Delta=2L$, $\angle\Delta AE'=\angle E'A\Delta'=\varphi'$, regarding the two pairs of triangles AΓΔ', ABΔ and AΔE', AΔ'E', we similarly find that: $$\frac{(A\Gamma\Delta')}{(AB\Delta)}=\frac{A\Gamma\bullet A\Delta'}{AB\bullet A\Delta},\quad \frac{(A\Delta E')}{(A\Delta' E')}=\frac{A\Delta\bullet AE'}{AE'\bullet A\Delta'}, \qquad (11).$$

respectively.

Thus, through relationships (10), (11)⇒(2).

Hence relation (2) is true in this case too.

<u>Therefore in all three cases above (Figures 1-3), relation (2) is true.</u>

B′. <u>Let us work in triangle AΔΔ′, with AB, AΓ isogonal:</u>

If we use triangle AΔΔ' with AB and AΓ isogonal and work just as we did above, we prove that relations (1) and (2) are still true. (It is possible to have the same result, if we depend on what is proven above).

(b). Converse.

A′. If we work in triangle ΑΒΓ, with ΑΔ, ΑΔ′ isogonal:

1st Case (Figures 1-3). **If relation (1) is true.**

In all three cases of the implication above, where relation (1) is true, we work in the following way: We consider that in triangle ΑΒΓ - ΑΔ, ΑΔ' are not isogonal, then we draw $ΑΔ_1$ isogonal to ΑΔ, hence:

$$\frac{(ΑΒΕ)}{(ΑΓΕ)}\cdot\frac{(ΑΓΔ_1)}{(ΑΔ_1Ε)}\cdot\frac{(ΑΔΕ)}{(ΑΒΔ)}=1. \qquad (12).$$

Thus, as relations (1) and (12) are true, it easily occurs through them that: $\frac{(ΑΓΔ_1)}{(ΑΔ_1Ε)}=\frac{(ΑΓΔ')}{(ΑΔ'Ε)}$. Hence, as all its four triangles, share the same altitude, it will also be $\frac{ΓΔ_1}{Δ_1Ε}=\frac{ΓΔ'}{Δ'Ε}\Rightarrow Δ_1\equiv Δ'$ and because $ΑΔ_1$ is isogonal to ΑΔ, ΑΔ' will be isogonal too.

2nd Case (Figures 1-3). **If relation (2) is true.**

In all three cases where relation (2) is true above, we work similarly to case 1 just above (of the converse) and easily prove that here as well ΑΔ' is in fact isogonal to ΑΔ, in triangle ΑΒΓ, using relation (2) and that its four triangles that will occur, are of the same altitude.

B′. If we work in triangle ΑΔΔ′, with ΑΒ, ΑΓ isogonal:

1st Case (Figures 1-3). **If relation (1) is true.**

In all three cases of the direct proof above, where relation (1) is true, we work in the following way: We consider that in triangle ΑΔΔ' - ΑΒ, ΑΓ are not isogonal, then draw $ΑΓ_1$ isogonal to ΑΒ, hence:

$$\frac{(ΑΒΕ)}{(ΑΓ_1Ε)} \cdot \frac{(ΑΓ_1Δ')}{(ΑΔ'Ε)} \cdot \frac{(ΑΔΕ)}{(ΑΒΔ)} = 1. \qquad (13).$$

Thus, as (1) and (13) are true, it easily occurs, through them that:

$\frac{(ΑΓΔ')}{(ΑΓΕ)} = \frac{(ΑΓ_1Δ')}{(ΑΓ_1Ε)}$. Therefore, as all its four triangles, have the same altitude it will also hold true that $\frac{Δ'Γ_1}{Γ_1Ε} = \frac{Δ'Γ}{ΓΕ} \Rightarrow Γ_1 \equiv Γ$ and because ΑΒ in triangle ΑΔΔ' is isogonal to $ΑΓ_1$, ΑΓ will be isogonal too.

2nd Case (Figures 1-3). If relation (2) is true.

In all three cases of the implication, where relation (2) is true, we work similarly to case 1 right above (of the converse) and easily prove that ΑΓ is in fact isogonal to ΑΒ, in triangle ΑΔΔ', using relation (2) and the fact that its four triangles, that will occur, are of the same altitude.

2.2. Isogonal Property of a Triangle.

This is a new Auxiliary Proposition. And regarding this Proposition we tried to perform its proof analytically and to include all possible cases, as the needs of this book demand.

Proposition 2. (Figures 1-3).

If two lines isogonal with respect to angle A of triangle ΑΒΓ, intersect triangle's side ΒΓ at points Δ, Δ' and if ΑΕ is the internal, and ΑΕ' the external bisector of angle A of the same triangle, then:

$$\frac{ΑΒ}{ΑΓ} = \frac{ΒΕ}{ΕΓ} = \frac{ΒΔ}{ΔΕ} \cdot \frac{ΕΔ'}{Δ'Γ} \text{ or } \frac{ΒΔ}{ΔΕ} \cdot \frac{ΕΔ'}{Δ'Γ} \cdot \frac{ΓΕ}{ΕΒ} = 1, \qquad (1).$$

$$\frac{A\Delta}{A\Delta'}=\frac{\Delta E}{E\Delta'}=\frac{\Delta B}{BE}\cdot\frac{E\Gamma}{\Gamma\Delta'} \text{ or } \frac{\Delta B}{BE}\cdot\frac{E\Gamma}{\Gamma\Delta'}\cdot\frac{\Delta'E}{E\Delta}=1, \qquad (1').$$

$$\frac{B\Delta}{\Delta E}\cdot\frac{E\Delta'}{\Delta'\Gamma}\cdot\frac{\Gamma E'}{E'B}=1, \quad \frac{\Delta B}{BE'}\cdot\frac{E'\Gamma}{\Gamma\Delta'}\cdot\frac{\Delta'E}{E\Delta}=1, \quad \frac{B\Delta}{\Delta E'}\cdot\frac{E'\Delta'}{\Delta'\Gamma}\cdot\frac{\Gamma E'}{E'B}=1, \qquad (1'').$$

and conversely:

Let AE be the internal bisector of angle BAΓ, in triangle ABΓ and AE' its external bisector and if one of the last equalities in the three relations above is true, then AE is also the bisector of angle ΔAΔ', or if AE is the internal bisector of angle ΔAΔ', in triangle AΔΔ' and AE' the external bisector of the same angle and if one of the last equalities in the three relations above is true, then AE is the bisector of angle BAΓ too.

Proofs.

1st Way. (Based on the proof of Auxiliary Proposition 1).

(a). Direct Proof.

Proof of relation (1).

1st Case (Figure 1). If points Δ, Δ' lie on side BΓ.

According to Auxiliary Proposition (AP) 1 (Relation 1), we will have:

$$\frac{(ABE)}{(A\Gamma E)}\cdot\frac{(A\Gamma\Delta')}{(A\Delta'E)}\cdot\frac{(A\Delta E)}{(AB\Delta)}=1. \qquad (2).$$

Because the six triangles in (2) have the same altitude then:

$$\frac{BE}{E\Gamma}\cdot\frac{\Gamma\Delta'}{\Delta'E}\cdot\frac{E\Delta}{B\Delta}=1 \text{ ή } \frac{B\Delta}{\Delta E}\cdot\frac{E\Delta'}{\Delta'\Gamma}\cdot\frac{\Gamma E}{EB}=1. \qquad (3).$$

This means that the second equality in (1) is true.

But, because AE is the bisector of BAΓ, it will be $\frac{AB}{A\Gamma}=\frac{BE}{E\Gamma}$, therefore if we take into consideration (3), we receive the triple equality in relation (1).

2nd Case (Figure 2). <u>If points Δ, Δ' are outside BΓ and lie on both its sides.</u>

In this case just like we saw in case 2 of Auxiliary Proposition 1, it is:

$$\frac{(A\Delta E)}{(AE\Delta')}\cdot\frac{(A\Gamma\Delta')}{(AB\Delta)}\cdot\frac{(ABE)}{(AE\Gamma)}=1. \qquad (4).$$

Because, the six triangles of (4), have the same altitude, it will be:

$$\frac{\Delta E}{E\Delta'}\cdot\frac{\Delta'\Gamma}{B\Delta}\cdot\frac{BE}{E\Gamma}=1 \text{ or } \frac{B\Delta}{\Delta E}\cdot\frac{E\Delta'}{\Delta'\Gamma}\cdot\frac{\Gamma E}{EB}=1. \qquad (5).$$

This means that the second equality of relation (1) is true here as well.

But, because AE is the bisector of triangle BAΓ, then $\frac{AB}{A\Gamma}=\frac{BE}{E\Gamma}$, thus if we take into consideration (5), we receive the triple equality of relation (1).

3rd Case (Figure 3). <u>If points Δ, Δ' are outside BΓ and towards the same side of M midpoint of segment EE'.</u>

This case will obviously occur, if $\varphi=\angle\delta AE=\angle EA\Delta'>\angle AEE'$ (where δ a point on the extension of AΔ and towards A). Then we easily find that also $\angle\Delta AE'=\angle E'A\Delta'=\varphi'$, $\varphi+\varphi'=\angle EAE'=1L$.

According to Auxiliary Proposition (AP) 1 (Case 3), relation (1) will be true here as well, this means that:

$$\frac{(ABE)}{(A\Gamma E)}\cdot\frac{(A\Gamma\Delta')}{(A\Delta'E)}\cdot\frac{(A\Delta E)}{(AB\Delta)}=1. \qquad (6).$$

Hence, because the six triangles in (6) have the same altitude it will be:

$$\frac{BE}{E\Gamma}\cdot\frac{\Gamma\Delta'}{\Delta'E}\cdot\frac{E\Delta}{B\Delta}=1 \text{ or } \frac{B\Delta}{\Delta E}\cdot\frac{E\Delta'}{\Delta'\Gamma}\cdot\frac{\Gamma E}{EB}=1. \qquad (7).$$

This means that the second equality of (1) holds true here as well.

But, because AE is the bisector of ΒΑΓ, it will be $\frac{AB}{A\Gamma}=\frac{BE}{E\Gamma}$, therefore if we take into consideration (7), we receive the triple equality of relation (1).

<u>Therefore, in all three cases above (Figures 1-3), the whole relation (1) is true.</u>

Proof of relation (1').

We observe that easily, from the second part of relation (1), we can derive the second part of (1'). This means that relation (1') (second part) is equivalent to the second part of (1). This means that in all three cases above where the second part of (1) is true, the second part of (1') will also be true.

After all, because for all three cases above (figures 1-3), AE is the bisector of angle ΔΑΔ' and AE' is the external bisector of the same angle, it will be $\frac{\Delta E}{E\Delta'}=\frac{\Delta A}{A\Delta'}$, we easily derive the first part of (1'), from the second part of (1').

<u>Therefore, in all three cases above (Figures 1-3), the whole relation (1') is true.</u>

Proof of relation (1").

To prove relation (1") for all three cases in figures 1 to 3, we work in the following way: According to Auxiliary Proposition 1 in all three cases mentioned there, it will be:

$$\frac{(\mathrm{A\Delta E'})}{(\mathrm{A\Delta' E'})}\cdot\frac{(\mathrm{A\Gamma\Delta'})}{(\mathrm{AB\Delta})}\cdot\frac{(\mathrm{ABE'})}{(\mathrm{A\Gamma E'})}=1.$$

Hence, from this, because its triangles, are of the same altitude, it will be:

$$\frac{\mathrm{\Delta E'}}{\mathrm{\Delta' E'}}\cdot\frac{\mathrm{\Gamma\Delta'}}{\mathrm{B\Delta}}\cdot\frac{\mathrm{BE'}}{\mathrm{\Gamma E'}}=1, \text{ or } \frac{\mathrm{B\Delta}}{\mathrm{\Delta E'}}\cdot\frac{\mathrm{E'\Delta'}}{\mathrm{\Delta'\Gamma}}\cdot\frac{\mathrm{\Gamma E'}}{\mathrm{E'B}}=1. \qquad (8).$$

Hence the third equality of (1'') is true.

But because AE is the internal bisector of angle BAΓ and AE' is the external bisector of the same angle, it will be: $\frac{\mathrm{AB}}{\mathrm{A\Gamma}}=\frac{\mathrm{BE}}{\mathrm{E\Gamma}}=\frac{\mathrm{BE'}}{\mathrm{E'\Gamma}}$.

Hence, from this and the triple equality in (1), we receive:

$$\frac{\mathrm{AB}}{\mathrm{A\Gamma}}=\frac{\mathrm{BE}}{\mathrm{E\Gamma}}=\frac{\mathrm{BE'}}{\mathrm{E'\Gamma}}=\frac{\mathrm{B\Delta}}{\mathrm{\Delta E}}\cdot\frac{\mathrm{E\Delta'}}{\mathrm{\Delta'\Gamma}}.$$

Thus, from this equality, we easily find that the first and the second equality of (1'') are true.

Thus all three equalities in (1'') are true (Figures 1-3).

(b). Converse.

1st Case (Figures 1-3). If the last equality in relation (1) is true or of its equivalent (1').

In this case we work in the following way:

From the second equality in (1) or in (1') and because the six triangles ABE, AΓE, AΓΔ', ABΔ, AΔE, AΔ'E have the same altitude it appears that:

$\frac{(AB E)}{(A\Gamma E)}\cdot\frac{(A\Gamma\Delta')}{(A\Delta' E)}\cdot\frac{(A\Delta E)}{(AB\Delta)}=1$. Thus, as this one true, according to the converse of Auxiliary Proposition 1, it occurs that AΔ' is isogonal to AΔ in triangle ABΓ as well.

2nd Case (Figures 1-3). **If the last equality in relation (1") is true.**

In this case we work similarly to case 1 (converse) above, taking into consideration that the third relation of (1") is true and that the six triangles ABE', AΓE', AΓΔ', ABΔ, AΔE', AΔ'E' have the same alttiude, relation (2) of Proposition 1 will be true. Thus, according to case 2 of the converse of Proposition 1, it appears that in triangle AΔΔ', AΓ is isogonal to AB.

2nd Way.

Proof of Relation (1) (Figure 1).

(a). Direct Proof.

It is obvious that pairs of lines AB-AΓ, AΔ- AΔ' are lines symmetrical with respect to AE and hence the flat pencils of four beams of opposite directions, A(BΔEΓ) and A(ΓΔ'EB), will be equal, as all of their homologous angles are equal one by one, (∠BAΔ=∠ΓAΔ'=ω, ∠ΔAE=∠Δ'AE=φ, ∠BAE=∠ΓAE, ∠BAΓ=∠ΓAB).

This means that: A(BΔEΓ)=A(ΓΔ'EB). (2).

Therefore, these pencils, when line BΓ intersects them, according to the known Pappus's Theorem {Geometry [6] § 8-1,3}, will have equal unharmonic ratios (BΔEΓ)=(ΓΔ'EB). This means that from (2) ⇒(BΔEΓ)=(ΓΔ'EB)=ρ ⇒(BEΔΓ)=(ΓEΔ'B)=1-ρ ⇒

$\frac{B\Delta}{\Delta E}:\frac{B\Gamma}{\Gamma E}=\frac{\Gamma\Delta'}{\Delta' E}:\frac{\Gamma B}{BE}\Rightarrow\frac{B\Delta}{\Delta E}\cdot\frac{E\Gamma}{\Gamma B}=\frac{\Gamma\Delta'}{\Delta' E}\cdot\frac{EB}{B\Gamma}\Rightarrow(1)$, as well as $\frac{BE}{E\Gamma}=\frac{AB}{A\Gamma}$.

Additionally, because from (1) we can easily derive (1'), (1') will also be true [Though it is possible to work analytically, as above and prove that (1') is also true].

(b). Conversely.

<u>If AE is the bisector of angle BAΓ in triangle ABΓ.</u>

Let us suppose that AΔ' is not isogonal to AΔ, but isogonal to $A\Delta_1$, then according to the implication above we will have:

$$\frac{BE}{E\Gamma}=\frac{B\Delta}{\Delta E}\cdot\frac{E\Delta_1}{\Delta_1\Gamma}. \qquad (3).$$

But, because, from the hypothesis, the first from (1) is true, then from (1) and (3) $\Rightarrow \frac{E\Delta'}{\Delta'\Gamma}=\frac{E\Delta_1}{\Delta_1\Gamma}\Rightarrow \Delta'\equiv\Delta_1\Rightarrow A\Delta'\equiv A\Delta_1$ and because $A\Delta_1$ was taken to be isogonal to AΔ, AΔ' will also be isogonal to AΔ.

<u>If AE is the bisector of angle ΔAΔ' in triangle AΔΔ'.</u>

We suppose that AE is not the bisector of angle BAΓ as well, but the bisector of angle BAΓ is AE_1, for which we suppose that $BE_1<BE$, hence as well

$$\Delta E_1<\Delta E,\ \Gamma E_1>\Gamma E,\ \Delta' E_1>\Delta' E. \qquad (4).$$

Therefore, regarding E_1, according to the implication above, it will be:

$$\frac{BE_1}{E_1\Gamma}\cdot\frac{\Gamma\Delta'}{\Delta' E_1}\cdot\frac{E_1\Delta}{\Delta B}=1. \qquad (5).$$

After all, according to the hypothesis, let relation (1) be true.

Hence, from (1) $\Rightarrow \frac{BE}{E\Gamma}.\frac{\Delta E}{E\Delta'} = \frac{B\Delta}{\Delta'\Gamma}$, (6).

while from (5) $\Rightarrow \frac{BE_1}{E_1\Gamma}.\frac{\Delta E_1}{E_1\Delta'} = \frac{B\Delta}{\Delta'\Gamma}$. (7).

Thus, from (6), (7) $\Rightarrow \frac{BE}{E\Gamma}.\frac{\Delta E}{E\Delta'} = \frac{BE_1}{E_1\Gamma}.\frac{\Delta E_1}{E_1\Delta'}$. (8).

Therefore, if we take into consideration (4) too, it will be: $\frac{BE}{E\Gamma} > \frac{BE_1}{E_1\Gamma}$ and $\frac{\Delta E}{E\Delta'} > \frac{\Delta E_1}{E_1\Delta'}$, thus we will also have $\frac{BE}{E\Gamma}.\frac{\Delta E}{E\Delta'} > \frac{BE_1}{E_1\Gamma}.\frac{\Delta E_1}{E_1\Delta'}$. (9).

We realize that we have reached a contradiction, because then (8) is not true, but (9) is. We have reached a contradiction, because we set as bisector line AE_1 and not AE.

We will also reach a contradiction if we suppose that for point E_1 it holds that $BE_1 > BE$, thus $\Delta E_1 > \Delta E$, etc.

Therefore, AE will in fact be the bisector of angle BAΓ.

3rd Way (Computational) (Figure 1-3).

(a). Direct Proof.

Because AΔ, AΔ', in triangle ABΓ, are isogonal, according to known Theorem [Geometry Book [6] § 11-12.1], it will be:

$$\frac{AB^2}{A\Gamma^2} = \frac{BE^2}{E\Gamma^2} = \frac{B\Delta}{\Delta\Gamma}.\frac{B\Delta'}{\Delta'\Gamma}. \quad (1).$$

After all because, in triangle AΔΔ', AB, AΓ are isogonal too, for the same reason, it will be: $\frac{A\Delta^2}{\Delta'A^2} = \frac{\Delta E^2}{\Delta'E^2} = \frac{\Delta B}{B\Delta'}.\frac{\Delta\Gamma}{\Gamma\Delta'}$. (2).

Hence, from relations (1), (2), we receive:

$$\frac{BE^2}{E\Gamma^2}\cdot\frac{\Delta E^2}{\Delta' E^2}=\frac{\Delta B}{\Delta\Gamma}\cdot\frac{B\Delta'}{\Gamma\Delta'}\cdot\frac{\Delta B}{B\Delta'}\cdot\frac{\Delta\Gamma}{\Gamma\Delta'}=\frac{B\Delta^2}{\Delta'\Gamma^2} \text{ or } \frac{B\Delta}{\Delta'\Gamma}=\frac{BE}{E\Gamma}\cdot\frac{\Delta E}{E\Delta'} \text{ or}$$

$$\frac{BE}{E\Gamma}=\frac{B\Delta}{\Delta'\Gamma}\cdot\frac{\Delta' E}{E\Delta} \text{ or } \frac{BE}{E\Gamma}=\frac{B\Delta}{\Delta E}\cdot\frac{E\Delta'}{\Delta'\Gamma}=\frac{AB}{A\Gamma}.$$ This means that relation (1) is proven true.

Proof of the Direct of Relation (1') of the 2nd and 3rd Way of Proof (Figure 1-3).

To prove relation (1'), we work similarly to the proof of relation (1) above, or in a more simple way, as shifting appropriately the segments in relation (1'), we receive relation (1), we conclude that relations (1) and (1') are equivalent and because (1) is true then (1') will also be true.

Auxiliary Proposition 2 is proven in analogous ways, and is true when for example AΓ, AB are symmetrical with respect to AE', which is perpendicular to AE at point A.

Proof of the Direct of Relation (1'') of the 2nd and 3rd Way of Proof of (1'') (Figure 1-3).

It is obvious that relations (1) and (1') above are true and for the Conjugate Harmonic E' of E, with respect to B, Γ and Δ, Δ', as it is:

$$\frac{BE}{E\Gamma}=\frac{BE'}{E'\Gamma} \text{ and } \frac{\Delta' E}{E\Delta}=\frac{\Delta' E'}{E'\Delta}. \qquad \text{(a).}$$

Thus, from relation (1) and (a) we can easily find that:

$$\frac{AB}{A\Gamma}=\frac{BE}{E\Gamma}=\frac{BE'}{E'\Gamma}=\frac{B\Delta}{\Delta E}\cdot\frac{E\Delta'}{\Delta'\Gamma}. \qquad \text{(b).}$$

From relation (b), we can easily find that:

$$\frac{B\Delta}{\Delta E}\cdot\frac{E\Delta'}{\Delta'\Gamma}\cdot\frac{\Gamma E'}{E'B}=1, \qquad \frac{\Delta E}{E\Delta'}\cdot\frac{\Delta'\Gamma}{\Gamma E'}\cdot\frac{E'B}{B\Delta}=1. \qquad (c).$$

This means that the first and the second equality in relation (1'') are true.

After all, from the second equality in (c) and (a), we can easily derive the following:

$$\frac{B\Delta}{\Delta E'}\cdot\frac{E'\Delta'}{\Delta'\Gamma}\cdot\frac{\Gamma E'}{E'B}=1.$$

Hence the third equality in (1'') is also true.

(b). Converses of the 3rd Way of Proof.

The proof of the converse here as well, can be performed, similarly to the converses of the 2nd Way above, or the other ways above.

Comment.

According to relevant definitions, in paragraph 3.3 (of Part A), pencil A(BΔEΔ'ΓE') is a «flat symmetrical pencil of six beams, with respect to an axis», as pencils A(BΔEΓ) and A(ΓΔ'EB) are symmetrical with respect to their beam AE and ΔE′, while the following four collinear points B, Δ, E, Δ', constitute a «Simple Harmonic Involution of Two Pairs of Conjugate Points with Double Points E, E' (paragraph 3.4, of the main part of this book). In paragraph 3.9 the properties of the above pencil and the involution are mentioned, in the form of Propositions (Criteria, with their proofs included).

Conclusion.

From the 1st Part of relation (1) we have: $\frac{BA}{A\Gamma}=\frac{B\Delta}{\Delta E}\cdot\frac{E\Delta'}{\Delta'\Gamma}$.

This means that the above Proposition 2 is true in triangle ABΓ. Additionally, from the 1st Part of relation (1) we have: $\frac{BE}{E\Gamma}=\frac{B\Delta}{\Delta E}\cdot\frac{E\Delta'}{\Delta'\Gamma}$.

This means that Proposition 2 is true in a triangle degenerate into a line. Thus it is also true in a special collinear series of points which we have defined as «Simple Collinear Harmonic Involution» (See Criterion 4, in Part A).

2.3. Isogonal Property of a Triangle.

(Auxiliary Proposition).

This Proposition exists in Bibliography [6] (§11-12.1), but only its implication. Here, besides its direct proof, we refer to its converse as well. Moreover, we mention some of its special cases, which are essential to this book.

Proposition 3. (Figures 1-3).

If two isogonal lines with respect to angle A in triangle ABΓ, intersect its side BΓ at points Δ, Δ' and if AE is the internal, AE' the external bisector of angle A in this triangle or in triangle AΔΔ', then we will have:

$$\frac{BA^2}{A\Gamma^2}=\frac{BE^2}{E\Gamma^2}=\frac{BE'^2}{E'\Gamma^2}=\frac{B\Delta}{\Delta\Gamma}\cdot\frac{B\Delta'}{\Delta'\Gamma}. \quad (1).$$

$$\frac{\Delta A^2}{A\Delta'^2}=\frac{\Delta E^2}{E\Delta'^2}=\frac{\Delta E'^2}{E'\Delta'^2}=\frac{\Delta B}{B\Delta'}\cdot\frac{\Delta\Gamma}{\Gamma\Delta'}. \quad (2).$$

And conversely:

If AE is the internal bisector of angle BAΓ, in triangle ABΓ and AE' the external bisector of the same angle and if relation (1) is true, then AE is the bisector of angle ΔAΔ' too, or if AE is the internal bisector of angle ΔAΔ', in triangle AΔΔ' and AE' the external bisector of the same angle and if relation (2) is true, then AE is the bisector of angle BAΓ too.

Proof.

(1). Direct Proof.

(a). Based on Triangle ABΓ.

Proof of relation (1).

1st Case (Figure 1). If points Δ, Δ' lie on side BΓ.

Because in triangles ABΔ and AΓΔ' angles BAΔ, ΓAΔ' are equal, it will be:

$$\frac{(AB\Delta)}{(A\Gamma\Delta')}=\frac{AB\bullet A\Delta}{A\Gamma\bullet A\Delta'}, \qquad (3).$$

Moreover because, $\angle BA\Delta'=\angle\Gamma A\Delta$, it will also hold that:

$$\frac{(AB\Delta')}{(A\Gamma\Delta)}=\frac{AB\bullet A\Delta'}{A\Gamma\bullet A\Delta}. \qquad (4).$$

Thus, through relations

$$(3),\ (4)\Rightarrow \frac{(AB\Delta)}{(A\Gamma\Delta')}\cdot\frac{(AB\Delta')}{(A\Gamma\Delta)}=\frac{AB\bullet A\Delta}{A\Gamma\bullet A\Delta'}\cdot\frac{AB\bullet A\Delta'}{A\Gamma\bullet A\Delta}=\frac{BA^2}{A\Gamma^2}. \qquad (5).$$

After all, because the four triangles in (5) are of the same altitude, it will be:

$$\frac{(AB\Delta)}{(A\Gamma\Delta')}\cdot\frac{(AB\Delta')}{(A\Gamma\Delta)}=\frac{B\Delta}{\Delta'\Gamma}\cdot\frac{B\Delta'}{\Delta\Gamma}=\frac{B\Delta}{\Delta\Gamma}\cdot\frac{B\Delta'}{\Delta'\Gamma}. \qquad (6).$$

Therefore, from (5), (6) and because AE is the internal and AE' the external bisector of angle BAΓ, it is: $\frac{BA}{A\Gamma}=\frac{BE}{E\Gamma}=\frac{BE'}{E'\Gamma}$, wanted relation (1) occurs easily.

2nd Case (Figure 2). <u>If points Δ, Δ' are outside BΓ and lie on both its sides..</u>

Again, because in triangles ABΔ and AΓΔ' angles BAΔ, ΓAΔ' are equal, it will be:

$$\frac{(AB\Delta)}{(A\Gamma\Delta')}=\frac{AB\cdot A\Delta}{A\Gamma\cdot A\Delta'}. \qquad (7).$$

Additionally because, ∠BAΔ'=∠ΓAΔ, it will also be:

$$\frac{(AB\Delta')}{(A\Gamma\Delta)}=\frac{AB\cdot A\Delta'}{A\Gamma\cdot A\Delta}. \qquad (8).$$

Hence, from relations

$$(7),(8)\Rightarrow \quad \frac{(AB\Delta)}{(A\Gamma\Delta')}\cdot\frac{(AB\Delta')}{(A\Gamma\Delta)}=\frac{AB\cdot A\Delta}{A\Gamma\cdot A\Delta'}\cdot\frac{AB\cdot A\Delta'}{A\Gamma\cdot A\Delta}=\frac{BA^2}{A\Gamma^2}. \qquad (9).$$

After all, because the four triangles in (9) are of the same altitude, it will be:

$$\frac{(AB\Delta)}{(A\Gamma\Delta')}\cdot\frac{(AB\Delta')}{(A\Gamma\Delta)}=\frac{B\Delta}{\Delta'\Gamma}\cdot\frac{B\Delta'}{\Delta\Gamma}=\frac{B\Delta}{\Delta\Gamma}\cdot\frac{B\Delta'}{\Delta'\Gamma}. \qquad (10).$$

Therefore, from (9), (10) and because AE is the internal and AE' the external bisector of angle BAΓ, it is: $\frac{BA}{A\Gamma}=\frac{BE}{E\Gamma}=\frac{BE'}{E'\Gamma}$, wanted relation (1) easily occurs.

3rd Case (Figure 3). <u>If points Δ, Δ' are outside BΓ and towards the same side of M midpoint of segment EE'.</u>

This case will obviously occur, if φ=∠δAE=∠EAΔ'>∠AEE' (where δ is a point on the extension of AΔ towards A). Then we can easily find that also ∠ΔAE' =∠E'AΔ'=φ', φ+φ'=∠EAE'=1L.

Additionally because in triangles ΑΒΔ and ΑΓΔ' we have: $\angle B A \Delta + \angle \Gamma A \Delta' = \angle B A \Delta + \omega = \angle B A \Delta + \angle B A \delta = 2L$,

it will be: $$\frac{(AB\Delta)}{(A\Gamma\Delta')} = \frac{AB \bullet A\Delta}{A\Gamma \bullet A\Delta'}. \quad (11).$$

Moreover because, $\angle \Gamma A \Delta + \angle B A \Delta' = (\omega + 2\varphi') + (\omega + \angle B A \Gamma) = 2\omega + 2\varphi' + 2\angle E A \Gamma = 2$ $[(\angle E A \Gamma + \omega) + \varphi'] = 2(\varphi + \varphi') = 2L$, it will also be: $\frac{(AB\Delta')}{(A\Gamma\Delta)} = \frac{AB \bullet A\Delta'}{A\Gamma \bullet A\Delta}$. (12).

Hence, through relations

$$(11), (12) \Rightarrow \frac{(AB\Delta)}{(A\Gamma\Delta')} \cdot \frac{(AB\Delta')}{(A\Gamma\Delta)} = \frac{AB \bullet A\Delta}{A\Gamma \bullet A\Delta'} \cdot \frac{AB \bullet A\Delta'}{A\Gamma \bullet A\Delta} = \frac{BA^2}{A\Gamma^2}. \quad (13).$$

After all, because the four triangles in (13) are of the same height, it 'd be:

$$\frac{(AB\Delta)}{(A\Gamma\Delta')} \cdot \frac{(AB\Delta')}{(A\Gamma\Delta)} = \frac{B\Delta}{\Delta'\Gamma} \cdot \frac{B\Delta'}{\Delta\Gamma} = \frac{B\Delta}{\Delta\Gamma} \cdot \frac{B\Delta'}{\Delta'\Gamma}. \quad (14)$$

Therefore, from (13), (14) and because AE is the internal and AE' the external bisector of angle ΒΑΓ, it is: $\frac{BA}{A\Gamma} = \frac{BE}{E\Gamma} = \frac{BE'}{E'\Gamma}$, wanted relation (1) occurs easily.

<u>As a result, in all three cases above (Figures 1-3), the whole relation (1) is true.</u>

Proof of relation (2).

1st Case (Figure 1). <u>If points Δ, Δ' lie on side ΒΓ.</u>

If in triangle ΑΔΔ', we apply the above 2nd case in figure 1, then: $\frac{\Delta A^2}{A\Delta'^2} = \frac{\Delta E^2}{E\Delta'^2} = \frac{\Delta E'^2}{E'\Delta'^2} = \frac{\Delta B}{B\Delta'} \cdot \frac{\Delta\Gamma}{\Gamma\Delta'}$.

Thus, for this case wanted relation (2) is true.

2nd Case (Figure 2). <u>If points Δ, Δ' are outside ΒΓ and lie on both its sides.</u>

If in triangle ΑΔΔ', we apply the above 1st case in figure 2, then: $\frac{\Delta A^2}{A\Delta'^2}=\frac{\Delta E^2}{E\Delta'^2}=\frac{\Delta E'^2}{E'\Delta'^2}=\frac{\Delta B}{B\Delta'}.\frac{\Delta\Gamma}{\Gamma\Delta'}$.

Thus, for this case wanted relation (2) is true.

3rd Case (Figure 3). <u>If points Δ, Δ' are outside ΒΓ and towards the same side of M midpoint of segment ΕΕ'.</u>

If in triangle ΑΔΔ', we apply the above 3rd case in figure 3, then:

$$\frac{\Delta A^2}{A\Delta'^2}=\frac{\Delta E^2}{E\Delta'^2}=\frac{\Delta E'^2}{E'\Delta'^2}=\frac{\Delta B}{B\Delta'}.\frac{\Delta\Gamma}{\Gamma\Delta'}.$$

Thus for this case as well relation (2) is true.

<u>Consequently, in all three cases above (Figures 1-3), the whole relation (2) is true.</u>

(b). <u>Based on triangle ΑΔΔ'.</u>

Here we work similarly to above too, but here ΑΒ, ΑΓ are isogonal.

(b). Converse.

1st Case (Figures 1-3). <u>If relation (1) is true.</u>

In the three cases above where relation (1) is true, we work in the following way: We consider ΑΔ, ΑΔ' to not be isogonal in triangle ΑΒΓ, then we draw $A\Delta_1$ isogonal to ΑΔ, hence:

$$\frac{BA^2}{A\Gamma^2} = \frac{BE^2}{E\Gamma^2} = \frac{BE'^2}{E'\Gamma^2} = \frac{B\Delta}{\Delta\Gamma} \cdot \frac{B\Delta_1}{\Delta_1\Gamma}. \qquad (15).$$

Thus, as (1) and (15) hold true, easily through them, it occurs that:. $\frac{B\Delta_1}{\Delta_1\Gamma} = \frac{B\Delta'}{\Delta'\Gamma} \Rightarrow \Delta_1 \equiv \Delta'$ and because from hypothesis $A\Delta_1$ is isogonal to AΔ in triangle ABΓ, AΔ' will also be isogonal.

2nd Case (Figures 1-3). **If relation (2) is true.**

In the three cases above where relation (2) is true, we work in the following way: We consider that AB, AΓ are not isogonal in triangle AΔΔ', then we draw $A\Gamma_1$ isogonal to AB, thus:

$$\frac{\Delta A^2}{A\Delta'^2} = \frac{\Delta E^2}{E\Delta'^2} = \frac{\Delta E'^2}{E'\Delta'^2} = \frac{\Delta B}{B\Delta'} \cdot \frac{\Delta\Gamma_1}{\Gamma_1\Delta'}. \qquad (16).$$

Hence, as (2) and (16) are true, easily through them, it occurs that:. $\frac{\Delta\Gamma_1}{\Gamma_1\Delta'} = \frac{\Delta\Gamma}{\Gamma\Delta'} \Rightarrow \Gamma_1 \equiv \Gamma$ and because from hypothesis $A\Gamma_1$ is isogonal to AB in triangle AΔΔ', AΓ will also be isogonal to AB.

Conclusion.

From relation (1) we have: $\frac{BA^2}{A\Gamma^2} = \frac{B\Delta}{\Delta\Gamma} \cdot \frac{B\Delta'}{\Delta'\Gamma}$.

This means that the above Proposition 3 is true in triangle ABΓ.

In addition, from (1) we have: $\frac{BE^2}{E\Gamma^2} = \frac{B\Delta}{\Delta\Gamma} \cdot \frac{B\Delta'}{\Delta'\Gamma}$.

This means that Proposition 3 is true in a triangle degenerate to a line, too. Thus is true in a special collinear series of points which we call «Simple Collinear Harmonic Envolution» (See Criterion 5 in Part A).

2.4. Important Auxiliary Proposition.

This Proposition was not found in bibliography, we introduce it here, as we will base the proof of other Propositions on it.

Proposition 4. (Figures 4).

From an arbitrary point Γ, on diameter AB of semicircle AΔB of a circle (O,ρ) or (κ), we draw a perpendicular (to this diameter), which intersects the semicircle at point Δ. On this semicircle and on both sides of Δ we set points E, Z such that ΓΔ is the bisector of angle EΓZ. Prove that, then and only then, we have: $\Gamma\Delta^2 = \Gamma E.\Gamma Z.$ **(1).**

Proof.

(a). Direct Proof.

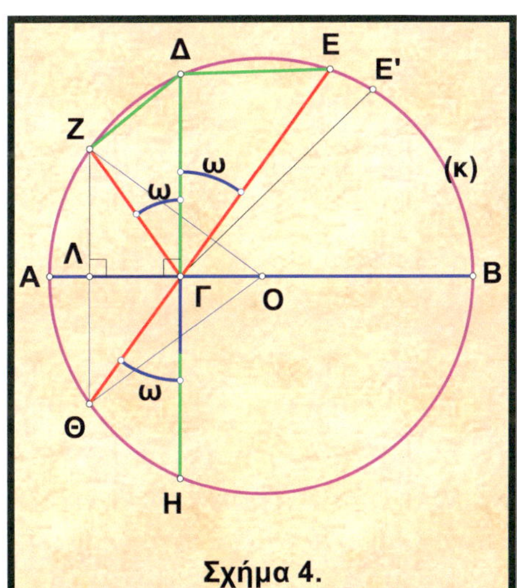

Σχήμα 4.

Hence, if ΓΔ⊥AB and ∠EΓΔ=∠ΔΓZ, we will prove that relation (1) is true.

If Θ is symmetrical to Z, with respect to AB, then Θ will lie on circle (κ) and if Λ is the intersection of ZΘ and AB, then right triangles ΛΓZ and ΛΓΘ, are easily proved to be equal, thus ∠ZΓΛ=∠ΛΓΘ, ΓZ=ΓΘ and hence ∠ΔΓZ=∠HΓΘ [where H≡ΓΔ∩(κ)].

Although we also have ∠ΔΓZ=∠ΔΓE. Therefore ∠ΔΓE=∠HΓΘ. This means that EΓΘ is straight. Thus :

$$E\Gamma.\Gamma\Theta=\Gamma\Delta.\Gamma H \text{ ή } \Gamma\Delta^2=\Gamma H^2=\Gamma E.\Gamma\Theta=\Gamma E.\Gamma Z \Rightarrow (1).$$

(b). Converse.

Hence, if (1) is true and $\Gamma\Delta \perp AB$, then we will prove that $\angle E\Gamma\Delta = \angle \Delta\Gamma Z$.

1st Way.

In fact, if ΔΓ, ΕΓ intersect circle (κ) at points Η, Θ respectively, we will have:

$$E\Gamma.\Gamma\Theta = \Gamma\Delta.\Gamma H = \Gamma\Delta^2 = \Gamma H^2. \qquad (2).$$

Thus, from (1), (2), we get: $E\Gamma.\Gamma Z = \Gamma E.\Gamma\Theta$ ή $\Gamma Z = \Gamma\Theta$. (3).

Consequently, triangles ΟΓΖ, ΟΓΘ are equal (have equal sides) and thus $\angle O\Gamma Z = \angle O\Gamma\Theta$ or $\angle \Delta\Gamma Z = \angle H\Gamma\Theta = \angle \Delta\Gamma E$ or $\angle E\Gamma\Delta = \angle \Delta\Gamma Z$. As a result ΓΖ is in fact the bisector of angle ΕΓΖ.

2nd Way.

If $\angle E\Gamma\Delta \neq \angle \Delta\Gamma Z$, then we can draw the line symmetrical to ΓΖ with respect to ΓΔ, that intersects the semicircle at point Ε', then according to the above direct proof, it will be: $\Gamma\Delta^2 = \Gamma E'.\Gamma Z$ and because (1) is true, then: $\Gamma\Delta^2 = \Gamma E'.\Gamma Z = \Gamma E.\Gamma Z$ or $\Gamma E'.\Gamma Z = \Gamma E.\Gamma Z$ or $\Gamma E' = \Gamma E$. This means that: $E' \equiv E$ or $\Gamma E' \equiv \Gamma E$, as points Ε, Ε', both lie on the intersection of the semicircle with circle (Γ, ΓΕ') or (Γ, ΓΕ), which intersection is only one (the other one is on the other semicircle). But ΓΔ was drawn to be the bisector of angle Ε'ΓΖ, thus ΓΔ will be the bisector of angle ΖΓΕ too, as $\Gamma E' \equiv \Gamma E$.

Remark.

In Proposition 11(33), of book [31], it is also proven that :

$\frac{E\Delta^2}{\Delta Z^2} = \frac{E\Gamma}{\Gamma Z} = -\frac{EH^2}{HZ^2}$, from which we get: $E\Delta.ZH = \Delta Z.EH$ and which means that the inscribed quadrilateral ΔEHZ is harmonic. This is directly proven if we apply Proposition 133, which is mentioned in Part B of this book.

2.5. Important Auxiliary Proposition with Various Applications.

We have come up with this Proposition since 1988 and we first published it in 1993 with our book [1] on page 1042, in 1996 with our book [31] § 1α(17) and in 1998 in the magazine «Dimension» of the HMS (issue 3-4/ 1998) with some of its applications.

We introduce it here, as we will be based on it to prove other Propositions (Lemma).

Proposition 5. (Figures 5 and 6).

In every hexagon ΑΒΓΔΕΖ inscribed in circle (o), the product of the three ratios of the lengths of its consecutive sides, is equal to the unity, if and only if its main diagonals are concurrent.

Hence, if $A\Delta \cap BE \cap \Gamma Z \equiv K \Leftrightarrow \frac{AB}{B\Gamma}\cdot\frac{\Gamma\Delta}{\Delta E}\cdot\frac{EZ}{ZA}=1.$

Proof.

1st Way [Figures 5(α) to 5(ζ)].

(a). Direct Proof.

For this hexagon we are given that: $A\Delta \cap BE \cap \Gamma Z \equiv K$ [We simultaneously refer to all six figures: 5 (α) to 5(ζ)]. We are asked to prove:

$$\frac{AB}{B\Gamma}\cdot\frac{\Gamma\Delta}{\Delta E}\cdot\frac{EZ}{ZA}=1. \qquad (1).$$

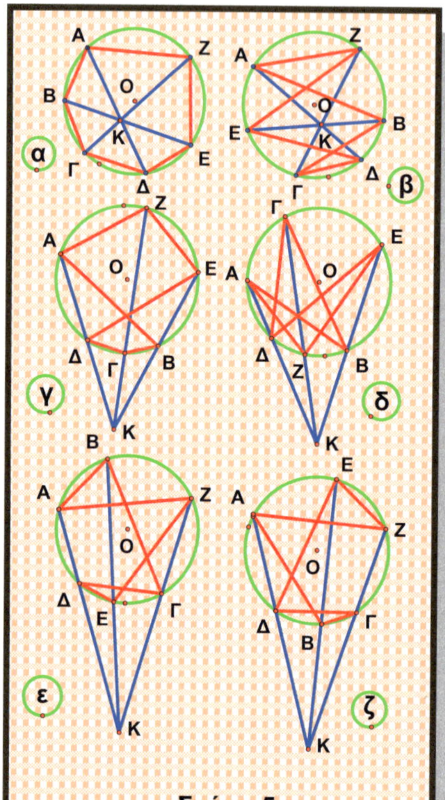

Σχήμα 5.

Actually, we can easily find that for example triangles ABK and EΔK, are similar, thus:

$$\frac{AB}{\Delta E}=\frac{KA}{KE}. \quad (2).$$

Similarly we find that:

$$\frac{\Gamma\Delta}{ZA}=\frac{K\Gamma}{KA}, \quad \frac{EZ}{B\Gamma}=\frac{KE}{K\Gamma}. \quad (3).$$

Hence, from relations (2), (3), if we multiply their analogous sides, we get:

$$\frac{AB}{\Delta E}\cdot\frac{\Gamma\Delta}{ZA}\cdot\frac{EZ}{B\Gamma}=\frac{KA}{KE}\cdot\frac{K\Gamma}{KA}\cdot\frac{KE}{K\Gamma}=1. \quad (4).$$

Consequently, from relation (4), we directly receive relation (1).

(b). Converse.

In the given simple hexagon ABΓΔEZ, if the above relation (1) is true, then we will also prove that: AΔ∩BE∩ΓZ≡K.

In fact, if diagonals AΔ, BE, ΓZ, are not concurrent at K (Figures 5) and it is true that AΔ∩BE≡α, BE∩ΓZ≡β, ΓZ∩ΔA≡γ, then if for example Aβ∩(O,ρ)≡Δ′ and not Δ, according to the direct proof of the Proposition, it will be:

$$\frac{AB}{B\Gamma}\cdot\frac{\Gamma\Delta'}{\Delta'E}\cdot\frac{EZ}{ZA}=1. \quad (5).$$

From relations (1), (5) it easily occurs that:

$$\frac{AB}{B\Gamma}\cdot\frac{\Gamma\Delta}{\Delta E}\cdot\frac{EZ}{ZA}=\frac{AB}{B\Gamma}\cdot\frac{\Gamma\Delta'}{\Delta' E}\cdot\frac{EZ}{ZA} \quad \text{or} \quad \frac{\Gamma\Delta}{\Delta E}=\frac{\Gamma\Delta'}{\Delta' E}. \qquad (6).$$

The last relation means that: Δ′ ≡Δ [see also annotation (a) below] and thus: $A\Delta\cap BE\cap\Gamma Z\equiv A\Delta'\cap BE\cap\Gamma Z\equiv K\equiv\alpha\equiv\beta\equiv\gamma$, or that the main diagonals (See terminology § 4) of the hexagon, are concurrent at K.

Remarks.

(a). Δ′ does in fact coincide with Δ, as these two points obviously belong to the same arc ΓΔE of circle (O, ρ), from the two arcs in which chord ΓE divides the circle, as it is about the same vertex Δ, of the simple hexagon for which the order of its vertexes on circle (O, ρ), is given (This means that the form of the hexagon is given), to begin with and then because triangles ΓΔ′E, ΓΔE are similar, as relation (6) is true and: ∠ΓΔE=∠ΓΔ′E. After all, these same triangles share side ΓE, thus they will be equal. Therefore, it will also be ΓΔ′=ΓΔ and thus Δ′≡Δ.

(b). In figure 5 we are given, beside the convex hexagon (a) and five more forms of concave hexagons (b) to (g), for which this Proposition is true and to which the above proof.

In book [1] we introduce twelve forms of hexagons in total. There, we mention forms of concave hexagons as well, for which this Proposition is not true [For example in figure 6, we are given two forms of hexagons for which the Proposition is not true.

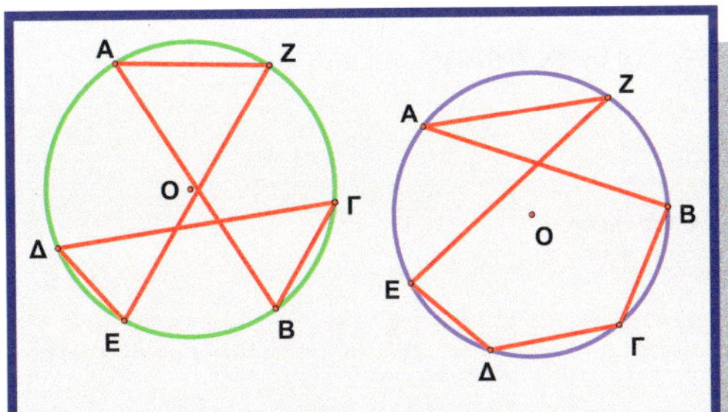

Σχήμα 6.

For these two forms it is obvious that the concurrency of their diagonals is not possible (Because their diagonals always form a triangle)].

Consequently, in order to apply the Proposition in another form of hexagon, we have to check whether for this form, the concurrency of the diagonals is possible or not (i.e. if the order of the hexagon's vertexes on the circle, allows the concurrency of its diagonals or not).

(c). The proof of this Proposition can be based on the application of Ceva's Theorem.

(d). In paragraph 1α(17) of book [31], we present alternative ways of expressing this Proposition. In the same paragraph of the same book, the reasons of the great value of this Proposition are mentioned as well.

Additionally in the same book generalizations of this Proposition are mentioned, many of its applications, its extensions and Propositions relevant to it.

(e). We have published Proposition 5, with many of its applications in issue 3-4/1998 of the magazine of the HMS in Central Macedonia «Dimension».

(f). We came up with this Proposition in 1988, published it in 1993 in our book [1] page 1042 and in 1996 in our book [31] § 1α(17).

(g). The great value of this Proposition, regarding its applications, is due to the fact that we proved its converse (Which we had not found in bibliography until today).

2nd Way.

Proof of Proposition 5 (From Architect Mr. Kostas Vittas):

LEMMA 1. – In every cyclic hexagon, the product of three non-consecutive sides of it, is equal to the product of the three remaining and conversely.

Proof.

Let $ABCDEF$ be the given hexagon, inscribed in circle (O) and

$AD \cap BE \cap CF \equiv P$.

In similar triangles $\triangle PAB$, $\triangle PED$ we have that $\frac{AB}{DE} = \frac{PA}{PE} = \frac{PB}{PD}$, (1).

From $(1) \Longrightarrow \frac{(AB)^2}{(DE)^2} = \frac{PA}{PE} \cdot \frac{PB}{PD}$, (2).

Similarly, in similar triangles $\triangle PCD$, $\triangle PAF$ and $\triangle PEF$, $\triangle PBC$,

we have that $\frac{(CD)^2}{(FA)^2} = \frac{PC}{PA} \cdot \frac{PD}{PF}$, (3) **and** $\frac{(EF)^2}{(BC)^2} = \frac{PE}{PC} \cdot \frac{PF}{PB}$, (4).

From $(2), (3), (4) \Longrightarrow (AB) \cdot (CD) \cdot (EF) = (BC) \cdot (DE) \cdot (FA)$, (5) **and we have proved what was wanted.**

•The converse (when relation (5) is true in an inscribed hexagon $ABCDEF \Longrightarrow AD \cap BE \cap CF \equiv P$) is easily proven, let us consider point $P \equiv AD \cap BD$ and let $F' \equiv (O) \cap CP$.

Then according to what we have previously said :

$(AB) \cdot (CD) \cdot (EF') = (BC) \cdot (DE) \cdot (F'A)$, (6).

From $(5), (6) \Longrightarrow \frac{EF}{EF'} = \frac{FA}{F'A}$, (7) **which is absurd if $F' \neq F$. Thus, it is concluded that $F' \equiv F$ and we complete the proof of the converse.**

To substantiate the contradiction in (7) if we suppose that $F' \neq F$, we can say that $F' \equiv (O) \cap CP$ is defined between E, A, as the extension of CP always lies in the internal of angle $\angle EPA$. If we now suppose, that F' lies between F, A, then it appears that $\frac{EF}{EF'} < 1$, (8) **and** $\frac{FA}{F'A} > 1$, (9).

From (8), (9) the contradiction about the hypothesis that F' lies between F, Δ occurs (as it is concluded that relation (7) is true).

Similarly we also end up in a contraction if we suppose that F' lies between E, F.

It is, thus, concluded that $F' \equiv F$ and we have proven what was wanted.

Kostas Vittas

2.6. Important Auxiliary Proposition with Various Applications.

This Proposition is mentioned here, as we will depend on it to prove other Propositions (Lemma).

Proposition 6. (Figures 7).

The symmedian of each side of a triangle, is conjugate [6] (§ 8-4.8) to this side, with respect to its circumscribed circle and conversely.

Proof.

(a). Direct Proof.

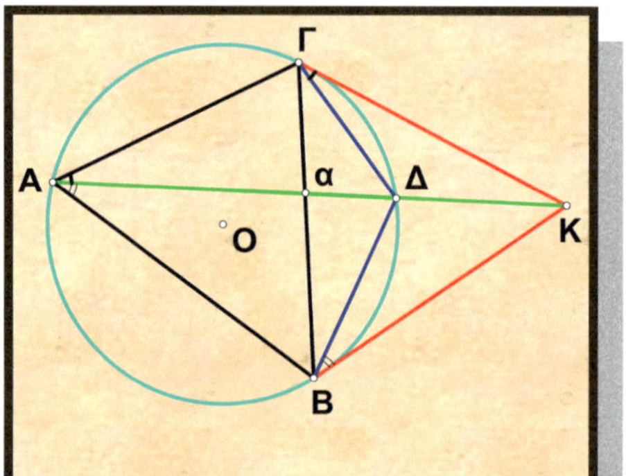

Σχήμα 7.

We consider triangle ΑΒΓ and its symmedian Αα (α∈ΒΓ). We will show that Αα and ΒΓ are conjugate with respect to the circumscribed circle (o) of triangle ΑΒΓ.

In fact, according to a known Theorem (§ 11-14.3) from Geometry [6], regarding the symmedian Αα, it will be:

$$\frac{AB^2}{A\Gamma^2}=\frac{\alpha B}{\alpha\Gamma}. \qquad (1).$$

If we prolong Aα and if $\Delta \equiv A\alpha \cap (o)$, then from the two pairs of similar triangles αAB, αΓΔ and αBΔ, αAΓ, we can easily get: $\frac{AB}{\Delta\Gamma}=\frac{\alpha B}{\alpha\Delta}$, $\frac{B\Delta}{A\Gamma}=\frac{\alpha\Delta}{\alpha\Gamma}$, thus it is concluded that:

$$\frac{AB\bullet B\Delta}{\Delta\Gamma\bullet\Gamma A}=\frac{\alpha B}{\alpha\Gamma}. \qquad (2).$$

Hence, from (1), (2) $\Rightarrow \frac{AB^2}{A\Gamma^2}=\frac{AB\bullet B\Delta}{\Delta\Gamma\bullet\Gamma A} \Rightarrow \frac{AB}{A\Gamma}=\frac{B\Delta}{\Delta\Gamma} \Rightarrow AB.\Delta\Gamma=A\Gamma.\Delta B.$

This relation, according to known Theorem [6] (§ 8-4.10), means that the convex quadrilateral ABΔΓ is harmonic, thus AΔ≡Aα and BΓ, are conjugate, with respect to circle (o) {definition of Harmonic Quadrilaterals [6] (§ 8-4.10}.

[This shows that Aα would pass through pole K of BΓ (definition of conjugate lines [6] (§ 8-4.8)].

(2). Converse.

Hence, if Aα and BΓ are conjugate, we will prove that Aα is symmedian in triangle ABΓ.

In fact, if $\Delta \equiv A\alpha \cap (o)$, then quadrilateral ABΔΓ would be harmonic (definition). Hence, according to known Theorem [6], it will be: AB.ΔΓ=AΓ.ΔB

$$\Rightarrow \qquad \frac{AB}{A\Gamma}=\frac{B\Delta}{\Delta\Gamma}\Rightarrow\frac{AB^2}{A\Gamma^2}=\frac{AB\bullet B\Delta}{\Delta\Gamma\bullet\Gamma A}. \qquad (3).$$

Moreover, relation (2) above will obviously be true as well.

Thus, from (2), (3) $\Rightarrow \frac{AB^2}{A\Gamma^2}=\frac{\alpha B}{\alpha\Gamma}$. This one, according to the converse of a known theorem (§ 11-14.3) in book [6], means that Aα is symmedian in triangle ABΓ.

Remarks.

(a). This Proposition has been established in our book [31] [§ 2α(43)]. Another proof of this Proposition is given in book [31] (paragraph 10ι(132) volume 10}.

(b). The value of this Proposition, regarding its applications, has to do with the fact that its converse is true as well.

2.7. Important Auxiliary Proposition with Various Applications.

Intersection of a Central Pencil of Three Beams by a Circle (Pappus' Analogous Theorem).

This Auxiliary Proposition (Lemma), is introduced here, as we will depend on it to prove other Propositions. It is about Proposition 2α(45) in book [31].

Criterion 7. (Figures 8α, β, γ).

If a flat central pencil of three beams KA, KB, KΓ, is intersected by circle (O, R), at six points A, B, Γ, Δ, E, Z, where Δ∈KA, E∈KB, Z∈KΓ, then and only then regarding the two groups of four points A, B, Γ, E and Δ, E, Z, B, it will be: (AΓBE)=(ΔZEB) or $\frac{AB}{B\Gamma}:\frac{AE}{E\Gamma}=\frac{\Delta E}{EZ}:\frac{\Delta B}{BZ}$. (1).

or $\frac{AB}{B\Gamma}\cdot\frac{\Gamma E}{EA}=\frac{\Delta E}{EZ}\cdot\frac{ZB}{B\Delta}$. (2).

Proof.

(1). Direct Proof (Figures 8α and 8β).

In fact, from the pairs of similar triangles ΑΒΚ~ΕΔΚ, ΒΓΚ ~ ΖΕΚ, ΓΕΚ ~ ΒΖΚ, ΕΑΚ ~ ΔΒΚ, we will have:

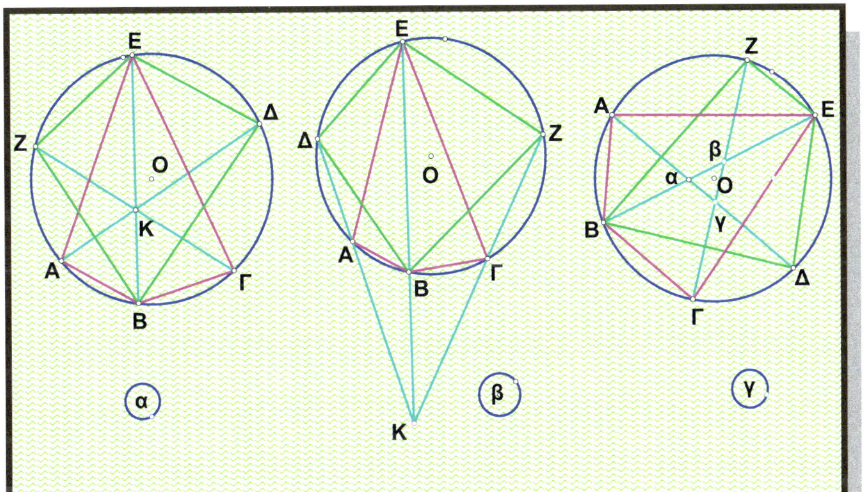

Σχήμα 8.

$$\frac{AB}{BK}=\frac{E\Delta}{\Delta K},\quad \frac{KB}{B\Gamma}=\frac{KZ}{ZE},\quad \frac{\Gamma E}{EK}=\frac{BZ}{ZK},\quad \frac{KE}{EA}=\frac{K\Delta}{\Delta B}. \qquad (3).$$

From relations (3), if we multiply their analogous sides, we get (2) and from that we get (1).

[Additionally, regarding the groups of four points Α, Β, Γ, Ε and Δ, Ε, Ζ, Β, because: (ΑΓΒΕ)=(ΒΕΑΓ)=(ΓΑΕΒ)=(ΕΒΓΑ), (ΔΖΕΒ)=(ΕΒΔΖ)=(ΖΔΒΕ)=(ΒΕΖΔ) {known Theorem [26](§ 332 § 3)} and because as proven above it is: (ΑΓΒΕ)=(ΔΖΕΒ)⇒

$$\frac{BA}{AE}\cdot\frac{E\Gamma}{\Gamma B}=\frac{E\Delta}{\Delta B}\cdot\frac{BZ}{ZE},\quad \frac{\Gamma E}{EA}\cdot\frac{AB}{B\Gamma}=\frac{ZB}{B\Delta}\cdot\frac{\Delta E}{EZ},\quad \frac{E\Gamma}{\Gamma B}\cdot\frac{BA}{AE}=\frac{BZ}{ZE}\cdot\frac{E\Delta}{\Delta B}, \text{ etc.}$$

It is obvious that the last relations can occur directly from (3) or even (2)].

(2). Converse (Figure 8γ).

If for example relation (2) is true, we will prove that : $A\Delta \cap BE \cap \Gamma Z \equiv K$.

In fact, if $A\Delta$, BE, ΓZ are not concurrent and we set: $A\Delta \cap BE \equiv \alpha$, $BE \cap \Gamma Z \equiv \beta$, $\Gamma Z \cap A\Delta \equiv \gamma$ (it is possible that α, β, γ, are outside of the circle), then from the pairs of similar triangles $AB\alpha \sim E\Delta\alpha$, $B\Gamma\beta \sim ZE\beta$, $\Gamma E\beta \sim BZ\beta$, $EA\alpha \sim \Delta B\alpha$, we get:

$$\frac{AB}{B\alpha} = \frac{E\Delta}{\Delta\alpha}, \quad \frac{\beta B}{B\Gamma} = \frac{\beta Z}{ZE}, \quad \frac{\Gamma E}{E\beta} = \frac{BZ}{Z\beta}, \quad \frac{\alpha E}{EA} = \frac{\alpha\Delta}{\Delta B}. \qquad (4).$$

From relations (4), if we multiply their analogous sides, it appears that:

$$\frac{AB.\Gamma E.B\beta.E\alpha}{B\Gamma.EA.B\alpha.E\beta} = \frac{\Delta E.ZB}{EZ.B\Delta}. \qquad (5).$$

Relation (5), because of (2), which is given to us, becomes: $\frac{B\alpha}{\alpha E} = \frac{B\beta}{\beta E} \Rightarrow \alpha \equiv \beta$.

This means that ΓZ, that passes through β, will also pass through α the intersection of $A\Delta$ and BE, hence: $A\Delta \cap BE \cap \Gamma Z \equiv K \equiv \alpha \equiv \beta \equiv \gamma$.

Remarks.

(a). This Proposition was established in our book [31] [§ 2α(45) which can also occur from Proposition 2α(44) in book [31], or Proposition 50IΓ (page 1582) in book [1]. Both these firstly appearing Propositions are important, as their converses are also true and mostly because they expand the known Pappus's Theorem, which has to do with intersections of a flat central pencil from lines, while here we have intersections of a flat central pencil of 3 and 4 beams respectively, intersected and re-intersected by a circle.

Both these Propositions, have also been published in the magazine «Mathematics in High School (issue 13 page 36), with some of their applications.

(b). The above Proposition 7 was proven valuable for the proof of Propositions and particularly for Propositions concerning Harmonic Hexagons.

2.8. Important Auxiliary Proposition.

Prolongation of Newton's Theorem in Hexagons {Special Case of Proposition 1δ(1) in Geometry [31]}.

This Auxiliary Proposition (Lemma) , is mentioned here, as we will depend on it to prove other Propositions.

Proposition 8 (Figures 9).

If the main diagonals of a hexagon inscribed in circle are concurrent at a point, then this point coincides with point Brianchon {[6] (§ 11-11)}, of the respective circumscribed (tangential).

Proof.

If i.e. we are given two respective hexagons ΑΒΓΔΕΖ inscribed in circle (O, R) and A'B'Γ'Δ'E'Z' circumscribed (tangential) in the same circle, as to $A \in A'Z'$, $B \in A'B'$, $\Gamma \in B'\Gamma'$, ...and it is: $A\Delta \cap BE \cap \Gamma Z \equiv K$, we will prove that also $A'\Delta' \cap B'E' \cap \Gamma'Z' \equiv K$. Hence K coincides with point Brianchon, of hexagon A'B'Γ'Δ'E'Z'.

In fact, according to known Newton's Theorem {Books [7] § 1274 and [31] Proposition 1β(2)}, where a new proof is given}, for the respective quadrilaterals ABΔΕ inscribed and A'H'Δ'Θ' circumscribed (where H'≡A'B'∩Γ'Δ', Θ'≡A'Z'∩Δ'E'), their diagonals will pass through the intersection of AΔ and BE, which is point

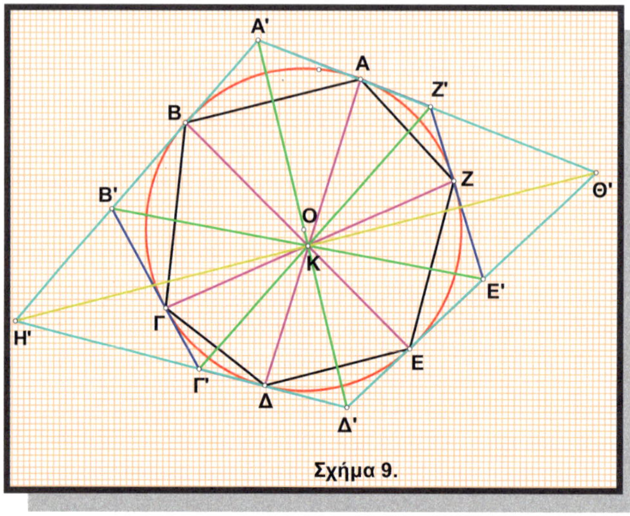

Σχήμα 9.

K. Hence, to state it otherwise A'Δ' goes through K (as does H'Θ'). In the exact same way we prove that B'E', Γ'Z' go through K.

Consequently we see that diagonals A'Δ', B'E', Γ'Z' go through K, which is known to be point Brianchon {[6](§ 11-11)}, og hexagon A'B'Γ'Δ'E'Z'.

Remarks.

(a). We can easily realize that the converse of Lemma 8 is not true, as it is known, that circumscribed in circle hexagons, always have concurrent diagonals at point Brianchon.

(b). Generalization of Proposition 8, for n-laterals, is given in § 1δ(1) of book [31]. Both these Propositions are obviously generalizations of Newton's Theorem, which is true only in quadrilaterals.

(c). Proposition 8 is also true in concave hexagons, (See Geometry [31], Structures 1δ(2) and 1δ(3).

(d). Lemma 8 above was proven valuable for the proof of Propositions concerning Harmonic Hexagons, whose diagonals are concurrent.

(e). Proposition 146 in Part B is also relevant, which is about Harmonic Hexagons and Proposition 206 about Harmonic Decagons.

ΕΠΙΛΟΓΟΣ

Here is the end of the "SKY" !!!

We started our research in the "sky" about four years ago with a high cubic capacity motorcycle [as it is shown in the picture at the beginning of the book (page 13)] and after crossing difficult paths and "Symplegades Rocks"("Clashing Rocks"), which I assure you were Exciting, enchanting, delightful and effective, I arrived happy and healthy at the end of the writing of my new book "ARMONIC GEOMETRY".

Everything has its end!!!

But that does not mean that the work is over.

There is still a lot of work to be done on complements, error corrections (where applicable) and improvements.

I thank God who gave me the strength, the ability and the days, to finish my new book "Harmonic Geometry".

I also thank all the friends who followed the entire or part of the journey I had and I believe they also enjoyed it.

I also would like to thank my own people, who tolerated and helped me morally.

Finally, I am extremely indebted to the site mathematica.gr, which hosted me and gave me the opportunity to present all the work to my friends.

We have been working methodically, with passion, patience and perseverance, and we believe that our realistic but not ambitious goal has been achieved. However, this is enough for us for the moment as we believe that the work is of HIGH QUALITY.

Thus, we believe that a new Euclidean geometry branch has emerged, or if you want a new Geometry, the Harmonic Geometry.

Anyone who will carefully study this book, I am convinced, that he will really enjoy the Magic, the Harmony and all the Greatness of Geometry, and will find in it Geometric Treasures that were hidden till today.

There are many and great challenges in this book, and we have given all the relevant tools to those who want to expand it.

Finally, we believe that we built (stone by stone) a building of a great Architectural value and of great Strength.

It is proved that way once again that Geometry is alive and grows, against the stubbornness of its enemies.

Whoever wants to really get involved seriously with the development of hard Geometry needs to do further research work but also to study the new elements of Geometry that come to light.

RESEARCH = Enjoyment + Discoveries + Intense Emotions ⇒ Creation ⇒ Progress ⇒ Civilization

If, with my previous frequent encouragement, I managed to persuade even a friend of Geometry to follow this difficult, but attractive and constructive way of research, it would be a great satisfaction for me.

We believe that Harmonic Geometry and more specifically the New Method of the Harmonic transformation, can be introduced to other fields of Mathematics as well, such as Complex Analysis, etc. This is stated as an idea for research, for those who are interested.

FIGURE INDEX

Equivalence of the Figure Number and the Page it is on.

A. Harmonic Geometry.

95↔531, 96↔535, 97↔546, 98↔549, 99↔555, 100↔566, 101↔573, 102↔575, 103↔578, 104↔582, 105↔586, 106↔591, 107↔596, 108↔604, 109↔605, 110↔607, 111↔609, 112↔614, 113↔616, 114↔622, 115↔625, 116↔627, 117↔660, 118↔668, 120↔677, 121↔680, 122↔685, 123↔692, 124↔693, 125↔695, 126↔698, 127↔704, 128↔707, 3.2.6.(4).(β).1/↔714.

B. Auxiliary Propositions.

1↔722, 2↔722, 3↔723, 4↔743, 5↔746, 6↔747, 7↔750, 8↔753, 9↔756.

CONTENTS

PART B

PART C

PART D

PART E

IT WAS SAID

(See at Google entry Nikos Kyriazis a distinguished member).

I would like to welcome to Mathematica the exceptional expert in geometry Nikos Kyriazis, whose proofs-solutions and suggestions we have read in several magazines / books both Greek and foreign!

Alexandros Kongelakis (Professor of Mathematics, Mathematica Site Administrator).

I would also like to officially welcome the eminent and multi award-winning expert in geometry Nikos Kyriazis. He is the author of the well known multi-volume *Geometry*, which should be acquired by all of us, in addition to a great number of articles in magazines.

His presence here is an honor for all of us. I wish him good health and to provide us with new and original solutions to the issues we are discussing.

Babis Stergiou (Professor of Mathematics - Author).

At this very moment, I really felt completely ignorant, as I did not know (even though I say that I am involved in Geometry) about this great work of Nikos Kyriazis! Thanks to Mathematica, I was informed and I thank above all Nikos for this. I hope to find and study thorough his work.

Nikos, even delayed, warm congratulations (please forgive me of my ignorance).

Andreas Varverakis (Professor of Mathematics).

I was completely ignorant about it before it was mentioned here, so when I visited the Mathematical Library (the bookshop of Haris Vafeiadis) a few days later, I asked for and saw some of the volumes of Nikos ... and I was really stunned. For me it is already too late to wander in such a lengthy work – even though I consider myself an expert in geometry (or rather *a man of symmetry *) – apart from the topics (and discussions) provoked by such endeavors regarding the possibility of further originality in sectors which are considered studied thoroughly are indeed provocative! (In addition of course, an additional big issue is when and why a mathematical sector could be considered as an "exhausted one"--deep waters…)

Giorgos Baloglou (Professor of Mathematics at the University of New York.

I am stunned facing the beauty of the problem and the sensational simplicity of Nikos' solution!

If this construction is universally unknown, I would suggest that the term "Kyriazis's hexagon" should be internationally adopted.

Giorgos Baloglou (Professor of Mathematics at the University of New York).

Many Happy returns to Nikos Kyriazis, a loyal lover of geometry, with a rich and original work.

Leonidas Tharralidis (Professor of Mathematics).

Nikos, a wonderful and absolutely basic solution!

It may very well be used in the classroom as a spontaneous adventure with the students as a part of the quadrilaterals course.

With a circle, everything is over .... and at the same time everything is restarted!

Thank you for the inspired solution.

Babis Stergiou (Professor of Mathematics - Author).

Nikos congratulations on the honorary distinction and thank you once more for honoring us with your membership at our Club.

Mihalis Lambrou, General Coordinator (Professor of Mathematics, University of Crete).

Nikos Kyriazis (whom I personally met during the summer and admired his liveliness, zeal, passion and love, I would say, he shows for Geometry). He does not deal with the foundation of Geometry. He has fully grasped and entirely respects the work of Euclid. What he does, based on this work, is to discover the qualities of the various shapes and then he is trying to prove them in as many ways as possible. Nowhere did he claim that these should be taught at high school. He has said several times that these are for the lovers of geometry or for spiritual satisfaction only. Is there anything better and purer? I am sure that, if there had been an opportunity for Euclid to study his work, he would have congratulated him.

Antonis Kyriakopoulos (Professor of Mathematics - Author).

Nikos, you are getting on our nerves with your criteria. We will all be suffering from complexes.

Andreas Poulios (Doctor of Pedagogy, Professor of Mathematics - Writer).

Nikos, although I think I do not know you personally, I admire indeed the tremendous will you possess to accomplish your work!! Good follow-up. I shall always follow the progress of your work.

Ioannis Dimitris (Mathematician, Istiaia Evia)

My friend Nikos,

I was deeply touched with your "GREAT THANK YOU". Your progress so far guarantees that you will offer us more, because our astronomical age is of no importance. The importance is in the age of our soul and you have proved to us that your soul is one of a teenager.

Yannis Kerassarides (Professor of Mathematics).

Dear Nikos Kyriazis, looking once more at the shapes in your article, I remember the despair (and the dizziness ...) I felt when I received them by fax from Nikos Iosifidis, when we had to type and layout the text!

Keep well and continue your EXCELLENT WORK.

Georgios Rizos, Coordinator (Professor of Mathematics).

Moreover, everybody bows in front of a **giant** of geometry such as Nikos Kyriazis.

Stathis Koutras (Professor of Mathematics).

My tireless friend Nikos

Your suggestions are inconceivable and your extensions are magnificent.

I would like to publicly confess that at the age of 52, the emotion I get from you concerning the greatness of geometry is unprecedented. Your new proposal is incredible; fantastic as well are the shapes that you have created the beauty of "symmetry" in all its grandeur.

I could not fail to react at your new challenge (and I confess that I have lost my sleep because of it).

Stathis Koutras (Professor of Mathematics).

Dear friend Nikos,

Your love and your efforts for "Hardcore Geometry" are worthy of admiration and respect, especially for your deeds at that beautiful but rather neglected branch of Mathematics.

Giorgos Apokis (Professor of Mathematics).

Niko,

Best wishes and keep well for the future, continue to impress us. You are an example to us (I will not stop saying it) we draw strength from your appetite and energy, I am 33 years old and I already feel tired and disappointed by many things in our field ... It is a good refuge for us and that's the reason we are here.

Concerning your books, I am ashamed to admit that I haven't bought them yet, but at the first opportunity I will acquire them and I will bring them to our first meeting, for your signing them all – we would need at least three hours!!

Keep well.

Makis Hatzopoulos (Professor of Mathematics).

Mr. Nikos, keep well and have in mind that we, the younger ones, admire you and have you as an example to follow.

Makis Hatzopoulos (Professor of Mathematics).

Mr. Nikos,

I very often follow your discoveries and admire you.

Andreas Poulos (Doctor of Pedagogy, Professor of Mathematics - Author).

Nikos, I do not know if you have realized, you have become the first of your league and the honored one.

Andreas Poulos (Doctor of Pedagogy, Professor of Mathematics - Author).

Nikos,

the right movement is not for us to accept your gratitude but for you to accept our admiration.

Continue to produce work on the Eternal Building of Euclid and let us admire you.

Nikos Mavroyiannis, General Coordinator (Professor of Mathematics).

I hope that you will continue your excellent work with the same passion, so that all the following generations (and mine amongst them) shall inherit something great in the field of Geometry.

Vassilis Papaspilotopoulos (Electrical Engineering Student at N.M.P.).

Mr Nikos,

I would simply like to personally thank you for the subjects, you tirelessly post at the site.

I hope to find the time to personally work with the problems you set, which are numerous and excellent so far.

Achilleas Sinefakopoulos

Mr. Nikos good morning

We have to voice a huge Thank You for a great number of reasons:

-for your being one of us - for giving us generously all that from your personal work-

-for teaching us so many things - for being able to teach us even more if we had more time to work upon it

-I wish you the best

-I am glad to see you through your publications – all the best for the future

-With love and unlimited respect

Fotini Kaldi (Professor of Mathematics).

Nikolas, although I do not know you personally, I wish you to be long lived and keep working on mathematics (although I have to admit, I do not always fully grasp the issues you deal with).

Omega Man

Many Happy returns from me to all those people having their birthday and above all to the tireless man of geometry Nikos Kyriazis whom I hope to meet personally in the future.

Marantidis Fotis (Professor of Mathematics).

My Friend Nikos, good evening.

It is a beautiful proposal indeed the one you have suggested.

I would like to meet you in person.

I feel the need to tell you that you are GREAT and please simply accept it.

Stathis Koutras (Professor of Mathematics).

I would like to present a nice problem by a friend of mine Nikos Kyriazis from Thessaloniki – Hellas.

Kostas Vittas (Architect at NMP).

File comment: Kyriazis's problem of concurrency.

Kostas Vittas (Architect at NMP).

Dear Nikos,

Be always well, strong, in good health, with inspirations and discoveries!

Rekoumis Konstantanos (Professor of Mathematics).

Dear Nikos,

I follow with great interest your research studies - so to speak - in geometry and I am not hesitating to voice my thoughts, I admire the methodology of your work and the depth of your research. Keep well!

With my best wishes for good health, every success, progress, prosperity and inspiration!

Rekoumis Konstantanos (Professor of Mathematics).

Dear Nikos.

I want to meet you in person so as to think together of our school years.

In one of your messages you mentioned that when we were pupils, you admired me. Now I have to tell you that I admire you for your excellent work in Geometry (Honestly, why you did not become a mathematician?)

Antonis Kyriakopoulos (Professor of Mathematics - Author)

My Friend Nikos.

I'm following what you write about geometry in “mathematica” and I'm proud of being your friend.

Antonis Kyriakopoulos (Professor of Mathematics - Author).

Dear friends and colleagues, the work dealing with this subject is excellent.

Of course, congratulations to Mr. Nikos Kyriazis - the tireless and always creative man - who started this project.

I too believe that all this informative material should be exploited in multiple ways

Andreas Poulos (Doctor of Pedagogy, Professor of Mathematics - Writer).

Thank you for sharing with us your thoughts and concerns on Geometry! Thanks to you we've learned a lot of new things!

Keep well and continue what you are doing!

I wish you all the best!

Nikos Katsipis (Professor of Mathematics).

My particular GEOMETRY wishes to the kind GIANT of Geometry NIKO KIRIAZI.

Nikos, be always strong and be next to us with your new and amazing discoveries.

Stathis Koutras (Professor of Mathematics).

Dear Nikos

.... You made me feel proud, when you said that I, even without knowing it, became the reason for you to work with mathematics and become the most important researcher in Euclidean Geometry and to gift us with your great and significant work (Over 30 volumes !!! with honorable distinctions from the Academy of Athens and awards obtained by many other important associations).

The torrent of the original proposals you articulate and prove in your books and in your articles about the Euclidean Geometry, place you among the great men of geometry of all time.

Nikos, keep well and continue to surprise us with the results of your research in this exceptional branch of mathematics.

Antonis Kyriakopoulos (Mathematician-Author).

I would like to thank Antonis Kyriakopoulos and Nikos Kyriazis for their contribution to Mathematics! The least we can do for them, we, the total number of the coordinators at mathematica.gr, have decided to add them in the team of the Distinguished Members as the New Regulation foresees! Both of them, each one from a different stand, are trying to improve and promote Mathematics in our country...

Thank you very much and we urge the younger generation to follow your example!

Alexandros Sygelakis (Professor of Mathematics-Administrator).

Antonis Kyriakopoulos, Nikos Kyriazis, allow us to nominate you as Distinguished Members of the Greek Mathematical, web community, as a minimum recognition of your contribution to the mathematical education in this country.

Grigoris Kostakos (Professor of Mathematics - Administrator).

Antonis Kyriakopoulos – Nikos Kyriazis thank you for your long-term contribution to the mathematical community. We've gained a lot of math knowledge from you and we hope to gain even more in the future. The education of Mathematics in our country is in need of people of your kind.

With respect and appreciation toward you, I ask for your permission to call you my masters.

Kardamitsis Spiros (Professor of Mathematics - Coordinator).

It is our honor, if anything else, that people of this magnitude at mathematics are close to us. Above all I wish you good health.

Christos Kyriazis (Professor of Mathematics).

I will also congratulate the excellent masters, Antonis Kyriakopoulos and Nikos Kyriazis, for their enormous contribution in the science of Mathematics.

Stavros Stavropoulos (Professor of Mathematics).

Mathematics with us, feel very proud of you. Congratulations from my part as well!

Tilegrafos Kostas (Professor of Mathematics).

Antonis and Nikos

I think your inclusion on that list is yours by right, not to say officially! We are proud to have you next to us, you beautify and ornament our mathematics, you make them simpler and more glamorous, consistent and accurate and with substance and purpose ... You are our masters in mathematica.gr.

Makis Hatzopoulos (Professor of Mathematics).

We thank our colleagues Antonis and Nikos, who are examples for us, the younger mathematicians.

Sukaras Efthimios (Professor of Mathematics).

Nikos Kyriazis.

I congratulate you for your contribution to Geometry. It is great honor to have you with us.

X.K. (Professor of Mathematics)

A QUESTION.

DOES ANY UNIVERSITY PROFESSOR VISIT THE MATEMATICA WEBSITE?

A GREAT NUMBER OF ORDINARY PEOPLE WERE NAMED HONORARY DOCTORS; NO ONE WAS MOVED BY THE GREATNESS OF THE EXCEPTIONAL WORK OF MISTER NIKOS?

I WONDER IF MISTER NIKOS WAS HONORED BY THE ACADEMIC WORLD.

Mr Nikos, I apologize for my message, the greatness of your modesty would not have wanted me to send it out ... but I would have exploded if I hadn't written it.

Finally, I'm sorry I do not have more time to work on it any longer... and I will not have because I simply do not hope to get a pension.

Leonidas 1

I have to congratulate you on your work or better on your valuable contribution. I hope you will be an example to others and especially Mathematicians. I also wish for the continuation of your efforts.

I offer you all of my Geometry books (six volumes), as well as my five research papers, because I'm sure they will be of a value to you.

Nikos Kiskyras (Mathematician).

. .. I congratulate you on your appeal to deal with such passion for Geometry, but also for your achievements, which are giving the impression not only for your complete knowledge of Geometry but, above all, a deeply complex and unified thought.

I wish you well and to keep working, with similar great success, on the most beautiful of sciences, on this the purely Greek science.

I also wish you to keep well and have every success in the divine paths of research and knowledge.

Michael Diakomanolis (Mathematician).

The main Theorem of your article is very interesting. I have not seen it anywhere and it can be a powerful tool for proving other Theorems like the ones of Steiner, Vecten, Morley, Brocard, Kariya, convergence of triangle heights, etc

Assessor in Geometry, the periodical Mathematics and Informatics Quarterly.

Your massive and multi-page work in Euclidean Geometry has been a pleasant surprise for me in terms of its content. This work is definitely the fruit of many years of long term, deeply successful or better intelligent research. Mr Nikolaos Kyriazis, a lover and admirer of Geometry, with this extraordinary written work, emerges as a brilliant man of geometry, a remarkable researcher in Geometric issues and in any case all praise is deserved.

Zissis Balpouzis (Mathematician).

THE ATHENS ACADEMY WAS OFFERED THE HONOR TO ATTRIBUTE TO NIKOLAO KYRIAZI CONGRATULATIONS CONCERNING HIS WORK AND SUCCESSFUL CONTRIBUTION ON THE SCIENCE OF BASIC GEOMETRY.

THE HONORS TOOK PLACE IN THE MEETING OF THE MONTH DECEMBER THE SEVENTH OF THE YEAR TWO THOUSAND TWO.

Athens Academy (This honor was obviously awarded for his work prior to 27-12-2002).

Nikos, a work of such importance will stay immortal and you as well with it!

Keep well, live long and enjoy the fruits and glories of this unique effort and contribution!

I hope that among all those that will honor you, there will be with additional love the Greek scientific Associations, the institutions and the Universities!

Babis Stergiou (Professor of Mathematics - Author).

Mr. Nikos Kyriazis, I would like you to allow me to thank you for this unforgettable gift that touches the spirit of Euclid and you generously offered it to us.

I sincerely wish you to continue such scientific interventions of progress and substantial advancement.

Best wishes to you and your family. (*) ...

Personally, I strongly believe that the message of this work is clearly wider and it is an essential response to the challenges of our time having several recipients....

Sotiris E. Louridas (Professor of Mathematics - Author).

Personally, as I have already written,

Nikos thank you again, I admire you for what you are doing even if I do not understand you.

Parmenides51 (Professor of Mathematics).

In certain research work a thank you is not enough !!!
We will "gulp" it, Nikos at every opportunity !!!
Sotiris D. Hasapis (Professor of Mathematics).

I admire Mr. Nikos Kyriazis for his courage to engage in Euclidean Geometry. When a person has got a professional career that is irrelevant to mathematics and manages to have so great results at an age when others lose interest in what was once their main activity, then this man is enviable. I really would like to meet and to him ... I have so many things to ask him ... I would like him so many things to tell me ...
What I'm writing is not exaggeration, it's from the bottom of my heart ...
Mr. Nikos, I do not know if we will ever meet, I want you to know that you are an example to be followed. Keep well and continue to work - with your own way - with the Euclidean Geometry. You offer and help ...
And you show what a retired person can do when he keeps in mind and heart something he loved from the time he was a student
Tilemachos Baltsavias. A simple Professor of Mathematics.

Nikos, this is my greatest honor ever. I feel like I do not deserve it, but because it is from you, I accept it humbly.
I hope for you the best, keep your spirit always in a creative mood and give to all of us these geometric diamonds. Good day and every happiness to you!
Babis Stergiou. (Professor of Mathematics, Mathematics Bookwriter).

Nikos, thank you very much for your enormous contribution to this field of geometry, which may be for the young people who shall wish to study the publications a real treasure. Keep always good and I think we will be amazed again with your new ideas.
DIMITRIS IOANNOU (Professor of Mathematics).

Important remark from page 12:

1. Greek alphabet, in which we can see the sequence (order) of the Greek letters, that interests the researcher.

(a). CAPITAL LETTERS.

A-alpha,	I-iota,	P-rho,
B-beta,	K-kappa,	Σ-sigma,
Γ-gamma,	L-lambda,	T-tau,
Δ-delta,	M-mu,	Y-upsilon,
E-epsilon,	N-nu,	Φ-phi,
Z-zeta,	Ξ-xi,	X-chi,
H-eta,	O-omicron,	Ψ-psi,
Θ-theta,	Π-pi,	Ω-omega

(b). LOWERCASE LETTERS.

α-alpha,	ι-iota,	ρ-rho,
β-beta,	κ-kappa,	σ-sigma,
γ-gamma,	λ-lambda,	τ-tau,
δ-delta,	μ-mu,	υ-upsilon,
ε-epsilon,	ν-nu,	φ-phi,
ζ-zeta,	ξ-xi,	χ-chi,
η-eta,	ο-omicron,	ψ-psi,
θ-theta,	π-pi,	ω-omega

2. Translation in English of some words that appear in the book in Greek.

Σχήμα – Figure, Μοίρες – Degrees, Γωνία – Angle, Προβολή – Projection,

Τόξο – Arc, Συνημμένο - Attached, Μέγεθος - Size.

This is followed by volume 3, with all the figures of the book assembled, to facilitate the reader.

NIKOS KYRIAZIS is a retired high-ranking officer of the Greek Army. Because of his great love for mathematics, he devoted himself to studies and research in Euclidean geometry after he retired.

The fruit of the above-mentioned research is the creation of many important new elements in geometry and he recently formulated Harmonic Geometry. All this work has been documented in five books, twenty-two (22) volumes so far, of which thirteen (13) have been released, available at cost price for geometry lovers. He has also written fifty-seven (57) articles in Greek and foreign magazines and newspapers, some of which have been published in two additional volumes (the 23rd and 24th).

In his works, he only records new elements, which have not been encountered before and which arose from his research.

The value of his books do not derive from their perfect appearance and their elegant presentation, but from their original content, which has significantly changed what we know about geometry, and will surprise the reader. A part of the work is secondary-school level, while the rest is of a higher level, suitable for the Olympiads, the Balkaniade, etc.

The comments and reviews of mathematics experts are excellent and favourable (Page 768).
The press wrote numerous glowing comments about him and his work.

All his books and articles prior to 27-12-2002 were honoured by the ACADEMY OF ATHENS at its annual celebratory session on 27-12-2002. The Academy hailed the whole work as "outstanding and useful" when presenting it with the award.

For his work and his Academy Award, he was honoured by many other institutions in Greece.

In 2011, he was unanimously named Distinguished Member in geometry by the website mathematica.gr.

In 1957, Nikos Kyriazis graduated from the Hellenic Army Cadet School as a Second Lieutenant of Engineering, in 1966, from the School of Civil Engineering in Athens, in 1975, from STAMAM (Military School, similar to the School of Civil Engineering of the Polytechnic School) and in 1978, from the Supreme War College.

He served his country continuously for 33 years and retired from his honourable service in 1987, with the rank of Brigadier. He has received numerous medals of honour during his military service.

Nikos Kyriazis was born in Thouria, Kalamata (Greece) in 1934. He permanently resides and creates in Kalamaria (Trapezoundos 9, tel. +30 2310-413.539). He is married and has two children and three grandchildren.

Made in the USA
Las Vegas, NV
03 November 2026

57424062R00245